Titles in This Series

16 **Sergei Gelfand and Simon Gindikin, editors,** I. M. Gelfand seminar, Parts 1 and 2, 1993
15 **A. T. Fomenko, editor,** Minimal surfaces, 1993
14 **Yu. S. Il′yashenko, editor,** Nonlinear Stokes phenomena, 1992
13 **V. P. Maslov and S. N. Samborskiĭ, editors,** Idempotent analysis, 1992
12 **R. Z. Khasminskiĭ, editor,** Topics in nonparametric estimation, 1992
11 **B. Ya. Levin, editor,** Entire and subharmonic functions, 1992
10 **A. V. Babin and M. I. Vishik, editors,** Properties of global attractors of partial differential equations, 1992
9 **A. M. Vershik, editor,** Representation theory and dynamical systems, 1992
8 **E. B. Vinberg, editor,** Lie groups, their discrete subgroups, and invariant theory, 1992
7 **M. Sh. Birman, editor,** Estimates and asymptotics for discrete spectra of integral and differential equations, 1991
6 **A. T. Fomenko, editor,** Topological classification of integrable systems, 1991
5 **R. A. Minlos, editor,** Many-particle Hamiltonians: spectra and scattering, 1991
4 **A. A. Suslin, editor,** Algebraic K-theory, 1991
3 **Ya. G. Sinaĭ, editor,** Dynamical systems and statistical mechanics, 1991
2 **A. A. Kirillov, editor,** Topics in representation theory, 1991
1 **V. I. Arnold, editor,** Theory of singularities and its applications, 1990

Part 1

Photograph courtesy of V. Minachin

Advances in
SOVIET MATHEMATICS

Volume 16, Part 1

I. M. Gelfand Seminar

Sergei Gelfand
Simon Gindikin
Editors

American Mathematical Society
Providence, Rhode Island

1991 *Mathematics Subject Classification*. Primary 00B15.

Library of Congress Catalog Card Number: 91-640741

ISBN 0-8218-4118-1 (Part 1)
ISBN 0-8218-4119-X (Part 2)
ISBN 0-8218-4117-3 (Set)

ISSN 1051-8037

The paper used in this book is acid-free and falls within the guidelines
established to ensure permanence and durability. ♾

This publication was typeset using AMS-TEX,
the American Mathematical Society's TEX macro system.

10 9 8 7 6 5 4 3 2 1   98 97 96 95 94 93

Contents

Photographs courtesy of C. MacPherson

Foreword

September 2, 1993 is Israel Moiseevich Gelfand's 80th birthday. This date practically coincides with the 50th anniversary of the Gelfand Seminar in Functional Analysis at Moscow University. The present volume consists of papers written by "young" participants of this seminar. One of the reasons for introducing an age limit was to keep the volume's size within reasonable bounds, another was Gelfand's constant orientation to the younger participants of his seminar. This collection is intended to be a surprise to the man whose birthday we are celebrating, and I hope that he will learn of the book's existence only after its publication. However, we have tried to imitate Gelfand's own preferences as much as possible. All of the invited authors were participants and welcome speakers at the seminar; if one imagines its Golden Jubilee Session, it may be safely conjectured that these mathematicians would have been invited to participate. I hope that Gelfand will approve our choice and will enjoy seeing articles written by these remarkable mathematicians. The invited authors were free to choose their topics and, if they so desired, their coauthors.

The problem of defining the notion of "young" scientist, in particular, of "young mathematician", is one of the most difficult unsolved problems, and is also of the utmost importance for the applications. In the given case its solution was formalized in the following way. For the upper bound of the age of an invited author to this collection, we chose the age of Serezha, Gelfand's older son, who incidentally played a key role in the appearance of this book. As paradoxal as this may sound, there are serious grounds for this choice. When Serezha first appeared at the seminar in 1961, significant changes in it took place. The orientation to the younger participants that had always been important, but concerned only Ph.D.-track students (occasionally younger graduate students) until then, was now drastically amended to include freshmen and sophomores, and later even some high school students. Among the students of this "first draft" one should note Dima Kazhdan and, somewhat later, Ossya Bernstein. Both became important participants in the seminar for many years. Progressively the other authors of this volume also appeared there. By 1962 the seminar left the relatively small auditorium 13-11 to move to the roomier 14-08. By then no less than half the participants were undergraduates. At that time Gelfand liked to repeat that the seminar was open to all students of the lower courses and to the most talented professors.

I think that it is only natural to include in this foreword some information about the seminar itself. I understand that it was well known in the West. Usually mathematical voyagers from Western Europe or the US felt compelled to visit it (to do that, they had to overcome the vigilance of the Moscow University guards, by no means an easy task even for Muscovites). It is difficult to explain to the Western reader what the seminar meant to "Soviet mathematical life". Surprisingly, that life, in many respects, was not at all so bad, despite the almost unwavering antisemitism and the constantly increasing control by the mathematical rabble with communist party background over the key positions in mathematics. Mathematics was an Oasis of sorts, very attractive to young people with strong interest in science and without career aspirations in the communist hierarchy. If one were lucky, you were able to live an intense intellectual life and write articles free of any references to the classics of Marxism-Leninism. Fortunately for mathematics, Stalin did not find time for the subject (unlike economics, biology, and linguistics).

The Gelfand seminar was always an important event in the very vivid mathematical life in Moscow, and, doubtless, one of its leading centers. A considerable number of the best Moscow mathematicians participated in it at one time or another. Mathematicians from other cities used all possible pretexts to visit it. I recall how a group of Leningrad students agreed to take turns to come to Moscow on Mondays (the day of the seminar, to which other events were linked), and then would retell their friends what they heard there. There were several excellent and very popular seminars in Moscow, but nevertheless the Gelfand seminar was always an event.

I would like to point out that, on the other hand, the seminar was very important in Gelfand's own personal mathematical life. Many of us witnessed how strongly his activities were focused on the seminar. When, in the early fifties, at the peak of antisemitism, Gelfand was chased out of Moscow University, he applied all his efforts to save the seminar. The absence of Gelfand at the seminar, even because of illness, was always something out of the ordinary.

One cannot avoid mentioning that the general attitute to the seminar was far from unanimous. Criticism mainly concerned its style, which was rather unusual for a scientific seminar. It was a kind of a theater with a unique stage director, simultaneously playing the leading role in the performance and organizing the supporting cast, most of whom had the highest qualifications. I use this metaphor with the utmost seriousness, without any intension to mean that the seminar was some sort of a spectacle. Gelfand had chosen the hardest and most dangerous genre: to demonstrate in public how he understood mathematics. It was an open lesson in the grasping of mathematics by one of the most amazing mathematicians of our time. This role could only be played under the most favorable conditions: the genre dictates the rules of the game, which are not always very convenient for the listeners. This means, for example, that the leader follows only his own intuition in the final choice of the topics of the talks, interrupts them with comments and questions (a priviledge not granted to other participants), organizes their "understanding", mainly by the younger participants (parts of the reports are repeated, participants are summoned to the blackboard). All this is done with extraordinary generosity, a true passion for mathematics.

Let me recall some of the stage director's stratagems. An important feature were improvisations of various kinds. The course of the seminar could change dramatically at any moment. Another important *mise en scène* involved the "trial listener" game, in which one of the participants (this could be a student as well as a professor) was instructed to keep informing the seminar of his understanding of the talk, and whenever the information was negative, that part of the report would be repeated. A well-qualified trial listener could usually feel when the head of the seminar wanted an occasion for such a repetition. Also, Gelfand himself had the faculty of being "unable to understand" in situations when everyone around was sure that everything is clear. What extraordinary vistas were opened to the listeners, and sometimes even to the mathematician giving the talk, by this ability not to understand. Gelfand liked that old story of the professor complaining about his students: "Fantastically stupid students—five times I repeat proof, already I understand it myself, and still they don't get it."

It has remained beyond my understanding how Gelfand could manage all that physically for so many hours. Formally the seminar was supposed to begin at 6 P.M., but usually started with about an hour's delay. I am convinced that the free conversations before the actual beginning of the seminar were part of the scenario. The seminar would continue without any break until 10 or 10:30 (I have heard that before my time it was even later). The end of the seminar was in constant conflict with the rules and regulations of Moscow State University. Usually at 10 P.M. the cleaning woman would make her appearance, trying to close the proceedings to do her job. After the seminar, people wishing to talk to Gelfand would hang around. The elevator would be turned off, and one would have to find the right staircase, so as not to find oneself stuck in front of a locked door on the ground floor, which meant walking back up to the 14th (where else but in Russia is the locking of doors so popular!). The next riddle was to find the only open exit from the building. Then the last problem (of different levels of difficulty for different participants)—how to get home on public transportation, at that time in the process of closing up. Seeing Gelfand home, the last mathematical conversations would conclude the seminar's ritual. Moscow at night was still safe and life seemed so unbelievably beautiful!

The Gelfand seminar exerted a great influence on many mathematicians, not only on his direct pupils. The seminar for all of us was a unique opportunity to witness Gelfand's informal understanding of mathematics, to learn the usually hidden mysteries of the craft. We are all glad that Israel Moiseevich approaches his 80th anniversary in fine health and spirits, has preserved his workaholic ways, and, as always, is bristling with new plans and ideas, on the seventh decade of his service to Mathematics.

Simon Gindikin

ADVANCES IN SOVIET MATHEMATICS
Volume 16, Part 1, 1993

A Proof of Jantzen Conjectures

A. BEILINSON AND J. BERNSTEIN

To our teacher Israel M. Gelfand

In this paper we will show that the localization functor sends the Jantzen filtration on Verma modules (or, more generally, standard Harish-Chandra modules) to the weight filtration on the corresponding perverse sheaves. This fact immediately implies a lot of remarkable properties of the Jantzen filtration: the hereditary property (conjectured by Jantzen), the socle and cosocle properties, and the Kazhdan-Lusztig algorithm for the computation of multiplicities in consecutive quotients (conjectured in [GJ1] and [GM]).

The paper is divided into two parts. The first part occupies Sections 1 and 2. In Section 1 we consider sheaves of noncommutative algebras on an algebraic variety which are of "local origin" (we call them D-algebras); a typical example is the algebra of differential operators. We list some basic functorial properties of such algebras. Section 2 deals with an important class of D-algebras: rings of twisted differential operators (tdo). We included in Section 2 more material than the minimum needed for the main part of our paper, since the language of tdo's is a convenient gadget in many situations (e.g., in algebraic versions of conformal field theory), and we thought it would be nice to have a review of the subject.

The second part deals with representation theory. It is the heart of this paper. In Section 3 we recall the localization construction [BB1] and write down some properties of K-orbits of flag varieties that will be of use. In Section 4 we define the Jantzen filtration in a geometric setting and describe its intimate relation with the monodromy filtration on vanishing cycles. In Section 5 mixed sheaves appear. We present a proof of Gabber's theorem about the weight filtration on vanishing cycles; since it is transmitted to representation theory, it provides, together with the constructions of Section 4, the Jantzen conjecture.

The main results of this paper were proved in the spring of 1981. The first draft of this paper, which followed notes of a spring 1982 seminar at

1991 *Mathematics Subject Classification*. Primary 22E45; Secondary 14F32.

Moscow University, appeared in 1986. The second part of the present paper is an abridged version of this draft (we just added a few recent references); therefore it is a bit archaic in style (we use l-adic mixed sheaves instead of mixed Hodge sheaves of Saito, etc.). We thank Robert Becker for carefully typing this manuscript.

In what follows (except for Section 5) "variety" = "scheme" = "separated scheme of finite type over $\mathbb{C}$", and "algebra" = "associative $\mathbb{C}$-algebra with 1". In fact, one can replace $\mathbb{C}$ by any field of characteristic 0 in any place that has nothing to do with the Riemann-Hilbert correspondence (we deal with arbitrary schemes in 1.1).

If $\pi: X \to Y$ is a morphism of varieties then π_*, π^{-1} denote the sheaf theoretic direct and inverse image functors. If A is an algebra then $\mathcal{M}(A)$ denotes the category of left A-modules, and ${}^r\mathcal{M}(A)$ denotes the category of right A-modules. A subcategory $\mathcal{B}$ of an abelian category $\mathcal{A}$ is a Serre subcategory if $\mathcal{B}$ is a strictly full subcategory closed under extensions and subquotients; in such a case we have the quotient abelian category $\mathcal{A}/\mathcal{B}$.

If M is a sheaf and m is a local section of M, we write $m \in M$.

§1. *D*-calculus

1.1. ***D*-algebras.** Let R be a commutative algebra and M be an R-bimodule. For $r \in R$ we have the endomorphism $\operatorname{ad} r$ of M, $\operatorname{ad} r(m) = rm - mr$. An increasing filtration $M_.$ on M is called a *D-filtration* if $M_{-1} = 0$ and $\operatorname{ad} r(M_i) \subset M_{i-1}$ for any i and $r \in R$ (i.e., on $\operatorname{gr}_. M$ the left and right R-module structures coincide). For example, one has the D-filtration $M_.^\vee$ defined by induction: $M_i^\vee = \{m \in M: \operatorname{ad} r(m) \in M_{i-1} \text{ for any } r \in R\}$ for $i \geqslant 0$; this D-filtration is maximal in an obvious sense. We call $M^\vee = \bigcup M_i^\vee$ the *differential part* of M; our M is a *differential bimodule* if $M^\vee = M_.$.

Another way to spell this out is to consider M as an $(R \otimes R)$-module; then $M_i^\vee = \{m \in M: I^{i+1} m = 0\}$, where I is the kernel of the multiplication map $R \otimes R \to R$ (i.e., the ideal of functions vanishing on the diagonal $\operatorname{Spec} R \hookrightarrow \operatorname{Spec} R \times \operatorname{Spec} R$). This description shows that our objects localize nicely: if $f \in R$, then for the localized R_f-bimodule $M_f := R_f \underset{R}{\otimes} M \underset{R}{\otimes} R_f$ one has $(M_f)_i^\vee = (M_i^\vee)_f = R_f \underset{R}{\otimes} M_i^\vee = M_i^\vee \underset{R}{\otimes} R_f$.

1.1.1. Remark. Let $M_.$ be a D-filtration on M. We have a canonical morphism of R-modules $\delta = \delta_i: \operatorname{gr}_i M \to \operatorname{Hom}_R(\Omega^1 R, \operatorname{gr}_{i-1} M)$, defined by $\delta(\overline{m})(adb) = (abm - amb) \bmod M_{i-2}$, where $a, b \in R$, $m \in M_i$, and $\overline{m}$ is the image of m in $\operatorname{gr}_i M$.

Using these morphisms we construct a complex

$$\operatorname{gr}_i M \to \operatorname{Hom}_R(\Omega^1 R, \operatorname{gr}_{i-1} M) \to \cdots \to \operatorname{Hom}_R(\Omega^j R, \operatorname{gr}_{i-j} M) \to \cdots$$

where the differential δ^i: $\operatorname{Hom}_R(\Omega^j R, \operatorname{gr}_{i-j} M) \to \operatorname{Hom}_R(\Omega^{j+1} R, \operatorname{gr}_{i-j-1} M)$

is given by

$$\delta^i(\varphi)(a_0 da_1 \wedge \cdots \wedge da_{j+1}) = \sum (-1)^l \delta_{i-j}(\varphi(a_0 da_1 \wedge \cdots \wedge \widehat{da_l} \wedge \cdots \wedge da_{j+1}))(da_l).$$

Note that $M_.$ is maximal iff $\bigcup M_i = M^\vee$ and each map δ_i is injective for $i > 0$.

1.1.2. Let A be an associative algebra equipped with a morphism of algebras $i: R \to A$. An increasing filtration $A_.$ on A is called a *D-ring filtration* if it is a ring filtration (i.e., $A_i A_j \subset A_{i+j}$), $A_{-1} = 0$, $i(R) \subset A_0$, and $i(R)$ lies in the center of the associated graded algebra $\mathrm{gr}_. A$. We can consider A as an R-bimodule; then such $A_.$ is a D-filtration. Note that $A_.^\vee$ is a D-ring filtration, so it is a maximal D-ring filtration. We will say that A is an *R-differential algebra* if $A = A^\vee$ (i.e., if A is a differential R-bimodule).

1.1.3. The above definitions easily globalize. Namely, let X be a scheme. A *differential $\mathscr{O}_X$-bimodule* M is a quasicoherent sheaf on $X \times X$ supported on the diagonal $X \subset X \times X$. We will consider M as a sheaf of $\mathscr{O}_X$-bimodules on X. It has the following properties:

(i) For any open $U \subset X$, $M(U)$ is a differential $\mathscr{O}(U)$-bimodule.

(ii) If U is affine, $U = \operatorname{Spec} R$, and $f \in R$, then $M(U_f) = M(U)_f$.

Conversely, any sheaf of $\mathscr{O}_X$-bimodules M with properties (i) and (ii) is a differential $\mathscr{O}_X$-bimodule.

Differential $\mathscr{O}_X$-bimodules are Zariski local objects: they form a stack on the Zariski topology of X. If X is affine, $X = \operatorname{Spec} R$, then differential $\mathscr{O}_X$-bimodules are the same as differential R-bimodules. Note that if M, N are differential $\mathscr{O}_X$-bimodules, then so is $M \underset{\mathscr{O}_X}{\otimes} N$.

1.1.4. An *$\mathscr{O}_X$-differential algebra,* or simply a *D-algebra* on X, is a sheaf of associative algebras on the Zariski topology of X equipped with a morphism of algebras $i: \mathscr{O}_X \to \mathscr{A}$ such that $\mathscr{A}$ is a differential $\mathscr{O}_X$-bimodule. One defines morphisms of D-algebras in an obvious manner. The D-algebras form a stack on the Zariski topology of X. If X is affine, $X = \operatorname{Spec} R$, then $\mathscr{O}_X$-differential algebras are the same as R-differential algebras.

1.1.5. For a D-algebra $\mathscr{A}$ on X an $\mathscr{A}$-module M is, by definition, a sheaf of $\mathscr{A}$-modules which is quasicoherent as an $\mathscr{O}_X$-module. Usually we will use left $\mathscr{A}$-modules, and call them simply $\mathscr{A}$-modules. They form an abelian category $\mathscr{M}(X, \mathscr{A}) = \mathscr{M}(\mathscr{A})$. The category of right $\mathscr{A}$-modules will be denoted ${}^r\mathscr{M}(\mathscr{A})$.

If X is affine and $A := \mathscr{A}(X)$, then $\mathscr{A}$-modules are the same as A-modules, since one has canonical equivalence of categories $\mathscr{M}(\mathscr{A}) = \mathscr{M}(A)$.

The $\mathscr{A}$-modules are local objects: if $j: U \hookrightarrow X$ is an open embedding then $\mathscr{A}_U$ is an $\mathscr{O}_U$-differential algebra and we have the obvious adjoint functors $\mathscr{M}(\mathscr{A}) \underset{j_*}{\overset{j^*}{\rightleftarrows}} \mathscr{M}(\mathscr{A}_U)$; the categories $\mathscr{M}(\mathscr{A}_U)$ form a stack over the Zariski topology of X.

If N is any (quasicoherent) $\mathcal{O}_X$-module then an $\mathcal{A}$*-action* on N is a structure of $\mathcal{A}$-modules on N compatible with the $\mathcal{O}$-module structure (i.e., $i(f)m = fm$ for $f \in \mathcal{O}_X$, $m \in N$).

1.1.6. First examples. (i) Let M, N be (quasicoherent) $\mathcal{O}_X$-modules. A $\mathbb{C}$-linear morphism $f: M \to N$ is called a *differential operator* if for any affine $U \subset X$ the corresponding morphism $f_U: M(U) \to N(U)$ lies in the differential part of the $\mathcal{O}(U)$-bimodule $\mathrm{Hom}_{\mathbb{C}}(M(U), N(U))$.

The differential operators form a sheaf of $\mathcal{O}_X$-bimodules $\mathrm{Diff}(M, N) \subset \mathrm{Hom}_{\mathbb{C}}(M, N)$. If M is coherent then $\mathrm{Diff}(M, N)$ is a differential $\mathcal{O}_X$-bimodule. In particular, for a coherent sheaf M we have an $\mathcal{O}_X$-differential algebra $D_M := \mathrm{Diff}(M, M)$. We put $D_X := D_{\mathcal{O}_X}$.

(ii) If $\mathcal{A}$, B are D-algebras on schemes X, Y respectively, then we have a D-algebra $\mathcal{A} \boxtimes B$ on $X \times Y$ such that for affine $U \subset X$, $V \subset Y$ one has $(\mathcal{A} \boxtimes B)(U \times V) = \mathcal{A}(U) \underset{\mathbb{C}}{\otimes} B(V)$.

1.2. Lie algebroids. For a scheme X a *Lie algebroid* L on X is a (quasi-coherent) $\mathcal{O}_X$-module equipped with a morphism of $\mathcal{O}_X$-modules $\sigma: L \to \mathcal{T}_X$ ($:= \mathrm{Der}\,\mathcal{O}_X =$ tangent sheaf of X) and a $\mathbb{C}$-linear pairing $[\cdot, \cdot]: L \underset{\mathbb{C}}{\otimes} L \to L$ such that

- $[\cdot, \cdot]$ is a Lie algebra bracket and σ commutes with brackets,
- for $l_1, l_2 \in L$, $f \in \mathcal{O}_X$ one has $[l_1, fl_2] = f[l_1, l_2] + \sigma(l_1)(f)l_2$.

For a Lie algebroid L we set $L^{(0)} := \mathrm{Ker}\,\sigma$. This is an $\mathcal{O}_X$-Lie algebra.

A Lie algebroid is called *smooth* if it is a locally free $\mathcal{O}$-module of finite rank. Lie algebroids form a category in an obvious way; this is a stack on the Zariski topology of X.

A *connection* on a Lie algebroid L is an $\mathcal{O}_X$-linear section $\nabla: \mathcal{T}_X \to L$ of σ (so $\sigma\nabla = \mathrm{id}_{\mathcal{T}_X}$). Such a ∇ is integrable if it commutes with brackets. For a connection ∇, its *curvature* $C(\nabla) \in \mathrm{Hom}_{\mathcal{O}_X}(\bigwedge^2 \mathcal{T}_X, L^{(0)})$ is defined by $C(\nabla)(\tau_1 \wedge \tau_2) = [\nabla(\tau_1), \nabla(\tau_2)]$. The connections on L form a $\mathrm{Hom}_{\mathcal{O}_X}(\mathcal{T}_X, L^{(0)})$-torsor $\mathcal{C}(L)$.

Let us describe some examples of Lie algebroids.

1.2.1. The tangent sheaf $\mathcal{T}_X$ is a Lie algebroid (with $\sigma = \mathrm{id}_{\mathcal{T}_X}$). For any Lie algebroid L there is a unique morphism $L \to \mathcal{T}_X$ and a morphism $\mathcal{T}_X \to L$ is the same as an integrable connection on L.

1.2.2. Assume that a Lie algebra $\mathfrak{g}$ acts on X, i.e., we have a morphism of Lie algebras $\alpha: \mathfrak{g} \to \mathcal{T}_X$. Then $\tilde{\mathfrak{g}}_X = \mathcal{O}_X \underset{\mathbb{C}}{\otimes} \mathfrak{g}$ becomes a Lie algebroid in a natural way: the map $\sigma: \mathcal{O}_X \underset{\mathbb{C}}{\otimes} \mathfrak{g} \to \mathcal{T}_X$ is $\sigma(f \otimes \gamma) = f\alpha(\gamma)$ and the bracket on $\mathcal{O}_X \underset{\mathbb{C}}{\otimes} \mathfrak{g}$ is defined by $[f_1 \otimes \gamma_1, f_2 \otimes \gamma_2] = f_1 f_2 \otimes [\gamma_1, \gamma_2] + f_1\alpha(\gamma_1)(f_2) \otimes \gamma_2 - f_2\alpha(\gamma_2)(f_1) \otimes \gamma_1$.

1.2.3. Let G be an algebraic group, $\mathfrak{g} = \mathrm{Lie}\,G$, and F be a G-torsor over X. We have the Lie algebroid $\widetilde{\mathcal{T}_F}$ of infinitesimal symmetries of (X, F):

a section of $\widetilde{\mathcal{T}_F}$ is a pair $(\tau, \widetilde{\tau})$, where τ is a vector field on X and $\widetilde{\tau}$ is a G-invariant vector field on F that lifts τ. We have $\sigma(\tau, \widetilde{\tau}) = \tau$, $[(\tau_1, \widetilde{\tau}_1), (\tau_2, \widetilde{\tau}_2)] = ([\tau_1, \tau_2], [\widetilde{\tau}_1, \widetilde{\tau}_2])$. Note that $\mathfrak{g}_F = \widetilde{\mathcal{T}}_F^{(0)}$ coincides with the F-twist of $\mathcal{O}_X \otimes \mathfrak{g}$ (with the usual $\mathcal{O}_X$-linear bracket) with respect to the adjoint action of G.

One can consider Lie algebroids of infinitesimal symmetries of any geometric object over X of a local nature. For example, for a vector bundle $\mathcal{E}$, we have a Lie algebroid $\widetilde{\mathcal{T}_{\mathcal{E}}}$, which coincides with the Lie algebroid of the corresponding GL-torsor.

1.2.4. Assume we have a smooth groupoid acting on X, i.e., we have a scheme Y equipped with two smooth projections $\pi_1, \pi_2 \colon Y \to X$, an embedding $e \colon X \hookrightarrow Y$, and a composition law $Y \underset{\pi_2, \pi_1}{\times} Y \to Y$ that satisfies the usual axioms (see, e.g., [D4]). The Lie algebroid L of our groupoid is the normal bundle for the embedding e; one defines the Lie bracket and projection σ by the usual formulas. If our groupoid is an ordinary group G acting on X (so $Y = G \times X$), then L coincides with the Lie algebroid from example 1.2.2 for $\mathfrak{g} = \operatorname{Lie} G$, α being the corresponding infinitesimal action of $\mathfrak{g}$.

1.2.5. Let $\mathcal{A}$ be a D-algebra on X. Put $\operatorname{Lie}\mathcal{A} := \{(\tau, a) \in \mathcal{T}_X \times \mathcal{A} : i\tau(f) = ai(f) - i(f)a \text{ for any } f \in \mathcal{O}_X\}$. This is a Lie algebroid on X in an obvious manner (one has $\sigma(\tau, a) = \tau$ and $[(\tau_1, a_1), (\tau_2, a_2)] = ([\tau_1, \tau_2], [a_1, a_2])$); we will call $\operatorname{Lie}\mathcal{A}$ the *Lie algebroid of* $\mathcal{A}$.

Clearly $\mathcal{A} \rightsquigarrow \operatorname{Lie}\mathcal{A}$ is a functor from the category of D-algebras to the category of Lie algebroids. It has a left adjoint functor that assigns to a Lie algebroid L its *universal enveloping D-algebra* $\mathcal{U}(L)$. Explicitly, $\mathcal{U}(L)$ is a sheaf of algebras equipped with the morphisms of sheaves $i \colon \mathcal{O}_X \to \mathcal{U}(L)$, $i_L \colon L \to \mathcal{U}(L)$; it is generated, as an algebra, by the images of these morphisms and the only relations are

(i) i is a morphism of algebras;
(ii) i_L is a morphism of Lie algebras;
(iii) for $f \in \mathcal{O}_X$, $l \in L$ one has $i_L(fl) = i(f)i_L(l)$, $[i_L(l), i(f)] = i(\sigma(l)f)$.

One checks easily that $\mathcal{U}(L)$ is actually a D-algebra.

Note that a $\mathcal{U}(L)$-module is the same as an $\mathcal{O}$-module with an L-action (i.e., an $\mathcal{O}_X$-module M equipped with a Lie algebra map $L \to \operatorname{End}_{\mathbb{C}} M$ such that $l(fm) = \sigma(l)(f)m + (fl)m$, $(fl)m = f(lm)$ for $f \in \mathcal{O}_X$, $l \in L$, $m \in M$). We will call $\mathcal{U}(L)$-modules simply L-modules.

1.3. Étale localization. All the above definitions are actually étale local. To be precise, let $\varphi \colon Y \to X$ be an étale morphism. Then the Y-diagonal $Y \hookrightarrow Y \times Y$ is a component (i.e., an open and closed subscheme) of the preimage of the X-diagonal $(\varphi \times \varphi)^{-1}(X) \hookrightarrow Y \times Y$. For a differential $\mathcal{O}_X$-bimodule M (which is an $\mathcal{O}_{X \times X}$-module supported on the diagonal) we define its pull-back

$M_Y = \varphi^* M$ as the restriction of $(\varphi \times \varphi)^* M = \mathscr{O}_{Y\times Y} \underset{(\varphi\times\varphi)^{-1}\mathscr{O}_{X\times X}}{\otimes} (\varphi \times \varphi)^{-1} M$ to the Y-diagonal. This is a differential $\mathscr{O}_Y$-bimodule.

We have a canonical embedding $\varphi^{-1}M \hookrightarrow \varphi^* M$ of $\varphi^{-1}\mathscr{O}_X$-bimodules that induces isomorphisms $\mathscr{O}_Y \underset{\varphi^{-1}\mathscr{O}_X}{\otimes} \varphi^{-1}M \xrightarrow{\sim} \varphi^* M \xleftarrow{\sim} \varphi^{-1}M \underset{\varphi^{-1}\mathscr{O}_X}{\otimes} \mathscr{O}_Y$. Therefore, the differential $\mathscr{O}_X$-bimodules are sheaves on the étale topology $X_{\text{ét}}$ of X; they form a stack on $X_{\text{ét}}$.

If M, N are differential $\mathscr{O}_X$-bimodules then $\varphi^*(M \underset{\mathscr{O}_X}{\otimes} N) = \varphi^* M \underset{\mathscr{O}_Y}{\otimes} \varphi^* N$. Therefore, if $\mathscr{A}$ is an $\mathscr{O}_X$-differential algebra, then $\mathscr{A}_Y = \varphi^*\mathscr{A}$ is an $\mathscr{O}_Y$-differential algebra (the product is the composition $\varphi^*\mathscr{A} \underset{\mathscr{O}_Y}{\otimes} \varphi^*\mathscr{A} \xrightarrow{\sim} \varphi^*(\mathscr{A}\otimes\mathscr{A}) \xrightarrow{\varphi^*(\cdot)} \varphi^*\mathscr{A}$); the embedding $\varphi^{-1}\mathscr{A} \hookrightarrow \varphi^*\mathscr{A}$ is a ring morphism.

We see that the D-algebras are also sheaves on $X_{\text{ét}}$, and they form a stack over $X_{\text{ét}}$. If M is an $\mathscr{A}$-module then $M_Y = \varphi^* M := \mathscr{O}_Y \underset{\varphi^{-1}\mathscr{O}_X}{\otimes} \varphi^{-1}M$ is an $\mathscr{A}_Y$-module. So we can consider M as a sheaf of $\mathscr{A}$-modules on $X_{\text{ét}}$; the categories $\mathscr{M}(\mathscr{A}_Y)$ form a stack over $X_{\text{ét}}$.

1.4. Functoriality. Let $\varphi\colon Y \to X$ be a morphism of schemes, and $\mathscr{A}$, $\mathscr{B}$ be D-algebras on X, Y respectively. Consider the sheaf $\varphi^*\mathscr{A} = \mathscr{O}_Y \underset{\varphi^{-1}\mathscr{O}_X}{\otimes} \varphi^{-1}\mathscr{A}$. This is a (quasicoherent) $\mathscr{O}_Y$-module equipped with a right $\varphi^{-1}\mathscr{A}$-action.

1.4.1. DEFINITION. A *φ-morphism* $\lambda\colon \mathscr{B} \to \mathscr{A}$ is an action of D-algebra $\mathscr{B}$ on $\varphi^*\mathscr{A}$ that commutes with the right $\mathscr{A}$-action.

In other words, it is a morphism of algebras $\lambda\colon \mathscr{B} \to \operatorname{End}_{\varphi^{-1}\mathscr{A}}(\varphi^*\mathscr{A})$ such that $\lambda(i(f))a = fa$ for $f \in \mathscr{O}_Y$, $a \in \varphi^*\mathscr{A}$.

REMARK. If φ is the identity morphism, then $\varphi^*\mathscr{A} = \mathscr{A}$, and λ is a usual morphism of D-algebras $\mathscr{B} \to \mathscr{A}$.

Note that a φ-morphism λ defines for any $\mathscr{A}$-module M a $\mathscr{B}$-action on $\varphi^* M := \mathscr{O}_Y \underset{\varphi^{-1}\mathscr{O}_X}{\otimes} \varphi^{-1}M$: one has $\varphi^* M = \varphi^*\mathscr{A} \underset{\varphi^{-1}\mathscr{A}}{\otimes} \varphi^{-1}M$, and we put $b(a \otimes m) := \lambda(b)a \otimes m$ for $b \in \mathscr{B}$, $a \in \varphi^*\mathscr{A}$, $m \in \varphi^{-1}M$. We get a canonical functor $(\varphi, \lambda)^*\colon \mathscr{M}(\mathscr{A}) \to \mathscr{M}(\mathscr{B})$.

If $\psi\colon Z \to Y$ is another morphism of schemes, $\mathscr{C}$ is a D-algebra on Z, and $\mu\colon \mathscr{C} \to \mathscr{B}$ is a ψ-morphism, we define the composition $\lambda\cdot\mu\colon \mathscr{C} \to \mathscr{A}$ as the $(\varphi\cdot\psi)$-morphism which is the action of $\mathscr{C}$ on $(\varphi\cdot\psi)^*\mathscr{A} = \psi^*(\varphi^*\mathscr{A})$ considered as $(\psi, \mu)^*(\varphi, \lambda)^*\mathscr{A}$. The functors $(\varphi\cdot\psi, \lambda\cdot\mu)^*$, $(\psi, \mu)^* \cdot (\varphi, \lambda)^*\colon \mathscr{M}(\mathscr{A}) \to \mathscr{M}(\mathscr{C})$ coincide.

1.4.2. A *D-scheme*, or *D-variety*, is a pair $(X, \mathscr{A})$, where X is a scheme, and $\mathscr{A}$ is a D-algebra on X. A morphism of D-schemes $(Y, \mathscr{B}) \to (X, \mathscr{A})$ is a pair (φ, λ) as above. We see that D-schemes form a category which we denote D-Sch; one has a canonical projection $D\text{-Sch} \to \text{Sch}$, $(X, \mathscr{A}) \mapsto X$, whose fiber over X is the category of D-algebras on X.

In fact this projection makes D-Sch a prefibered category over Sch (see [SGA1] for the terminology). This means that for any morphism of schemes $\varphi: Y \to X$ and any D-algebra $\mathscr{A}$ on X, there is a (canonical) D-algebra $\varphi^{\cdot}\mathscr{A}$ on Y equipped with φ-morphisms $\varphi^{\cdot}\mathscr{A} \to \mathscr{A}$ such that for a D-algebra $\mathscr{B}$ on Y the φ-morphisms $\mathscr{B} \to \mathscr{A}$ are the same as morphisms $\mathscr{B} \to \varphi^{\cdot}\mathscr{A}$. Indeed, $\varphi^{\cdot}\mathscr{A}$ is the algebra $\mathrm{Diff}_{\varphi^{-1}(\mathscr{A})}(\varphi^{*}\mathscr{A}, \varphi^{*}\mathscr{A})$ of all $(\mathscr{O}_Y)$-differential operators on $\varphi^{*}\mathscr{A}$ that commute with the right $\varphi^{-1}\mathscr{A}$-action. The universality property implies that $\varphi^{\cdot}$ is a functor from the category of D-algebras on X to that on Y. It is compatible with Zariski localization; if $X = \operatorname{Spec} R$, $Y = \operatorname{Spec} S$ are affine schemes and $A = \Gamma(X, \mathscr{A})$ then $\Gamma(Y, \varphi^{\cdot}\mathscr{A})$ is the ring of all S-differential operators acting on $S \underset{R}{\otimes} A$ that commute with the right A-action.

1.4.3. Remark. For a differential bimodule M on X we define its *pull-back* $\varphi^{\cdot}M$ as $\mathrm{Diff}_{\varphi^{-1}\mathscr{O}_X}(\mathscr{O}_Y, \varphi^{*}M) = \{l \in \mathrm{Diff}(\mathscr{O}_Y, \mathscr{O}_Y \underset{\varphi^{-1}\mathscr{O}_X}{\otimes} \varphi^{-1}M) : l(f\varphi^{*}(g)) = l(f)g \text{ for } f \in \mathscr{O}_Y, g \in \mathscr{O}_X\}$; this is a differential $\mathscr{O}_Y$-bimodule. This notation is compatible with the earlier one: if $\mathscr{A}$ is a D-algebra on X then we have an isomorphism of $\mathscr{O}_Y$-bimodules

$$\mathrm{Diff}_{\varphi^{-1}\mathscr{A}}(\varphi^{*}\mathscr{A}, \varphi^{*}\mathscr{A}) \xrightarrow{\sim} \mathrm{Diff}_{\varphi^{-1}\mathscr{O}_X}(\varphi^{*}\mathscr{O}_Y, \varphi^{*}\mathscr{A}), \; n \mapsto n \circ \varphi^{*}(i).$$

Note that the functor $\varphi^{\cdot}$ is, in general, neither left nor right exact (being the composition of the right exact functor φ^{*} and the left exact one $\mathrm{Diff}_{\varphi^{-1}\mathscr{O}_X}(\mathscr{O}_Y, \cdot))$, and it behaves badly with respect to composition of φ's. To recover the φ-functoriality one should, perhaps, work with the derived categories from the very beginning.

1.4.4. Let $Z \xrightarrow{\psi} Y$ be another morphism of schemes. For a D-algebra $\mathscr{A}$ on X the composition of the canonical ψ- and φ-morphisms $\psi^{\cdot}\varphi^{\cdot}\mathscr{A} \to \varphi^{\cdot}\mathscr{A} \to \mathscr{A}$ defines, by the universality property, a canonical morphism $c_{\varphi,\psi}: \psi^{\cdot}\varphi^{\cdot}\mathscr{A} \to (\varphi\psi)^{\cdot}\mathscr{A}$. If $\chi: W \to Z$ is the third morphism of schemes then the compositions $c_{\varphi\psi,\chi}\chi^{\cdot}(c_{\varphi,\psi})$, $c_{\varphi,\psi\chi}c_{\psi,\chi}: \chi^{\cdot}\psi^{\cdot}\varphi^{\cdot}\mathscr{A} \to (\varphi\psi\chi)^{\cdot}\mathscr{A}$ coincide. In general, $c_{\varphi,\psi}$ are not isomorphisms (a possible remedy would be to change the definition of $\varphi^{\cdot}$ so that the right $\varphi^{\cdot}$ would send the D-algebras on X to the differential graded D-algebras on Y; we will not pursue this line further).

1.4.5. In certain cases one can describe $\varphi^{\cdot}$ quite explicitly.

(i) φ is étale. Then, by 1.3, $\varphi^{*}\mathscr{A}$ is an $\mathscr{O}_Y$-differential algebra. One has a canonical isomorphism $\varphi^{*}\mathscr{A} \xrightarrow{\sim} \varphi^{\cdot}\mathscr{A}$ of D-algebras that assigns to $\alpha \in \varphi^{*}\mathscr{A}$ the operator of the left multiplication by α on $\varphi^{*}\mathscr{A}$. The inverse map $\varphi^{\cdot}\mathscr{A} \to \varphi^{*}\mathscr{A}$ is $l \mapsto l(1 \otimes 1)$.

(ii) φ is a projection $Y = T \times X \to X$. Then $\varphi^*\mathscr{A} = \mathscr{O}_T \boxtimes \mathscr{A}$, and one has an isomorphism of D-algebras $D_T \boxtimes \mathscr{A} \xrightarrow{\sim} \varphi^{\cdot}\mathscr{A}$ that sends $\partial \otimes a \in D_T \boxtimes \mathscr{A}$ to the operator $t \otimes b \mapsto \partial(t) \otimes ab$.

(iii) $\varphi\colon Y \hookrightarrow X$ is a closed embedding defined by an ideal $I \subset \mathscr{O}_X$. Then $\varphi^*\mathscr{A} = \mathscr{A}/I\mathscr{A}$. Let $N(I\mathscr{A}) := \{\alpha \in \mathscr{A} : \alpha I \subset I\mathscr{A}\}$ be the normalizer of the ideal $I\mathscr{A}$ in the algebra $\mathscr{A}$. One has an isomorphism of D-algebras $N(I\mathscr{A})/I\mathscr{A} \xrightarrow{\sim} \varphi^{\cdot}\mathscr{A}$ that sends $\alpha \in N(I\mathscr{A})$ to the operator of left multiplication by α on $\mathscr{A}/I\mathscr{A}$; the inverse map is $\gamma \mapsto \gamma(1 \otimes 1)$.

1.4.6. Let $\varphi\colon Y \to X$ be a morphism of schemes, and L, N be Lie algebroids on X, Y respectively. A *φ-morphism* $\gamma\colon N \to L$ is an $\mathscr{O}_Y$-linear map $\gamma\colon N \to \varphi^*L$ which satisfies the following conditions. Take $n \in N$; write $\gamma(n) = \sum f^i \otimes l^i$, $f^i \in \mathscr{O}_Y$, $l^i \in L$. Then one requires that $\gamma([n_1, n_2]) = \sum f_1^i f_2^j \otimes [l_1^i, l_2^j] + \sigma(n_1)(f_2^j) \otimes l_2^j - \sigma(n_2)(f_1^i) \otimes l_1^i$, and $\sigma(n)(\varphi^* g) = \sum f^i \varphi^*(\sigma(l^i)(g))$ for $g \in \mathscr{O}_X$.

If $\psi\colon Z \to Y$ is another morphism of schemes, K is a Lie algebroid on Z, and $\delta\colon K \to N$ is a ψ-morphism, then the composition $\gamma\delta := \psi^*(\gamma)\delta$ is a $\varphi\psi$-morphism $K \to L$.

We will call a pair (X, L), where X is a scheme and L is a Lie algebroid on X, a *D-Lie scheme.* We see that D-Lie schemes form a category D-Lie (the morphisms are pairs φ, γ as above).

The projection $(X, L) \mapsto X$ makes D-Lie a prefibered category over the category of schemes (one easily constructs the pull-back functor $\varphi^{\cdot}$: (Lie algebroids on X) $\to$ (Lie algebroids on Y)).

If, in the above situation, P is an L-module on X, then φ^*P is naturally an N-module: the N-action is $n(f \otimes p) := \sigma(n)(f) \otimes p + \sum f f^i \otimes l^i p$, where $\gamma(n) = \sum f^i \otimes l^i$. Therefore, we have the pull-back functor $(\varphi, \gamma)^*$: (L-modules) $\to$ (N-modules), $(\varphi, \gamma)^*P = \varphi^*P$; clearly $(\varphi\psi, \gamma\delta)^* = (\psi, \delta)^*(\varphi, \gamma)^*$. In particular, $\varphi^*\mathscr{U}(L)$ is an N-module, hence γ defines a φ-morphism of D-algebras $\mathscr{U}(N) \to \mathscr{U}(L)$. Therefore, we have the universal enveloping algebra functor D-Lie $\to$ D-Sch, $(X, L) \mapsto (X, \mathscr{U}(L))$.

1.4.7. Let $(\varphi, \lambda)\colon (Y, \mathscr{B}) \to (X, \mathscr{A})$ be a morphism of D-schemes. It defines the *push-forward functor* between the derived categories of right modules $R(\varphi, \lambda)_*\colon D^{-r}\mathscr{M}(Y, \mathscr{B}) \to D^{-r}\mathscr{M}(X, \mathscr{A})$, $R(\varphi, \lambda)_*(P) := R\varphi_*(P \overset{L}{\underset{\mathscr{B}}{\otimes}} \varphi^*\mathscr{A})$. Here we consider $\varphi^*\mathscr{A}$ as a $(\mathscr{B}, \varphi^{-1}\mathscr{A})$-bimodule (so $P \underset{\mathscr{B}}{\otimes} \varphi^*\mathscr{A}$ is a right $\varphi^{-1}\mathscr{A}$-module) and φ_* is the sheaf theoretic direct image.

If $X = \operatorname{Spec} R$, $Y = \operatorname{Spec} S$ are affine schemes, $A = \Gamma(X, \mathscr{A})$, and $B = \Gamma(Y, \mathscr{B})$, then the functor $R(\varphi, \lambda)_*$ sends a right B-module P to the complex of A-modules $P \overset{L}{\underset{B}{\otimes}} (S \underset{R}{\otimes} A)$.

1.5. Smooth localization. Assume we are in the situation of 1.4.4.

1.5.1. LEMMA. *If either* φ *or* ψ *is a smooth morphism of schemes, then* $c_{\varphi,\psi}$ *is an isomorphism.*

PROOF. The statement is Zariski local. We assume that $X = \operatorname{Spec} R$ is affine; put $A = \Gamma(X, \mathscr{A})$. Since locally any smooth morphism $U \to V$ is a composition $U \xrightarrow{\alpha} V \times \mathbb{C}^n \xrightarrow{\pi} V$, where α is étale and π is a projection, it suffices to check our lemma when the smooth morphism in question is either étale or the projection.

(i) ψ is étale. Then $\psi^{\cdot}\varphi^{\cdot}\mathscr{A} = \psi^{*}\varphi^{\cdot}\mathscr{A}$ by 1.4.5(i) and $c_{\varphi,\psi}$ is the action of $\psi^{*}\varphi^{\cdot}\mathscr{A}$ on $\psi^{*}\varphi^{*}\mathscr{A} = \mathscr{O}_Z \underset{R}{\otimes} A$. Since $\psi^{*}\operatorname{Diff}(\mathscr{O}_Y, \varphi^{*}\mathscr{A}) = \operatorname{Diff}(\mathscr{O}_Z, \psi^{*}\varphi^{*}\mathscr{A})$ the flatness of ψ implies that $c_{\varphi,\psi}$ is an isomorphism.

(ii) φ is étale. One has $\varphi^{\cdot}\mathscr{A} = \varphi^{*}\mathscr{A}$, $\psi^{*}\varphi^{\cdot}\mathscr{A} = (\varphi\psi)^{*}\mathscr{A}$, and $\psi^{\cdot}\varphi^{\cdot}\mathscr{A}$, respectively $(\varphi\psi)^{\cdot}\mathscr{A}$, is the sheaf of all operators $l \in \operatorname{Diff}(\mathscr{O}_Z, \mathscr{O}_Z \underset{R}{\otimes} A)$ that commute with the right action of $\psi^{-1}\mathscr{O}_Y$, respectively R. Since for any $f \in \mathscr{O}_Z$ the map $\mathscr{O}_Y \to \mathscr{O}_Z \underset{R}{\otimes} A = \psi^{*}\varphi^{\cdot}\mathscr{A}$, $g \mapsto l(f\psi^{*}(g)) - l(f)i(g)$, is a differential operator, it vanishes if and only if it vanishes on R. Hence $\psi^{\cdot}\varphi^{\cdot}\mathscr{A} = (\varphi\psi)^{\cdot}\mathscr{A}$.

(iii) $\psi\colon Z = \mathbb{C}^n \times Y \to Y$ is a projection. One has (see 1.4.5(ii)) $\psi^{\cdot}\varphi^{\cdot}\mathscr{A} = D_{\mathbb{C}^n} \boxtimes \varphi^{\cdot}\mathscr{A} = \{l \in \operatorname{Diff}(\mathscr{O}_Z, \mathscr{O}_Z \underset{R}{\otimes} A)\colon l(f(\varphi\psi)^{*}(g)) = l(f)i(g) \text{ for } f \in \mathscr{O}_Z, g \in R\} = (\varphi\psi)^{\cdot}\mathscr{A}$.

(iv) $\varphi\colon Y = \mathbb{C}^n \times X \to X$ is a projection. Let t_j, $j = 1, \ldots, n$, be the coordinates on $\mathbb{C}^n$, $\theta = \varphi\psi$, $q_j = \psi^{*}(t_j) \in \mathscr{O}_Z$, so

$$\psi = (q\,;\theta) = (q_1, \ldots, q_n\,;\theta).$$

One has $(\varphi\psi)^{\cdot}\mathscr{A} = \{l \in \operatorname{Diff}(\mathscr{O}_Z, \mathscr{O}_Z \underset{R}{\otimes} A)\colon l(fr) = l(f)i(r) \text{ for } f \in \mathscr{O}_Z, r \in R\}$, $\varphi^{\cdot}\mathscr{A} = D_{\mathbb{C}^n} \boxtimes A$, $\psi^{\cdot}\varphi^{\cdot}\mathscr{A} = \{m \in \operatorname{Diff}(\mathscr{O}_Z, q^{*}D_{\mathbb{C}^n} \underset{R}{\otimes} A)\colon m(fr) = m(f)i(r), m(fq_j) = m(f)i(t_j) \text{ for } f \in \mathscr{O}_Z, r \in R, j = 1, \ldots, n\}$.

Write $D_{\mathbb{C}^n} = \bigoplus \mathscr{O}_{\mathbb{C}^n}\partial_{t_1}^{i_1}\cdots\partial_{t_n}^{i_n}$, so that in the above formula $m = \sum m_{i_1\cdots i_n}\partial_{t_1}^{i_1}\cdots\partial_{t_n}^{i_n}$, where $m_{i_1\cdots i_n} \in \operatorname{Diff}(\mathscr{O}_Z, \mathscr{O}_Z \underset{R}{\otimes} A)$. Conditions on $m \in \psi^{\cdot}\varphi^{\cdot}\mathscr{A}$ imply $m_{i_1\cdots i_n} \in (\varphi\psi)^{\cdot}\mathscr{A}$ and

$$m_{i_1\cdots i_n}(fq_j) = q_j m_{i_1\cdots i_n}(f) + m_{i_1,\ldots,i_j+1,\ldots,i_n}(f).$$

The map $c_{\varphi,\psi}$ sends m to $m_{0\cdots 0}$; clearly this is an isomorphism.

1.5.2. Let $\pi_P\colon P \to X$ be a smooth X-scheme (this means that π_P is a smooth morphism). For a D-algebra $\mathscr{A}$ on X we will call $\mathscr{A}_P := \pi^{\cdot}\mathscr{A}$ the *(smooth) localization of* $\mathscr{A}$ *at* P. The above lemma claims that the pull-back functors are compatible with smooth localizations: for a commutative diagram

$$\begin{array}{ccc} Q & \xrightarrow{\widetilde{\varphi}} & P \\ {\scriptstyle \pi_Q}\downarrow & & \downarrow{\scriptstyle \pi_P} \\ Y & \xrightarrow{\varphi} & X \end{array}$$

where π_Q is also smooth, one has a canonical isomorphism $\widetilde{\varphi}^{\cdot}\mathscr{A}_P = (\varphi^{\cdot}\mathscr{A})_Q$ (both equal to $(\varphi\pi_Q)^{\cdot}\mathscr{A}$). In particular, for any morphism $\alpha\colon P' \to P$ of smooth X-schemes one has $\mathscr{A}_{P'} = \alpha^{\cdot}\mathscr{A}$; i.e., $\mathscr{A}_P$ forms a Cartesian section of D-Sch over the category X_{sm} of smooth X-schemes.

1.5.3. The category X_{sm} has a Grothendieck topology structure (smooth site of X; a covering in X_{sm} is a smooth surjective morphism of X-schemes). Note that a morphism between D-algebras on X is an isomorphism if and only if it is an isomorphism locally in smooth topology X_{sm} (use 1.4.5(i), (ii)). In fact D-algebras themselves are local objects with respect to the smooth topology: they satisfy the smooth descent property.

To be precise, let $\pi_P\colon P \to X$ be a smooth covering (i.e., a smooth surjective morphism). Consider the corresponding Čech system

$$\mathscr{P} := \cdots P \underset{X}{\times} P \underset{X}{\times} P \overset{\pi_{12},\pi_{23},\pi_{13}}{\rightrightarrows} P \underset{X}{\times} P \overset{\pi_1,\pi_2}{\rightrightarrows} P \xrightarrow{\pi_P} X.$$

For a D-algebra $\mathscr{A}$ on X denote by $g_{\mathscr{A}}$ the composition $\pi_1^{\cdot}\mathscr{A}_P \xrightarrow{\sim} \mathscr{A}_{P\underset{X}{\times}P} \xrightarrow{\sim} \pi_2^{\cdot}\mathscr{A}_P$.

A *D-algebra on $\mathscr{P}$* is a pair $(\mathscr{B}, g)$, where $\mathscr{B}$ is a D-algebra on P and $g\colon \pi_1^{\cdot}\mathscr{B} \xrightarrow{\sim} \pi_2^{\cdot}\mathscr{B}$ is an isomorphism that satisfies the cocycle property $\pi_{13}^{\cdot}(g) = \pi_{23}^{\cdot}(g)\pi_{12}^{\cdot}(g)$. The D-algebras on $\mathscr{P}$ form a category in an obvious way, and we have a canonical $\mathscr{P}$-localization functor (D-algebras on X) $\to$ (D-algebras on $\mathscr{P}$), $\mathscr{A} \mapsto (\mathscr{A}_P, g_A)$.

1.5.4. LEMMA. *This functor is an equivalence of categories.*

PROOF. Let us construct the inverse functor. Denote by $X_{\mathrm{sm}}^{(P)} \subset X_{\mathrm{sm}}$ the full subcategory of P-small objects (those T for which $\mathrm{Hom}_{X_{\mathrm{sm}}}(T, P)$ is not empty).

A D-algebra $(\mathscr{B}, g)$ on $\mathscr{P}$ defines a Cartesian section $T \mapsto \mathscr{B}_T$ of D-Sch over $X_{\mathrm{sm}}^{(P)}$ as follows. For a P-small T and a morphism $\gamma\colon T \to P$ put $\mathscr{B}_{T\gamma} := \gamma^{\cdot}\mathscr{B}$. If $\gamma'\colon T \to P$ is another morphism then, by 1.5.1, $\mathscr{B}_{T\gamma} = (\gamma, \gamma')^{\cdot}\pi_1^{\cdot}\mathscr{B}$ and $\mathscr{B}_{T\gamma'} = (\gamma, \gamma')^{\cdot}\pi_2^{\cdot}\mathscr{B}$, so we have a canonical isomorphism $g_{\gamma,\gamma'} = (\gamma, \gamma')^{\cdot}(g)\colon \mathscr{B}_{T\gamma} \xrightarrow{\sim} \mathscr{B}_{T\gamma'}$; one has $g_{\gamma,\gamma''} = g_{\gamma'\gamma''}g_{\gamma\gamma'}$ by the cocycle property of g. Therefore $\mathscr{B}_{T\gamma}$ does not depend on γ: this is our $\mathscr{B}_T$. If $T' \overset{\alpha}{\to} T$ is a morphism in $X_{\mathrm{sm}}^{(P)}$, then we have a canonical morphism

$c_\alpha : \alpha^{\bullet}\mathscr{B}_T \to \mathscr{B}_{T'}$ defined as a composition $\alpha^{\bullet}\mathscr{B}_{T\gamma} = \alpha^{\bullet}\gamma^{\bullet}\mathscr{B} \overset{c_{\gamma,\alpha}}{\to} (\gamma\alpha)^{\bullet}\mathscr{B} = \mathscr{B}_{T'\gamma\alpha}$ (this definition does not depend on the choice of $\gamma \in \mathrm{Hom}(T, P)$).

This c_α is actually an isomorphism (this is clear from 1.5.1 if either α is smooth or there exists a smooth $\gamma: T \to P$. The general case reduces to this, since c_α is compatible with smooth localizations: replace T and T' by $T \underset{X}{\times} P$ and $T' \underset{X}{\times} P$ respectively).

We can replace $X_{\mathrm{sm}}^{(P)}$ by the subcategory $X_{\text{ét}}^{(P)}$ that consists of P-small T's étale over X. By the étale descent (see 1.3) $\mathscr{B}_T$ defines a D-algebra $\mathscr{A}$ on X. Clearly $\mathscr{A}$ depends on $(\mathscr{B}, g)$ in a functorial way. One checks immediately that this functor (D-algebras on $\mathscr{P}$) $\to$ (D-algebras on X), $(\mathscr{B}, g) \mapsto \mathscr{A}$, is inverse to the $\mathscr{P}$-localization functor.

1.5.5. For a D-algebra $\mathscr{A}$ on X the categories $\mathscr{M}(\mathscr{A}_P)$ form a fibered category $\mathscr{M}(X_{\mathrm{sm}})$ over X_{sm} (with respect to the pull-back functor φ^*). The usual flat descent property for $\mathscr{O}$-modules implies that $\mathscr{M}(X_{\mathrm{sm}})$ is a stack over X_{sm}.

1.5.6. One easily checks that the analogs of 1.5.1 and 1.5.4 are valid in the Lie algebroid situation (see 1.4.6).

1.6. Affine D-schemes. Let $(X, \mathscr{A})$ be a D-scheme. Put $A := \Gamma(X, \mathscr{A})$. We have a pair (Δ, Γ) of adjoint functors $\mathscr{M}(X, \mathscr{A}) \overset{\Gamma}{\underset{\Delta}{\rightleftarrows}} \mathscr{M}(A)$; $\Gamma(M) = \Gamma(X, M)$ is the global sections functor, $\Delta(N) := \mathscr{A} \underset{A}{\otimes} N$.

1.6.1. LEMMA. *The following conditions on $(X, \mathscr{A})$ are equivalent:*

(i) *For any $M \in \mathscr{M}(X, \mathscr{A})$ one has $H^i(X, M) = 0$ for $i > 0$ and M is generated, as an $\mathscr{A}$-module, by its global sections.*

(ii) *The functors Γ, Δ are (mutually inverse) equivalences of categories.*

If the conditions of 1.6.1 hold, then we will call $(X, \mathscr{A})$ an *affine D-scheme*, and X an *$\mathscr{A}$-affine variety.*

EXAMPLES. (i) If X itself is an affine scheme, then it is $\mathscr{A}$-affine for any $\mathscr{A}$.

(ii) Let X be any quasiprojective variety. Then there exists a smooth surjective morphism $\pi: \widetilde{X} \to X$ such that $\widetilde{X}$ is affine (actually there exists π which is a torsor with respect to an action of some vector bundle over X). Then X is $\pi_*\mathscr{O}_{\widetilde{X}}$-affine. For a nontrivial noncommutative example, see Section 3.

1.7. D-stacks. The smooth descent property shows that we can repeat all the above notions and constructions in the context of algebraic stacks. Here is a brief sketch of the first definitions.

Let $\mathscr{X}$ be an algebraic stack (for the smooth topology, see [Lau]). A *D-algebra $\mathscr{A}$ on $\mathscr{X}$* consists of the following data:

(i) For any scheme X and a smooth 1-morphism $\pi: X \to \mathscr{X}$ one has a D-algebra $\mathscr{A}_{(X, \pi)}$ on X.

(ii) For any (X, π), (X', π') as above, a morphism $\alpha\colon X' \to X$, and a 2-morphism $\pi' \xrightarrow{\widetilde{\alpha}} \pi\alpha$, one has an isomorphism of D-algebras

$$\widetilde{\alpha}_{\mathscr{A}}\colon \mathscr{A}_{(X', \pi')} \to \alpha^{\cdot}\mathscr{A}_{(X, \pi)}.$$

We demand that $\widetilde{\alpha}_{\mathscr{A}}$ behave naturally with respect to compositions of $(\alpha, \widetilde{\alpha})$'s.

1.7.1. Example. If $\mathscr{X}$ is smooth then we have the D-algebra of differential operators $D_{\mathscr{X}}$ on $\mathscr{X}$. Namely, for any (X, π) as in (i) above the scheme X is regular; we put $(D_{\mathscr{X}})_{(X, \pi)} := D_X$, and take for $\widetilde{\alpha}$ in (ii) the canonical isomorphism $D_{X'} = \alpha^{\cdot} D_X$.

1.7.2. D-algebras on $\mathscr{X}$ are local objects: one can describe them as follows. Choose a smooth covering $\pi\colon X \to \mathscr{X}$ (so π is a smooth surjective 1-morphism); put $Y := X \times_{\mathscr{X}} X$. Then Y is a smooth groupoid acting on X, and $\mathscr{X}$ coincides with the quotient stack $Y \backslash X$. A D-algebra $\mathscr{A}$ on $\mathscr{X}$ yields a D-algebra $\mathscr{A}_X = \mathscr{A}_{(X, \pi)}$ on X equipped with the Y-action (which is an isomorphism $g_{\mathscr{A}}\colon \pi_1^{\cdot}\mathscr{A}_X \xrightarrow{\sim} \pi_2^{\cdot}\mathscr{A}_X$ that satisfies the cocycle condition). As follows from 1.5.3 the functor $\mathscr{A} \mapsto (\mathscr{A}_X, g_{\mathscr{A}})$ from the category of D-algebras on $\mathscr{X}$ to the category of Y-equivariant D-algebras on X is an equivalence of categories.

For a D-algebra $\mathscr{A}$ on $\mathscr{X}$ an *$\mathscr{A}$-module* M is a collection of $\mathscr{A}_{(X, \pi)}$-modules $M_{(X, \pi)}$ together with $\widetilde{\alpha}_{\mathscr{A}}$-isomorphisms $M_{(X', \pi')} \xrightarrow{\sim} \alpha^* M_{(X, \pi)}$ compatible with the composition of $(\alpha, \widetilde{\alpha})$'s for (X, π), $(\alpha, \widetilde{\alpha})$ as above. In terms of a smooth covering $X \xrightarrow{\pi} \mathscr{X}$ such an M is the same as a Y-equivariant $\mathscr{A}_X$-module (i.e., an $\mathscr{A}_X$-module equipped with a $g_{\mathscr{A}}$-isomorphism $g_M\colon \pi_1^* M \xrightarrow{\sim} \pi_2^* M$ that satisfies the cocycle condition). The $\mathscr{A}$-modules form an abelian category $\mathscr{M}(\mathscr{X}, \mathscr{A})$.

We leave to the reader further translations of 1.1–1.6 to the stack setting. The only delicate point here is the construction of the derived category of $\mathscr{A}$-modules (and, consequently, the construction of the push-forward functors from 1.4.7). It turns out that for the usual derived category $D\mathscr{M}(\mathscr{X}, \mathscr{A})$ the local to global spectral sequence for Ext's no longer holds in the stack case (the first example: take $\mathscr{X} = G_m \backslash \text{point}$, $\mathscr{A} = D_{\mathscr{X}}$). One must replace it by a certain canonical t-category $D(\mathscr{X}, \mathscr{A})$ with the heart $\mathscr{M}(\mathscr{X}, \mathscr{A})$; we hope to present a construction of $D(\mathscr{X}, \mathscr{A})$ elsewhere.

1.8. Equivariant setting. Let G be an algebraic group and X be a G-variety, i.e., a scheme equipped with a G-action $\mu\colon G \times X \to X$. In this subsection we will give an explicit description of D-algebras on the quotient stack $G\backslash X$ and of corresponding modules.

1.8.1. Let $\mathfrak{g}$ be the Lie algebra of G. Our μ defines the infinitesimal action $\alpha\colon \mathfrak{g} \to \mathscr{T}_X$; by 1.2.2 we have the Lie algebroid $\widetilde{\mathfrak{g}}_X$. For a (quasicoherent) $\mathscr{O}_X$-module P a G-action on P (which is an isomorphism $\mu^* P \xrightarrow{\sim} p_X^* P$ that satisfies the cocycle condition) defines a $\mathfrak{g}$-action α_P on P that lifts α (i.e., α_P is a Lie algebra morphism $\mathfrak{g} \to \operatorname{End}_{\mathbb{C}} P$ such that $\alpha_P(\gamma)(fp) = f\alpha_P(\gamma)(p) + \alpha(\gamma)(f)p$, $\gamma \in \mathfrak{g}$, $f \in \mathscr{O}_X$, $p \in P$; equivalently, we have a morphism of Lie algebroids $\alpha_P \colon \tilde{\mathfrak{g}}_X \to \widetilde{\mathscr{T}_P}$ (see 1.2.2–1.2.3)).

1.8.2. Let M be an $\mathscr{O}_X$-differential bimodule. The group $G \times G$ acts on $X \times X$, and the diagonal subgroup $G \hookrightarrow G \times G$ preserves the diagonal. Let $G^\wedge$ be the formal completion of $G \times G$ along the diagonal G; this is a formal subgroup of $G \times G$ that preserves the formal neighborhood $X^\wedge$ of the diagonal. We define a *G-action on M* as a $G^\wedge$-action on M considered as an $\mathscr{O}_{X \times X}$-module supported on the diagonal.

One can spell out this definition without explicitly mentioning $G^\wedge$ as follows. Consider the "complex Harish-Chandra pair" $(\mathfrak{g} \times \mathfrak{g}, G)$. It acts on $X \times X$, and a G-action on an $\mathscr{O}_X$-differential bimodule is the lifting of this action to M considered as an $\mathscr{O}_{X \times X}$-module. Explicitly, a $(\mathfrak{g} \times \mathfrak{g}, G)$-action on M is a pair (μ_M, α_M) where μ_M is a G-action on M considered as an $\mathscr{O}_{X \times X}$-module (G acts on $X \times X$ diagonally), and α_M is a $(\mathfrak{g} \times \mathfrak{g})$-action on M (in other words α_M is a Lie algebra map $\mathfrak{g} \times \mathfrak{g} \to \operatorname{End}_{\mathbb{C}} M$ such that one has $\alpha_M(\gamma_1, \gamma_2)(f_1 m f_2) = \alpha(\gamma_1)(f_1) m f_2 + f_1(\alpha_M(\gamma_1, \gamma_2)m) f_2 + f_1 m \alpha(\gamma_2)(f_2)$ for $\gamma_1, \gamma_2 \in \mathfrak{g}$, $f_1, f_2 \in \mathscr{O}_X$, $m \in M$). This pair should be compatible in the sense that the action of $\mathfrak{g}$ that comes from μ_M coincides with the α_M-action of the diagonal $\mathfrak{g} \subset \mathfrak{g} \times \mathfrak{g}$, and one has $g^*(\alpha_M(\gamma_1, \gamma_2)m) = \alpha_M(\operatorname{ad}_g \gamma_1, \operatorname{ad}_g \gamma_2) g^*(m)$ for $g \in G$, $m \in M$, $\gamma_1, \gamma_2 \in \mathfrak{g}$. Here g^* stands for the μ_M-action of g on M.

REMARK. A G-action on M as an $\mathscr{O}_{X \times X}$-module defines (and is completely determined by) a G-action on M as an $\mathscr{O}_X$-module (with respect to either only left or only right $\mathscr{O}_X$-action). We can formally weaken the above definition by demanding only that in the pair (μ_M, α_M) our μ_M should be the G-action on M as on a left (or right) $\mathscr{O}_X$-module. It is easy to check that nothing changes: the compatibility with α_M immediately implies that μ_M is actually a G-action on M as an $\mathscr{O}_{X \times X}$-module.

1.8.3. Let $\mathscr{A}$ be a D-algebra on X. A *G-action on* $\mathscr{A}$ is a G-action $(\mu_{\mathscr{A}}, \alpha_{\mathscr{A}})$ on $\mathscr{A}$ as an $\mathscr{O}_X$-differential bimodule such that

(i) $\mu_{\mathscr{A}}$ is compatible with the ring structure on $\mathscr{A}$, i.e., $g^*(a_1 a_2) = (g^* a_1)(g^* a_2)$, $g^*(1) = 1$ for $g \in G$, $a_1, a_2 \in \mathscr{A}$.
(ii) For $\gamma \in \mathfrak{g}$, $a_1, a_2 \in \mathscr{A}$ one has $\alpha_{\mathscr{A}}(\gamma, 0)(a_1 a_2) = (\alpha_{\mathscr{A}}(\gamma, 0)a_1)a_2$, $\alpha_{\mathscr{A}}(0, \gamma)(a_1 a_2) = a_1 \alpha_{\mathscr{A}}(0, \gamma)(a_2)$.

Note that (i) implies that the structure morphism $i \colon \mathscr{O}_X \to \mathscr{A}$ commutes with the G-action. For $\gamma \in \mathfrak{g}$ put $i_{\mathfrak{g}}(\gamma) := \alpha_{\mathscr{A}}(\gamma, 0)(1) \in \mathscr{A}$; then $\alpha_{\mathscr{A}}(\gamma_1, \gamma_2)a = i_{\mathfrak{g}}(\gamma_1)a - a i_{\mathfrak{g}}(\gamma_2)$. We can rewrite the above definition in terms of $i_{\mathfrak{g}}$. Namely, a G-action on $\mathscr{A}$ is the same as a pair $(\mu_{\mathscr{A}}, i_{\mathfrak{g}})$, where $\mu_{\mathscr{A}}$ is a G-action on $\mathscr{A}$ as on a left (or right) $\mathscr{O}_X$-module and $i_{\mathfrak{g}} \colon \mathfrak{g} \to \mathscr{A}$ is a

Lie algebra map such that

(i) $\mu_{\mathscr{A}}$ is compatible with the ring structure on $\mathscr{A}$.
(ii) $i_{\mathfrak{g}}$ commutes with the G-action (where G acts on $\mathfrak{g}$ by the adjoint representation).
(iii) The $\mathfrak{g}$-action on $\mathscr{A}$ that comes from $\mu_{\mathscr{A}}$ coincides with $\mathrm{ad}_{i_{\mathfrak{g}}}$.

We call a D-algebra equipped with a G-action a *Harish-Chandra algebra* or an *$(\mathscr{O}_X, G)$-differential algebra.*

EXAMPLE. If G acts on a coherent $\mathscr{O}_X$-module P then D_P is a Harish-Chandra algebra in an obvious way (here $i_{\mathfrak{g}} = \alpha_P$).

1.8.4. Now let P be a Lie algebroid on X. A *G-action on P* is a G-action μ_P on P as an $\mathscr{O}_X$-module together with a morphism of Lie algebras $i_{\mathfrak{g}}\colon \mathfrak{g} \to P$ such that

(i) One has $g^*[\zeta_1, \zeta_2] = [g^*\zeta_1, g^*\zeta_2]$, $g^*(\sigma\zeta) = \sigma(g^*\zeta)$ for $g \in G$, $\zeta_i \in P$.
(ii) $i_{\mathfrak{g}}(\mathrm{ad}_g \gamma) = g^* i_{\mathfrak{g}}\gamma$ for $g \in G$, $\gamma \in \mathfrak{g}$.
(iii) The $\mathfrak{g}$-action on P that comes from μ_P coincides with $\mathrm{ad}_{i_{\mathfrak{g}}}$.

We call a Lie algebroid equipped with a G-action a *Harish-Chandra Lie algebroid*. If X is a point, then this is the same as a Harish-Chandra pair.

For a Harish-Chandra Lie algebroid P its universal enveloping algebra $\mathscr{U}(P)$ is a Harish-Chandra algebra in an obvious manner. If $\mathscr{A}$ is a Harish-Chandra algebra on X, then $\mathrm{Lie}\mathscr{A}$ (see 1.2.5) is a Harish-Chandra Lie algebroid.

One defines morphisms of Harish-Chandra algebras and Lie algebroids in an obvious way.

EXAMPLE. $\widetilde{\mathfrak{g}}_X$ is a Harish-Chandra algebroid in an obvious manner (the G-action on $\widetilde{\mathfrak{g}}_X$ comes from the adjoint action on $\mathfrak{g} \subset \widetilde{\mathfrak{g}}_X$). For any Harish-Chandra Lie algebroid P there exists a unique morphism $\widetilde{\mathfrak{g}}_X \to P$. We have an obvious ring homomorphism $\mathscr{U}(\mathfrak{g}) \to \mathscr{U}(\widetilde{\mathfrak{g}}_X)$; the corresponding morphism of $\mathscr{O}_X$-modules $\mathscr{O}_X \otimes \mathscr{U}(\mathfrak{g}) \to \mathscr{U}(\widetilde{\mathfrak{g}}_X)$ is an isomorphism.

1.8.5. Let $\mathscr{A}$ be a Harish-Chandra algebra on X. A *weak $(\mathscr{A}, G)$-module* is an $\mathscr{A}$-module M equipped with a G-action (as an $\mathscr{O}_X$-module) compatible with the G-action on $\mathscr{A}$ (i.e., for $g \in G$, $a \in \mathscr{A}$, $m \in M$ one has $g^*(am) = g^*(a)g^*(m)$). An *$(\mathscr{A}, G)$-module* is a weak $(\mathscr{A}, G)$-module M such that the action of $\mathfrak{g}$ on M that comes from the G-action coincides with the one that comes from $i_{\mathfrak{g}}\colon \mathfrak{g} \to \mathscr{A}$ and the $\mathscr{A}$-module structure on M. We will also call $(\mathscr{A}, G)$-modules *Harish-Chandra modules.*

To see the difference between weak $(\mathscr{A}, G)$-modules and Harish-Chandra modules consider the tensor product algebra $\mathscr{A} \underset{\mathbb{C}}{\otimes} \mathscr{U}(\mathfrak{g})$. It carries the Harish-Chandra structure: the G-action is $g(a \otimes u) = g(a) \otimes \mathrm{Ad}_g(u)$, and the $i_{\mathfrak{g}}$-map is $i_{\mathfrak{g}}(\gamma) = i^A_{\mathfrak{g}}(\gamma) \otimes 1 + 1 \otimes \gamma$.

1.8.6. LEMMA. *Weak* $(\mathscr{A}, G)$*-modules are the same as* $(\mathscr{A} \underset{\mathbb{C}}{\otimes} \mathscr{U}(\mathfrak{g}), G)$*-modules.*

PROOF. An $(\mathscr{A} \underset{\mathbb{C}}{\otimes} \mathscr{U}(\mathfrak{g}), G)$-module is a weak $(\mathscr{A}, G)$-module (where $\mathscr{A}$ acts via $\mathscr{A} \hookrightarrow \mathscr{A} \otimes \mathscr{U}(\mathfrak{g})$, $a \mapsto a \otimes 1$). Conversely, let M be a weak $(\mathscr{A}, G)$-module. For $\gamma \in \mathfrak{g}$ consider the $\mathbb{C}$-linear endomorphism $w(\gamma)$ of M, $w(\gamma)m := \gamma^{(1)}m - \gamma^{(2)}m$, where $\gamma^{(1)}$ is the action of γ that comes from the action of G on M (as on $\mathscr{O}_X$-module) and $\gamma^{(2)}$ is the action of $i_{\mathfrak{g}}(\gamma) \in \mathscr{A}$. One checks that $w(\gamma)$ commutes with $\mathscr{A}$-action and the map $w: \mathfrak{g} \to \mathrm{End}_{\mathscr{A}} M$ is a Lie algebra homomorphism. Therefore the $\mathscr{A}$-module structure and w define a $(\mathscr{A} \underset{\mathbb{C}}{\otimes} \mathscr{U}(\mathfrak{g}))$-module structure on M, and the G-action makes M a Harish-Chandra $(\mathscr{A} \underset{\mathbb{C}}{\otimes} \mathscr{U}(\mathfrak{g}))$-module.

The $(\mathscr{A}, G)$-modules form an abelian category $\mathscr{M}(\mathscr{A}, G)$, the weak ones form an abelian category $\mathscr{M}(\mathscr{A}, G)_{\mathrm{weak}}$. Clearly $\mathscr{M}(\mathscr{A}, G)$ is a full subcategory of $\mathscr{M}(\mathscr{A}, G)_{\mathrm{weak}}$, which is closed under subquotients. We have the faithful forgetting of the G-action functor $o: \mathscr{M}(\mathscr{A}, G)_{\mathrm{weak}} \to \mathscr{M}(\mathscr{A})$.

EXAMPLES. (i) If $\mathscr{P}$ is a G-equivariant Lie algebroid on X then an $\mathscr{U}(\mathscr{P})$-Harish-Chandra module M is a $\mathscr{P}$-module equipped with a G-action (as an $\mathscr{O}_X$-module) such that the corresponding infinitesimal action of $\mathfrak{g}$ coincides with the action defined by the morphism $i_{\mathfrak{g}}: \mathfrak{g} \to \mathscr{P}$. If X is a point we recover the usual notion of a $(\mathscr{P}, G)$-module.

(ii) For any Harish-Chandra algebra $\mathscr{A}$ the free left $\mathscr{A}$-module $\mathscr{A}$ is a weak $(\mathscr{A}, G)$-module in an obvious manner. The corresponding $\mathfrak{g}$-action w is $w(\zeta)a = -ai_{\mathfrak{g}}(\zeta)$.

1.8.7. LEMMA. (i) *Harish-Chandra algebras on* X *are the same as* D*-algebras on the quotient stack* $G\backslash X$.

(ii) *For a Harish-Chandra algebra* $\mathscr{A}$, $(\mathscr{A}, G)$*-modules are the same as modules over corresponding algebras on* $G\backslash X$.

PROOF. (i) Denote by $p_G, p_X: G \times X \to G, X$ the projections. Consider a D-algebra on $G\backslash X$. According to 1.7 this is the same as a pair $(\mathscr{A}, g)$, where $\mathscr{A}$ is a D-algebra on X and $g: \mu^{\cdot}\mathscr{A} \xrightarrow{\sim} p_X^{\cdot}\mathscr{A}$ is an isomorphism of D-algebras that satisfies the cocycle condition. Let us define on $\mathscr{A}$ a Harish-Chandra structure $(\mu_{\mathscr{A}}, i_{\mathfrak{g}})$, i.e., a G-action on $\mathscr{A}$ in the sense of 1.8.3.

Recall that $p_X^{\cdot}\mathscr{A} = D_G \boxtimes \mathscr{A}$ (see 1.4.5(ii)), hence $p_X^*\mathscr{A} = \mathscr{O}_G \boxtimes \mathscr{A}$ is the centralizer of $p_G^{-1}\mathscr{O}_G \subset \mathscr{O}_{G\times X}$ in $p_X^{\cdot}\mathscr{A}$. Since $\mu^{\cdot}\mathscr{A}$ is the pull-back of $p_X^{\cdot}\mathscr{A}$ via the isomorphism $(p_G, \mu): G \times X \xrightarrow{\sim} G \times X$ we see that $\mu^*\mathscr{A}$ coincides with the centralizer of $p_G^{-1}\mathscr{O}_G$ in $\mu^{\cdot}\mathscr{A}$. Therefore our $g: \mu^{\cdot}\mathscr{A} \to p_X^{\cdot}\mathscr{A}$ defines an isomorphism between the $p_G^{-1}\mathscr{O}_G$-centralizers $\mu_{\mathscr{A}}: \mu^*\mathscr{A} \xrightarrow{\sim} p_X^*\mathscr{A}$. This is the desired action of G on $\mathscr{A}$ as an $\mathscr{O}_X$-module. To define $i_{\mathfrak{g}}$ consider the embedding $\nu: \mathfrak{g} \to D_G \subset p_X^{\cdot}\mathscr{A}$ that sends $\gamma \in \mathfrak{g}$ to the corresponding left invariant vector field. For $\gamma \in \mathfrak{g}$ put $i_g(\gamma) := \nu(\gamma) - g(p_G, \mu)^*(\nu(\gamma)) \in p_X^{\cdot}\mathscr{A}$; this element commutes with $p^{-1}\mathscr{O}_G$ and $\nu(\mathfrak{g})$ and, therefore, lies in $p_X^{-1}\mathscr{A} \subset p_X^{\cdot}\mathscr{A}$. We get $i_{\mathfrak{g}}: \mathfrak{g} \to \mathscr{A}$; it is easy to check that $(\mu_{\mathscr{A}}, i_{\mathfrak{g}})$ is a Harish-Chandra structure on $\mathscr{A}$.

Conversely, for a Harish-Chandra algebra $\mathscr{A}$ we define the isomorphism $g: \mu^{\cdot}\mathscr{A} \xrightarrow{\sim} p_X^{\cdot}\mathscr{A}$ as $\mu_{\mathscr{A}}$ on the centralizers of $p_G^{-1}\mathscr{O}_G$ and extend it to the entire algebra using the above formula relating $i_{\mathfrak{g}}$ and ν . One easily checks the cocycle property for g .

(ii) Clear.

1.8.8. Here is a simple corollary of 1.8.7. Assume we have an embedding of algebraic groups $G \subset G'$. We have the induced G'-variety $X' = G' \underset{G}{\times} X = G' \times X/\{(g', x) = (g'g^{-1}, gx)\}$. The embedding $X \hookrightarrow X'$ induces an isomorphism between the quotient stacks $G\backslash X \xrightarrow{\sim} G'\backslash X'$, hence the pull-back to X is an equivalence between the categories of G'-Harish-Chandra algebras on X' and G-Harish-Chandra algebras on X (and similarly for the corresponding Harish-Chandra modules). We leave to the reader an explicit construction of the inverse functors.

1.8.9. Assume that G acts on X in a free way, so we have a morphism of schemes $\pi: X \to Z$ that identifies Z with $G\backslash X$ (i.e., X is a G-torsor over Z). According to 1.8.7 the functor $\pi^{\cdot}$: (D-algebras on Z) $\to$ (Harish-Chandra algebras on X) is an equivalence of categories. Here is an explicit construction of the inverse functor. Denote by $\mathfrak{g}_Z^{\sim}$ the X-twist of $\mathscr{O}_Z \otimes \mathfrak{g}$; explicitly, $\mathfrak{g}_Z^{\sim} = (\pi_*\mathscr{O}_X \otimes \mathfrak{g})^G$, G acts on $\mathfrak{g}$ by the adjoint action. For a Harish-Chandra algebra $\mathscr{A}$ on X put $\mathscr{A}_Z^{\sim} := (\pi_*\mathscr{A})^G$; this is a D-algebra on Z. The map $i_{\mathfrak{g}}$ sends $\mathfrak{g}_Z^{\sim}$ to $\mathscr{A}_Z^{\sim}$. One can see that $\mathscr{A}_Z^{\sim} \cdot i_{\mathfrak{g}}(\mathfrak{g}_Z^{\sim})$ is actually a 2-sided ideal in $\mathscr{A}_Z^{\sim}$; put $\mathscr{A}_Z := \mathscr{A}_Z^{\sim}/\mathscr{A}_Z^{\sim} i_{\mathfrak{g}}(\mathfrak{g}_Z^{\sim})$. The left action of $\mathscr{A}$ on $\pi^*\mathscr{A}_Z = \pi^*\mathscr{A}_Z^{\sim}/\pi^*(\mathscr{A}_Z^{\sim})i_{\mathfrak{g}}(\mathfrak{g}_Z^{\sim}) = \mathscr{A}/\mathscr{A} i_{\mathfrak{g}}(\mathfrak{g})$ defines the isomorphism $\mathscr{A} = \pi^{\cdot}\mathscr{A}_Z$.

For any D-algebra $\mathscr{A}_Z$ on Z consider the corresponding algebras $\mathscr{A} = \pi^{\cdot}\mathscr{A}_Z$, $\mathscr{A}_Z^{\sim}$, so $\mathscr{A}_Z = \mathscr{A}_Z^{\sim}/\mathscr{A}_Z^{\sim} i_{\mathfrak{g}}(\mathfrak{g}_Z^{\sim})$. Note that $\pi_*\mathscr{A} = \pi_*\mathscr{O}_X \underset{\mathscr{O}_Z}{\otimes} \mathscr{A}_Z^{\sim}$ and any (local) section of π (i.e., any G-isomorphism $X = G \times Z$) induces an isomorphism $\mathscr{A}_Z^{\sim} = \mathscr{A}_Z \underset{\mathbb{C}}{\otimes} \mathscr{U}(\mathfrak{g})$. Assume that G is an affine group. Then the functor $\pi_*: \mathscr{M}(X, \mathscr{A}) \to \mathscr{M}(Z, \pi_*\mathscr{A})$ is an equivalence of categories. We have the adjoint functors $\mathscr{M}(\mathscr{A}, G)_{\text{weak}} \underset{\pi^{\cdot}}{\overset{\pi_{\cdot}}{\rightleftarrows}} \mathscr{M}(Z, \mathscr{A}_Z^{\sim})$,

$$\pi_{\cdot}M = (\pi_*M)^G,$$

$$\pi^{\cdot}N = \pi^{-1}\Big(\pi_*\mathscr{A} \underset{\mathscr{A}_Z^{\sim}}{\otimes} N\Big) = \mathscr{A} \underset{\pi^{-1}}{\otimes} \mathscr{A}_Z^{\sim}\pi^{-1}N = \mathscr{O}_X \underset{\pi^{-1}}{\otimes} \mathscr{O}_Y\pi^{-1}N.$$

1.8.10. LEMMA. (i) *The functors* $\pi^{\cdot}$, $\pi_{\cdot}$ *are mutually inverse equivalences of categories.*

(ii) $\pi_{\cdot}$ *identifies* $\mathcal{M}(\mathcal{A}, G)$ *with* $\mathcal{M}(Z, \mathcal{A}_Z)$, $\pi^{\cdot}$ *coincides on* $\mathcal{M}(Z, \mathcal{A}_Z)$ *with* $\pi_{\cdot}^{*}$.

§2. Twisted differential operators

2.1. First definitions and equivalences. ([1]) Let X be a smooth algebraic or analytic variety over $\mathbb{C}$.

2.1.1. DEFINITION. An algebra of twisted differential operators, or simply a tdo, on X is a sheaf of associative algebras D on X equipped with a morphism of algebras $i: \mathcal{O}_X \to D$ such that there exists an increasing filtration $D_{\cdot}$ on D with the following properties:

(i) $D_{\cdot}$ is a ring filtration (i.e., $D_iD_j \subset D_{i+j}$) such that the associated graded algebra $\operatorname{gr}_{\cdot} D$ is commutative; one has $D_{-1} = 0$, $\bigcup D_i = D$.

(ii) The morphism i identifies $\mathcal{O}_X$ with D_0, and the obvious morphism of the symmetric algebra $S^{\cdot}(D_1/D_0)$ into $\operatorname{gr}_{\cdot} D$ is an isomorphism of $\mathcal{O}_X$-algebras.

(iii) The morphism $\sigma: D_1/D_0 \to \mathcal{T}_X$, $\sigma(\partial)(f) := \partial f - f\partial$, where $\partial \in D_1$, $f \in \mathcal{O}_X = D_0$, is an isomorphism.

EXAMPLE. If $\mathcal{L}$ is a line bundle on X then $D_{\mathcal{L}}$ is a tdo.

REMARK. Let D be a tdo; according to (ii) and (iii) above we have a canonical isomorphism of $\mathcal{O}_X$-algebras $\operatorname{gr}_{\cdot} D = S^{\cdot}(\mathcal{T}_X) =$ functions on the cotangent bundle to X. The algebra $\operatorname{gr}_{\cdot} D$ carries a standard Poisson bracket $\{\cdot, \cdot\}: \operatorname{gr}_a D \times \operatorname{gr}_b D \to \operatorname{gr}_{a+b-1} D$, $\{f, g\} = fg - gf \bmod D_{a+b-2}$. The above isomorphism identifies $\{\cdot, \cdot\}$ with the Poisson bracket that comes from the standard symplectic structure on the cotangent bundle.

For a tdo D the filtration $D_{\cdot}$ is a D-filtration on D. Therefore D is an $\mathcal{O}_X$-differential algebra. In fact one checks immediately that $D_{\cdot}$ is the maximal D-filtration $D_{\cdot}^{\vee}$; in particular it is uniquely determined.

It is easy to see that the canonical complex from 1.1.1 for $M = D$ is the Koszul complex. To be precise, one has the following characterization of tdo's.

2.1.2. LEMMA. *An* $\mathcal{O}_X$*-differential algebra* $\mathcal{A}$ *is a tdo iff for the maximal D-filtration* $\mathcal{A}_{\cdot} = \mathcal{A}_{\cdot}^{\vee}$ *the following conditions hold:*

(i) $i: \mathcal{O}_X \to \mathcal{A}_0$ *is an isomorphism.*

(ii) *The morphism* $\sigma: \mathcal{A}_1/\mathcal{A}_0 \to \operatorname{Der}\mathcal{A}_0 = \mathcal{T}_X$, $\sigma(\partial)(f) = \partial f - f\partial$, *is surjective.*

PROOF. The above remarks show that any tdo satisfies the conditions in 2.1.2. Conversely, assume $\mathcal{A}$ satisfies the conditions 2.1.2. Since $\mathcal{A}_0$ is

([1]) We borrowed 2.1–2.4 below from an unpublished manuscript of A. Beilinson and D. Kazhdan.

the centralizer of $i(\mathcal{O}_X)$, we see that σ is injective, and hence σ is an isomorphism. The term $\mathcal{A}_1$ is closed under commutators (for $\partial_1, \partial_2 \in \mathcal{A}_1$, $f \in \mathcal{O}_X$ one has $\operatorname{ad} f([\partial_1, \partial_2]) = -\{[\partial_1, \partial_2], f\} = [\operatorname{ad}\partial_1, \operatorname{ad}\partial_2](f) = -[\sigma(\partial_1), \sigma(\partial_2)](f)$), therefore the $\mathcal{O}_X$-subalgebra of $\operatorname{gr}_\cdot\mathcal{A}$ generated by $\mathcal{T}_X = \mathcal{A}_1/\mathcal{A}_0$ is commutative. It remains to show that the corresponding map $S^i(\mathcal{T}_X) \to \operatorname{gr}_i\mathcal{A}$ is an isomorphism. This follows by induction on i. Namely, assume that we know this for any $i < j$. It is easy to check that for $i < j$ the maps $\delta_i\colon \operatorname{gr}_i\mathcal{A} = S^i(\mathcal{T}_X) \to \operatorname{Hom}(\Omega^1_X, \operatorname{gr}_{i-1}\mathcal{A}) = \mathcal{T}_X \otimes S^{i-1}(\mathcal{T}_X)$ from 1.1.1 are Koszul differentials, as well as the composition $S^j(\mathcal{T}_X) \to \operatorname{gr}_j\mathcal{A} \xrightarrow{\delta_j} \operatorname{Hom}(\Omega^1_X, \operatorname{gr}_{j-1}\mathcal{A}) = \mathcal{T}_X \otimes S^{j-1}(\mathcal{T}_X)$. Since δ_j is injective (see 1.1.1) the exactness of the Koszul complex implies that $S^j(\mathcal{T}_X) \xrightarrow{\sim} \operatorname{gr}_j\mathcal{A}$.

It is easy to see that any morphism between tdo's is an isomorphism, i.e., tdo's form a groupoid $\mathcal{TDO}(X)$. Below we give several equivalent descriptions of this groupoid.

2.1.3. Definition. A Picard Lie, or simply Picard, algebroid on X is a Lie algebroid $\widetilde{\mathcal{T}}$ equipped with a central section $1_{\widetilde{\mathcal{T}}}$ of $\widetilde{\mathcal{T}}^{(0)} = \operatorname{Ker}\sigma$, such that the sequence $0 \to \mathcal{O}_X \xrightarrow{i} \widetilde{\mathcal{T}} \xrightarrow{\sigma} \mathcal{T}_X \to 0$, $i(f) = f_{\widetilde{\mathcal{T}}} := f1_{\widetilde{\mathcal{T}}}$, is exact.

Recall that σ is the standard morphism of Lie algebroids defined in 1.2; we will identify $\mathcal{O}_X$ with $\widetilde{\mathcal{T}}^0$ using the isomorphism i.

Let $\widetilde{\mathcal{T}}$ be a Picard algebroid. Since $\widetilde{\mathcal{T}}^{(0)} = \mathcal{O}_X$, the sheaf $\mathcal{C}(\widetilde{\mathcal{T}})$ of connections is an Ω^1_X-torsor. The curvature map $c\colon \mathcal{C}(\widetilde{\mathcal{T}}) \to \Omega^2_X$ has the following property: $c(\nu + \nabla) = d\nu + c(\nabla)$ for $\nu \in \Omega^1_X$, $\nabla \in \mathcal{C}(\widetilde{\mathcal{T}})$ (see 1.2).

A morphism of Picard algebroids is a morphism of Lie algebroids that preserves $1_{\widetilde{\mathcal{T}}}$'s. Picard algebroids form a groupoid $\mathcal{PA}(X)$. The Baer sum construction defines on $\mathcal{PA}(X)$ a structure of "$\mathbb{C}$-vector space in categories". Namely, for $\widetilde{\mathcal{T}}_i \in \mathcal{PA}(X)$, $\lambda_i \in \mathbb{C}$ the linear combination $\widetilde{\mathcal{T}} = \lambda_1\widetilde{\mathcal{T}}_1 + \lambda_2\widetilde{\mathcal{T}}_2$ is a Picard algebroid $\widetilde{\mathcal{T}}$ equipped with a morphism of Lie algebroids $s_{\lambda_1,\lambda_2}\colon \widetilde{\mathcal{T}}_1 \times_{\mathcal{T}_X} \widetilde{\mathcal{T}}_2 \to \widetilde{\mathcal{T}}$ such that $s_{\lambda_1,\lambda_2}(f_1, f_2) = (\lambda_1 f_1 + \lambda_2 f_2)$.

For a tdo D consider the Lie algebroid $\widetilde{\mathcal{T}}_D := \operatorname{Lie} D$ (see 1.2.5). Clearly $\widetilde{\mathcal{T}}_D = D_1$, and $1_{\widetilde{\mathcal{T}}_D} = 1 \in \mathcal{O}_X \subset D_1$ defines on $\widetilde{\mathcal{T}}_D$ the structure of a Picard algebroid. Conversely, for a Picard algebroid $\widetilde{\mathcal{T}}$ denote by $D_{\widetilde{\mathcal{T}}}$ the quotient of $\mathcal{U}(\widetilde{\mathcal{T}})$ modulo the ideal generated by the central element $1 - 1_{\widetilde{\mathcal{T}}}$. One checks immediately that $D_{\widetilde{\mathcal{T}}}$ is a tdo.

2.1.4. Lemma. *The functors $\mathcal{T}D\mathcal{O}(X) \rightleftarrows \mathcal{PA}(X)$, $D \mapsto \widetilde{\mathcal{T}}_D$, $\widetilde{\mathcal{T}} \mapsto D_{\widetilde{\mathcal{T}}}$, are mutually inverse equivalences of categories.*

2.1.5. Let $d\colon A^n \to A^{n+1}$ be a morphism of sheaves of abelian groups on X, considered as a length 2 complex $A^\cdot$ supported in degrees n and $n+1$.

An $A^\cdot$-*torsor* is a pair $(\mathscr{F}, c)$, where $\mathscr{F}$ is an A^n-torsor and $c: \mathscr{F} \to A^{n+1}$ is a map such that $c(a+\varphi) = d(a)+c(\varphi)$ for $a \in A^n$, $\varphi \in \mathscr{F}$ (in other words, c is a trivialization of the induced A^{n+1}-torsor $d(\mathscr{F})$). The $A^\cdot$-torsors form a groupoid $A^\cdot$-tors. One has $\operatorname{Aut}\mathscr{F} = \Gamma(X, \operatorname{Ker} d) = H^n(X, A^\cdot)$, and isomorphism classes of $A^\cdot$-torsors are in a natural 1-1 correspondence with $H^{n+1}(X, A^\cdot)$.

REMARK. $A^\cdot$-tors is a stack in Picard categories on X; if $A^\cdot$ is a complex of $\mathbb{C}$-vector spaces, then $A^\cdot$-tors is a $\mathbb{C}$-vector space in categories (one forms $\mathbb{C}$-linear combinations of torsors in an obvious way). If D is surjective, then $A^\cdot$-tors $=$ $(\operatorname{Ker} d)$-tors.

Consider now the truncated de Rham complex $\Omega_X^{\geqslant 1} := (\Omega_X^1 \to \Omega_X^{2\,\mathrm{cl}})$, where $\Omega_X^{2\,\mathrm{cl}}$ are closed 2-forms. By 2.1.3 we have the functor $\mathscr{C}: \mathscr{PA}(X) \to \Omega_X^{\geqslant 1}$-tors, $\widetilde{\mathscr{T}} \mapsto (\mathscr{C}(\widetilde{\mathscr{T}}), c)$.

2.1.6. LEMMA. *$\mathscr{C}: \mathscr{PA}(X) \to \Omega_X^{\geqslant 1}$-tors is an equivalence of $\mathbb{C}$-vector spaces in categories.*

By 2.1.5 we can identify the set of isomorphism classes of tdo's with $H^2(X, \Omega_X^{\geqslant 1})$. For a tdo D we will denote by $c_1(D) \in H^2(X, \Omega_X^{\geqslant 1})$ the corresponding cohomology class.

2.1.7. For a tdo D a *connection* ∇ on D is a connection on the corresponding Picard algebroid $\widetilde{\mathscr{T}_D}$. Note that pairs (D, ∇), ∇ is a connection on a tdo D, are rigid: the only automorphism of D that preserves ∇ is the identity. The pairs (D, ∇) are in 1-1 correspondence with closed 2-forms; for $\omega \in \Omega^{2\,\mathrm{cl}}(X)$ we will denote by $(D_\omega, \nabla_\omega)$ the (unique up to a canonical isomorphism) tdo with $c(\nabla) = \omega$. A corresponding $\Omega_X^{\geqslant 1}$-torsor $(\mathscr{F}_\omega, c_\omega)$ is $\mathscr{F}_\omega = \Omega_X^1$, $c_\omega(\nu) = d\nu + \omega$.

2.1.8. Now consider the cotangent bundle $T^* = T^*(X) \xrightarrow{\pi} X$. This is a vector bundle over X; also T^* carries a canonical symplectic 2-form ω such that π is a polarization (which is a smooth projection with Lagrangian fibers). If ν is a 1-form on X and $t_\nu: T^* \to T^*$, $t_\nu(a) = a + \nu_{\pi(a)}$, is the translation by ν, then $t_\nu^*(\omega) = \pi^*(d\nu) + \omega$.

DEFINITION. A twisted cotangent bundle is a T^*-torsor $\psi \xrightarrow{\pi_\psi} X$ (i.e., π_ψ is a fibration equipped with a simple transitive action of T^* along the fibers) together with a symplectic form ω_ψ on ψ such that π_ψ is a polarization for ω_ψ, and for any 1-form ν one has $t_\nu^*(\omega_\psi) = \pi_\psi^* d\nu + \omega$.

For a twisted cotangent bundle ψ we will denote by A_ψ the $\mathscr{O}_X$-algebra $\pi_{\psi *}\mathscr{O}_\psi$. Then A_ψ carries a Poisson bracket $\{\cdot, \cdot\}$ (defined by ω_ψ) and a filtration $A_{\psi i} =$ functions of degree $\leqslant i$ along the fibers of π_ψ. Clearly one has $A_{\psi i} = \{a \in A_\psi : \{a, \mathscr{O}_X\} \subset A_{\psi(i-1)}\}$ and the associated graded algebra $\operatorname{gr.} A_\psi$ is naturally isomorphic to the symmetric algebra $A_{T^*} = S^\cdot\mathscr{T}_X$ as a Poisson algebra.

2.1.9. REMARKS. (i) A T^*-torsor structure on ψ is uniquely determined by the symplectic structure ω_ψ and the polarization π_ψ (since the infinitesimal action of a 1-form $\nu \in \Omega^1(X)$ is given by the vector field $\zeta(\nu)$ defined by $\zeta(\nu) \lrcorner \omega_\varphi = \pi_\psi^*(\nu)$).

(ii) Twisted cotangent bundles over X form a groupoid $\mathcal{TCB}(X)$. According to (i), $\mathcal{TCB}(X)$ is a full subcategory of the category of triples (Y, ω_Y, π_Y) where (Y, ω_Y) is a symplectic manifold and $\pi_Y: Y \to X$ is a polarization (for the symplectic structure).

2.1.10. LEMMA. *One has a canonical equivalence of categories* $\Gamma: \mathcal{TCB}(X) \xrightarrow{\sim} \Omega_X^{\geqslant 1}$*-tors.*

PROOF. Put $\Gamma(\psi) = \Omega^1$-torsor of a section of ψ; the map $c: \Gamma(\psi) \to \Omega_X^{2\,\mathrm{cl}}$ is $c(\gamma) := \gamma^*(\omega_\psi)$. Note that the corresponding Picard algebroid $\mathcal{T}_\psi$ is $A_{\psi 1}$ equipped with the bracket $\{\cdot, \cdot\}$.

The inverse functor Γ^{-1} sends an $\Omega_X^{\geqslant 1}$-torsor $(\mathcal{F}, c)$ to $(\psi, \pi_\psi, \omega_\psi)$, where $\pi_\psi: \psi \to X$ is the space of the torsor $\mathcal{F}$, and the symplectic form ω_ψ is the unique form such that for a section $\gamma \in \mathcal{F}$ of π_ψ the corresponding isomorphism $T^*X \xrightarrow{\sim} \psi$, $0 \mapsto \gamma$, identifies ω_ψ with $\omega + \pi^* c(\gamma)$.

2.1.11. REMARK. Let D be a tdo, and ψ be the corresponding twisted cotangent bundle. Then D is a "quantization" of ψ in the sense that D is a deformation of a commutative algebra A_ψ. To be precise, one has a canonical family $\mathbb{D} = \{D_t\}$ of sheaves of filtered rings on X parametrized by points $t \in \mathbb{P}^1$ (i.e., $\mathbb{D}$ is a flat $\mathcal{O}_{\mathbb{P}^1}$-algebra) such that:

(i) For $t \neq \infty$ one has $D_t = D_{t\widetilde{\mathcal{T}}}$ (here $\widetilde{\mathcal{T}} = \widetilde{\mathcal{T}}_D$; in particular, $D_1 = D$, $D_0 = D_{\mathcal{O}_X}$).

(ii) $D_\infty = A_\psi$, and the ω_ψ-Poisson bracket on A_ψ is given by the usual formula $\{\varphi_1, \varphi_2\} = [t(\widetilde{\varphi}_1\widetilde{\varphi}_2 - \widetilde{\varphi}_2\widetilde{\varphi}_1)] \bmod t^{-1}$ (here $\varphi_i \in D_\infty$, and $\widetilde{\varphi}_i$ are arbitrary local sections of D at $t = \infty$ such that $\widetilde{\varphi}_i(\infty) = \varphi_i$).

(iii) $\mathrm{gr}_a D = (S^a \mathcal{T}_X)(-a)$.

2.1.12. Let us see what the above constructions mean in the case $D = D_{\mathcal{L}}$, where $\mathcal{L}$ is a line bundle. The corresponding Picard algebroid $\widetilde{\mathcal{T}}_{\mathcal{L}} = \widetilde{\mathcal{T}}_{D_{\mathcal{L}}}$ is the Lie algebroid of infinitesimal symmetries of $(X, \mathcal{L})$, see 1.2.3. The $\Omega_X^{\geqslant 1}$-torsor $(\mathcal{F}_{\mathcal{L}}, c_{\mathcal{L}}) := \mathcal{C}(\widetilde{\mathcal{T}}_{\mathcal{L}})$ is the sheaf of connections on $\mathcal{L}$, and $c_{\mathcal{L}}$ is the usual curvature. Note that the functor $\mathcal{O}_X^*$-tors $\to \Omega_X^{\geqslant 1}$-tors is precisely the push-out functor for the morphism $d\log: \mathcal{O}_X^* \to \Omega_X^{1\,\mathrm{cl}} \subset \Omega_X^{\geqslant 1}[1]$. In particular it transforms $\otimes$ to the sum of torsors. One has $c_1(D_{\mathcal{L}}) = c_1(\mathcal{L}) \in H^2(X, \Omega_X^{\geqslant 1})$.

2.1.13. A tdo D is called locally trivial if locally it is isomorphic to $D_X = D_{\mathcal{O}_X}$; according to 2.1.6, locally trivial tdo's are the same as $\Omega_X^{1\,\mathrm{cl}}$-torsors. Note that in analytic situations every tdo is locally trivial. In algebraic situations this is not true in general. For example, let X be a compact algebraic variety. The space of isomorphism classes of tdo's $H^2(X, \Omega_X^{\geq 1})$ coincides with the Hodge filtration subspace $F^1 H^2_{DR}$, and the locally trivial tdo's correspond to those classes that vanish on some Zariski open subspace of X, i.e., precisely to $\mathbb{C}$-linear combinations of the algebraic cycles classes.

2.2. Functoriality. Let $\varphi: Y \to X$ be a morphism of smooth varieties. We have the corresponding morphism $\varphi^*: \varphi^{-1}\Omega_X^{\cdot} \to \Omega_Y^{\cdot}$ between the de Rham complexes, hence the morphism of groupoids $\varphi^{\cdot}: \Omega_X^{\geq 1}$-tors $\to \Omega_Y^{\geq 1}$-tors. If $\psi: Z \to Y$ is another morphism of varieties then we have a canonical isomorphism of functors $\psi^{\cdot}\varphi^{\cdot} = (\varphi\psi)^{\cdot}: \Omega_X^{\geq 1}$-tors $\to \Omega_Z^{\geq 1}$-tors, therefore $\Omega^{\geq 1}$-tors form a fibered category over the category of schemes. It is easy to see that $\varphi^{\cdot}$ satisfies the descent property if φ is a smooth surjective map, i.e., $\Omega^{\geq 1}$-tors form a stack on the category of smooth schemes equipped with the smooth topology.

The equivalent versions $\mathcal{TDO}$, $\mathcal{PA}$, and $\mathcal{TCB}$ of $\Omega^{\geq 1}$-tors therefore are also stacks. The pull-back functors for tdo's and Picard algebroids coincide with those defined in 1.4.

To be precise, for a Picard algebroid $\widetilde{\mathcal{T}_X}$ on X the Lie algebroid $\varphi^{\cdot}\widetilde{\mathcal{T}_X}$ (see 1.4.6) is a Picard algebroid. Namely, as an $\mathcal{O}_Y$-module $\varphi^{\cdot}\widetilde{\mathcal{T}_X}$ coincides with the fibered product $\varphi^*\widetilde{\mathcal{T}_X} \underset{\varphi^*\mathcal{T}_X}{\times} \mathcal{T}_Y$ (with respect to the projections $\varphi^*\widetilde{\mathcal{T}_X} \xrightarrow{\varphi^*(\sigma)} \varphi^*\mathcal{T}_X \xleftarrow{d\varphi} \mathcal{T}_Y$); and we put $1_{\varphi^{\cdot}\widetilde{\mathcal{T}_X}} = (\varphi^* 1_{\widetilde{\mathcal{T}_X}}, 0)$. One defines the pull-back for connections in an obvious way; this identifies $\mathcal{C}(\varphi^{\cdot}\widetilde{\mathcal{T}_X})$ with $\varphi^{\cdot}\mathcal{C}(\widetilde{\mathcal{T}_X})$.

Let us turn to tdo's:

LEMMA. (i) *Let $\varphi: Y \to X$ be a morphism of smooth schemes and D be a tdo on X. Then the $\mathcal{O}_Y$-differential algebra $\varphi^{\cdot}D$ (defined in* (1.4.1)) *is a tdo.*

(ii) *One has a canonical isomorphism of Picard algebroids $\widetilde{\mathcal{T}}_{\varphi^{\cdot}D} = \varphi^{\cdot}\widetilde{\mathcal{T}}_D$.*

PROOF. (i) The morphism φ can be written as a composition of a closed imbedding and a projection. By Lemma 1.5.1 it is enough to prove the statement for these two cases. The case of projection immediately follows from 1.4.5(ii). In the case of a closed imbedding one proves the statement by direct local computations using 1.4.5(iii).

(ii) easily follows from (i).

Note that for $\psi: Z \to Y$ the canonical morphism is compatible with the corresponding isomorphism of Picard algebroids.

According to 1.4 we have the pull-back functors $\varphi^*: \mathcal{M}(D) \to \mathcal{M}(\varphi^{\cdot}D)$ between the categories of D-modules that make the $\mathcal{M}(D)$'s a stack.

2.3. Twisted *D*-modules and projective connections. Let D be a tdo. We say that a D-module M is *lissé* if it is coherent as an $\mathcal{O}_X$-module.

2.3.1. LEMMA. *Let M be a lissé D-module. Then*

(i) *M is a vector bundle;*

(ii) *one has a canonical isomorphism of tdo's $D \xrightarrow{\sim} D_{(\det M)^{1/d}}$ where $d = \operatorname{rk} M$. In particular, D is locally trivial.*

PROOF. (i) Repeat a proof for ordinary D-modules (see [B]).

(ii) This canonical isomorphism comes from the morphism of Picard algebroids $\mathscr{T}_D \to \mathscr{T}_{\det M}$ that sends $1_{\mathscr{T}_D}$ to $d1_{\mathscr{T}_{\det M}}$. Formula: $\tilde{\tau} \mapsto \tilde{\tau}_{\det M}$, $\tilde{\tau}_{\det M}(m_1 \wedge \cdots \wedge m_d) = \tilde{\tau}(m_1) \wedge m_2 \wedge \cdots \wedge m_d + \cdots + m_1 \wedge \cdots \wedge \tilde{\tau}(m_d)$.

Let $\mathscr{E}$ be a coherent $\mathscr{O}_X$-module; consider the Lie algebroid $\widetilde{\mathscr{T}_{\mathscr{E}}}$ from 1.2.3. Note that $\mathscr{O}_X \cdot \mathrm{id}_{\mathscr{E}}$ is an ideal in $\widetilde{\mathscr{T}_{\mathscr{E}}}$; put $\overline{\mathscr{T}}_{\mathscr{E}} := \widetilde{\mathscr{T}_{\mathscr{E}}}/\mathscr{O}_X \cdot \mathrm{id}_{\mathscr{E}}$. This is a Lie algebroid. A *projective connection* on $\mathscr{E}$ is a connection on $\overline{\mathscr{T}}_{\mathscr{E}}$.

For a tdo D a D-action on $\mathscr{E}$ is the same as a morphism of Lie algebroids $\alpha \colon \widetilde{\mathscr{T}_D} \to \widetilde{\mathscr{T}_{\mathscr{E}}}$ that sends $1_{\widetilde{\mathscr{T}_D}}$ to $\mathrm{id}_{\mathscr{E}}$. Such an α defines an integrable projective connection $\overline{\nabla}_\alpha$ on $\mathscr{E}$ by the formula $\overline{\nabla}_\alpha(\sigma(\tau)) = \alpha(\tau) \bmod \mathscr{O}_X \cdot \mathrm{id}_{\mathscr{E}}$.

2.3.2. LEMMA. *Assume that the map $\mathscr{O}_X \to \operatorname{End}\mathscr{E}$, $f \mapsto f\,\mathrm{id}_{\mathscr{E}}$, is injective. Then the above map $\alpha \mapsto \overline{\nabla}_\alpha$ from the set of pairs (D, α), where D is a tdo and α is a D-action on $\mathscr{E}$, to the set of projective integrable connections on $\mathscr{E}$ is bijective.*

PROOF. One constructs the inverse map as follows. Let $\overline{\nabla} \colon \mathscr{T}_X \to \overline{\mathscr{T}}_{\mathscr{E}}$ be an integrable projective connection. Then $\widetilde{\mathscr{T}_{\overline{\nabla}}} := \mathscr{T}_X \times_{\overline{\mathscr{T}_{\mathscr{E}}}} \widetilde{\mathscr{T}_{\mathscr{E}}}$ is a Picard algebroid, and the projection $\alpha_{\overline{\nabla}} \colon \widetilde{\mathscr{T}_{\overline{\nabla}}} \to \widetilde{\mathscr{T}_{\mathscr{E}}}$ defines the $D_{\widetilde{\mathscr{T}_{\overline{\nabla}}}}$-action on $\mathscr{E}$.

2.4. Subprincipal symbols.(2) Let $\Omega = \det \Omega^1_X$ be the sheaf of volume forms on X, and $\widetilde{\mathscr{T}_\Omega}$ be the corresponding Picard algebroid. One has a canonical section $\ell \colon \mathscr{T}_X \to \widetilde{\mathscr{T}_\Omega}$ which assigns to $\partial \in \mathscr{T}_X$ its Lie derivative $\ell(\partial)$. Clearly ℓ commutes with the Lie bracket and for $f \in \mathscr{O}_X$ one has $f\ell(\partial) = \ell(f\partial) - \partial(f)$.

2.4.1. Let D be any tdo on X. The *dual* to the tdo D is a tdo D° equipped with an isomorphism $* \colon D \xrightarrow{\sim} D^\circ$ of filtered sheaves such that $*(ab) = *(a) * (b)$, $*f = f$ for any sections $a, b \in D$, $f \in \mathscr{O}_X = D_0 = D_0^\circ$. Clearly one has $\mathrm{gr}_i(*) = (-1)^i \mathrm{id}_{S^i\mathscr{T}_X}$.

Consider the corresponding Picard algebroids $\widetilde{\mathscr{T}} = \widetilde{\mathscr{T}_D}$, $\widetilde{\mathscr{T}}^\circ = \widetilde{\mathscr{T}_{D^\circ}}$: one has an isomorphism of sheaves $* \colon \widetilde{\mathscr{T}} \xrightarrow{\sim} \widetilde{\mathscr{T}}^\circ$ such that $*([\partial_1, \partial_2]) = -[*\partial_1, *\partial_2]$, $*(f\partial) = f{*\partial} - \sigma(\partial)(f)$, $*i_{\widetilde{\mathscr{T}}}(f) = i_{\widetilde{\mathscr{T}}^\circ}(f)$, and $\sigma(*\partial) = -\sigma(\partial)$ for $\partial_1, \partial_2 \in \widetilde{\mathscr{T}}$, $f \in \mathscr{O}_X$. The Picard algebroid $\widetilde{\mathscr{T}}^\circ$ is canonically isomorphic to $\widetilde{\mathscr{T}_\Omega} - \widetilde{\mathscr{T}}$. Namely, $*\partial \mapsto s_{1,-1}(-\ell\sigma(\partial), -\partial)$ (see 2.1.3 for notation).

(2) We are grateful to V. Drinfeld whose remarks made the exposition of this section much clearer.

2.4.2. EXAMPLE. Consider the tdo $D_{\Omega^{1/2}}$. The above isomorphism identifies $\widetilde{\mathscr{T}}^{\circ}_{\Omega^{1/2}}$ with $\widetilde{\mathscr{T}}_{\Omega} - \widetilde{\mathscr{T}}_{\Omega^{1/2}} = \widetilde{\mathscr{T}}_{\Omega^{1/2}}$, therefore $D_{\Omega^{1/2}}$ is a self-dual tdo (i.e., we have the involution $*: D_{\Omega^{1/2}} \to D_{\Omega^{1/2}}$, $*^2 = \mathrm{id}$, $*(ab) = *(b) * (a)$). Denote by $D^{\pm}_{\Omega^{1/2}}$ the ± 1 eigenspaces of $*$ on $D_{\Omega^{1/2}}$. One has $\mathrm{gr}\, D^{+}_{\Omega^{1/2}} = \bigoplus S^{2i}\mathscr{T}_X$, $\mathrm{gr}\, D^{-}_{\Omega^{1/2}} = \bigoplus S^{2i+1}\mathscr{T}_X$. The $\pm$-grading is not compatible with the product, but it is compatible (in different ways) with the brackets $[\partial_1, \partial_2] = \partial_1\partial_2 - \partial_2\partial_1$ and the symmetrized product $\partial_1 \bullet \partial_2 = (1/2)(\partial_1\partial_2 + \partial_2\partial_1)$.

2.4.3. For a tdo D consider the graded $\mathbb{C}[t]$-algebra $\widehat{D}_{\cdot} := \bigoplus D_i$ (for $\partial_1 \in \widehat{D}_i = D_i$ and $\partial_2 \in \widehat{D}_j = D_j$ their product is $\partial_1\partial_2 \in D_{i+j}$, and $t\partial_1 = \partial_1 \in D_{i+1} = \widehat{D}_{i+1}$). For any nonzero $\lambda \in \mathbb{C}$ we have the isomorphism $\widehat{D}/(t-\lambda)\widehat{D} \xrightarrow{\sim} D$, $\partial \mapsto \lambda^i \partial$ for $\partial \in \widehat{D}_i = D_i$, and $\widehat{D}/t\widehat{D} = \mathrm{gr}\, D = S^{\cdot}\mathscr{T}_X$.

Consider the operations $\bullet$, $\{\cdot, \cdot\}$ on $\widehat{D}$ defined by the formulas $\partial_1 \bullet \partial_2 := (1/2)(\partial_1\partial_2 + \partial_2\partial_1)$, $\{\partial_1, \partial_2\} := (1/t)(\partial_1\partial_2 - \partial_2\partial_1)$. Then $\{\cdot, \cdot\}$ is a Lie algebra bracket, $\bullet$ is a commutative (nonassociative) product, and one has $\{\partial_1, \partial_2 \bullet \partial_3\} = \{\partial_1, \partial_2\} \bullet \partial_3 + \partial_2 \bullet \{\partial_1, \partial_3\}$. Put $\widetilde{\mathrm{gr}}\, D = \widehat{D}/t^2\widehat{D} = \bigoplus D_i/D_{i-2}$; the operations $\bullet$, $\{\cdot, \cdot\}$ induce the corresponding operations on the quotient $\widetilde{\mathrm{gr}}\, D$.

2.4.4. LEMMA. *The operation* $\bullet$ *on* $\widetilde{\mathrm{gr}}\, D$ *is associative. Therefore* $\widetilde{\mathrm{gr}}\, D$ *is a Poisson* $\mathbb{C}[t]/t^2$*-algebra (with respect to the product* $\bullet$ *and the Poisson bracket* $\{\cdot, \cdot\}$*).*

PROOF. For $\partial_j \in D_{i_j}$, $j = 1, 2, 3$, one has $(\partial_1 \bullet \partial_2) \bullet \partial_3 - \partial_1 \bullet (\partial_2 \bullet \partial_3) = (1/4)[\partial_2, [\partial_1, \partial_3]] \in D_{i_1+i_2+i_3-2}$.

We can describe this Poisson algebra as follows. Consider the Picard algebroid $\mathscr{T}^{\vee}_D := \widetilde{\mathscr{T}}_D - \widetilde{\mathscr{T}}_{\Omega^{1/2}}$. Let $(\psi, \pi_\psi, \omega_\psi)$ be its twisted cotangent bundle, and $A_{\cdot} := \mathscr{O}_\psi$ be the corresponding filtered commutative algebra with Poisson bracket $\{\cdot, \cdot\}$ (see 2.1.8). As above we have the graded Poisson $\mathbb{C}[t]$-algebra $\widehat{A} := \bigoplus A_i$ (one has $\{A_i, A_j\} \subset A_{i+j-1}$), and the quotient $\widetilde{\mathrm{gr}}\, A = \widehat{A}/t^2\widehat{A}$ which is a $\mathbb{C}[t]/t^2$-Poisson algebra.

2.4.5. PROPOSITION. *One has a canonical isomorphism of graded Poisson* $\mathbb{C}[t]/t^2$*-algebras* $\tilde{\sigma}: \widetilde{\mathrm{gr}}\, D \xrightarrow{\sim} \widetilde{\mathrm{gr}}\, A$ *that lifts the isomorphism* $\sigma: \widetilde{\mathrm{gr}}\, D/t\widetilde{\mathrm{gr}}\, D = \mathrm{gr}\, D \xrightarrow{\sim} S^{\cdot}\mathscr{T}_X = \widetilde{\mathrm{gr}}\, A/t\widetilde{\mathrm{gr}}\, A$.

PROOF. For $\tau \in \mathscr{T}_X$ denote by τ^- the unique element of $\widetilde{\mathscr{T}}_{\Omega^{1/2}} = (D^{-}_{\Omega^{1/2}})_1$ such that $\sigma(\tau^-) = \tau$; for $f \in \mathscr{O}_X$ one has $(f\tau)^- = f\tau^- + \frac{\tau(f)}{2}$. Consider the isomorphism of sheaves $\theta_1: \widetilde{\mathscr{T}}_D \xrightarrow{\sim} \mathscr{T}^{\vee}_D$, $\theta_1(\partial) = s_{1,-1}(\partial, \sigma(\partial)^-)$ (see 2.1.3 for $s_{1,-1}$). Then θ_1 is an isomorphism of extensions of $\mathscr{T}_X$ by $\mathscr{O}_X$, it commutes with brackets and for $f \in \mathscr{O}_X$, $\partial \in \widetilde{\mathscr{T}}_D$ one has $\theta_1(f \bullet \partial) = \theta_1(f\partial + 1/2\sigma(\partial)(f)) = f\theta_1(\partial)$. Therefore, for any $i \geqslant 0$ we have the isomorphism $\theta_1 = S^i_\bullet(\theta_1)\colon \mathscr{S}^i_\bullet\widetilde{\mathscr{T}_D} \xrightarrow{\sim} A_i$ where $S^i_\bullet\widetilde{\mathscr{T}_D}$ is the symmetric power with respect to the $\mathscr{O}_X$-module structure on $\widetilde{\mathscr{T}_D}$, $(f, \partial) \mapsto f \bullet \partial$. The graded ring $\widehat{\mathscr{S}} := \bigoplus S^i_\bullet\widetilde{\mathscr{T}_D}$ is a graded $\mathbb{C}[t]$-algebra in the usual manner (the multiplication by t is the multiplication by $1 \in \mathscr{O}_X \subset \widetilde{\mathscr{T}_D}$, which is an embedding $S^i_\bullet\widetilde{\mathscr{T}_D} \subset S^{i+1}_\bullet\widetilde{\mathscr{T}_D}$), and the bracket on $\widetilde{\mathscr{T}_D}$ defines the Poisson structure on $\mathscr{S}$. Clearly $\theta = \bigoplus \theta_i\colon \widehat{\mathscr{S}} \to \widehat{A}$ is an isomorphism of the graded Poisson $\mathbb{C}[t]$-algebras. On the other hand, the identity map $\mathrm{id}_{\widetilde{\mathscr{T}_D}}$ extends to a morphism $\psi\colon \widehat{\mathscr{S}} \to \widetilde{\mathrm{gr}}\, D$ of graded $\mathbb{C}[t]$-algebras. It commutes with Poisson brackets, and the induced map $\widehat{\mathscr{S}}/t^2\widehat{\mathscr{S}} \to \widetilde{\mathrm{gr}}\, D$ is an isomorphism of $\mathbb{C}[t]/t^2$-Poisson algebras. We put $\tilde{\sigma} := \theta\psi^{-1}\colon \widetilde{\mathrm{gr}}\, D \to \widehat{A}/t^2\widehat{A} = \widetilde{\mathrm{gr}}\, A$: this is the desired isomorphism.

REMARK. An explicit formula for the inverse isomorphism is $\tilde{\sigma}^{-1}\colon \widetilde{\mathrm{gr}}\, A \xrightarrow{\sim} \widetilde{\mathrm{gr}}\, D$. We have $\tilde{\sigma}_0^{-1} = \mathrm{id}_{\mathscr{O}_X}$, $\tilde{\sigma}_1^{-1}(a) = s_{1,1}(a, \sigma(a)^-)$ (here $a \in \widetilde{\mathrm{gr}}_1 A = A_1 = \widetilde{\mathscr{T}}^\vee_D$, and we identify $\widetilde{\mathrm{gr}}_1 D = D_1 = \widetilde{\mathscr{T}_D}$ with $\widetilde{\mathscr{T}}^\vee_D + \widetilde{\mathscr{T}}_{\Omega^{1/2}}$).

If $a = a_1 \cdots a_i \in \widetilde{\mathrm{gr}}_i A = A_i/A_{i-2}$, where $a_j \in A_1$, one has $\tilde{\sigma}_1^{-1}(a) = 1/i! \sum_g \tilde{\sigma}_1^{-1}(a_{g(1)}) \cdots \tilde{\sigma}_1^{-1}(a_{g(i)})$, where the sum is taken over all permutations g of i indices.

2.4.6. COROLLARY. *The boundary map*

$$\delta_D\colon H^i(X, S^j\mathscr{T}_X) \to H^{i+1}(X, S^{j-1}\mathscr{T}_X)$$

for the short exact sequence $0 \to S^{j-1}\mathscr{T}_X \to D_j/D_{j-2} \to S^j\mathscr{T}_X \to 0$ *is convolution with the class* $c_1(D) - \frac{1}{2}c_1(\Omega) \in H^1(X, \Omega^1_X)$.

2.5. Monodromic D-modules. Let H be a torus (i.e., an algebraic group isomorphic to a product of $\mathbb{C}^*$'s), $\mathfrak{h} = \mathrm{Lie}\, H$. For a variety X we will call an H-torsor $\pi\colon \widetilde{X} \to X$ over X an *H-monodromic structure* on X; we call a pair $(X, \widetilde{X})$ an *H-monodromic variety.*

2.5.1. Assume that X is smooth. An H-monodromic structure $\widetilde{X}$ on X defines, by 1.2.3, a Lie algebroid $\widetilde{\mathscr{T}} := \mathscr{T}_{\widetilde{X}}$ on X. The $\mathscr{O}_X$-Lie algebra $\widetilde{\mathscr{T}}^{(0)}$ coincides with $\tilde{\mathfrak{h}} = \mathscr{O}_X \otimes \mathfrak{h}$ identified with the (commutative) Lie algebra of vertical H-invariant vector fields; we have the short exact sequence $0 \to \tilde{\mathfrak{h}} \to \widetilde{\mathscr{T}} \xrightarrow{\sigma} \mathscr{T}_X \to 0$. The group H acts on $D_{\widetilde{X}}$ in a usual way. On X we have the D-algebras $\pi_* D_{\widetilde{X}} \supset \widetilde{D} := [\pi_* D_{\widetilde{X}}]^H$; clearly $\pi_* D_{\widetilde{X}} = \pi_*\mathscr{O}_{\widetilde{X}} \underset{\mathscr{O}_X}{\otimes} \widetilde{D}$ as an $\mathscr{O}_X$-module.

The embedding $\widetilde{\mathscr{T}} \subset \widetilde{D}$ induces an isomorphism $\mathscr{U}(\widetilde{\mathscr{T}}) \xrightarrow{\sim} \widetilde{D}$; in particular $S(\mathfrak{h})$ coincides with the center of $\widetilde{D}$ and $\widetilde{D}/\mathfrak{h}\widetilde{D} = D_X$ (since any (local) section of π, i.e., an isomorphism $\widetilde{X} = H \times X$, identifies $\widetilde{D}$ with $D_X \otimes \mathscr{U}(\mathfrak{h}) = D_X \otimes S(\mathfrak{h})$). For $\chi \in \mathfrak{h}^*$ denote by m_χ the corresponding maximal ideal of $S(\mathfrak{h})$. Then $D_\chi := \widetilde{D}/m_\chi \widetilde{D}$ is a (locally trivial) tdo on X; clearly D_χ depends on χ in a $\mathbb{C}$-linear way. Therefore, one can consider $\widetilde{D}$ as a "linear" family of tdo's on X parametrized by $\mathfrak{h}^*$.

2.5.2. A *monodromic D-module on* X is a weak $(D_{\widetilde{X}}, H)$-module (see 1.8); denote by $\widetilde{\mathscr{M}(X)} = \mathscr{M}(D_{\widetilde{X}}, H)_{\text{weak}}$ the corresponding category. According to 1.8.9, we have the mutually inverse equivalences of categories $\widetilde{\mathscr{M}(X)} \overset{\pi^\cdot}{\underset{\pi_\cdot}{\rightleftarrows}} \mathscr{M}(\widetilde{D})$ compatible with $\mathscr{O}$-tensor products of modules.

Let M be a $\widetilde{D}$-module. For an ideal $I \subset S(\mathfrak{h})$ put $M^I := \{m \in M : \mathrm{Im} = 0\}$; this is a $\widetilde{D}$-submodule of M. In particular for $\chi \in \mathfrak{h}^*$ we have the submodules $M_\chi := M^{m_\chi} \subset M_{\widetilde{\chi}} := \bigcup_{n\geqslant 1} M^{m_\chi^n}$; let M_{fin} be the union of the M^I's, where the I's are ideals of finite codimension in $S(\mathfrak{h})$. Clearly $M_{\text{fin}} = \bigoplus_{\chi\in\mathfrak{h}^*} M_{\widetilde{\chi}}$. Denote by $\mathscr{M}(\widetilde{D})_{\text{fin}} \subset \mathscr{M}(\widetilde{D})$ the full subcategory of $\mathfrak{h}$-*finite* modules, i.e., such M that $M_{\text{fin}} = M$; we also have the full subcategories $\mathscr{M}(X)_\chi := \mathscr{M}(D_\chi) = \{M \in \mathscr{M}(\widetilde{D}) : M_\chi = M\} \subset \mathscr{M}(X)_{\widetilde{\chi}} := \{M \in \mathscr{M}(\widetilde{D}) : M_{\widetilde{\chi}} = M\}$, so $\mathscr{M}(\widetilde{D})_{\text{fin}} = \prod_{\chi\in\mathfrak{h}^*} \mathscr{M}(X)_{\widetilde{\chi}}$. The equivalence $\pi^\cdot$ sends $\mathscr{M}(\widetilde{D})_{\text{fin}}$ to the subcategory $\widetilde{\mathscr{M}(X)}_{\text{fin}} \subset \widetilde{\mathscr{M}(X)}$; we will identify $\mathscr{M}(X)_\chi$, $\mathscr{M}(X)_{\widetilde{\chi}}$ with the corresponding subcategories of $\mathscr{M}(X)$.

2.5.3. Consider now the category $\mathscr{M}(\widetilde{X})$ of $D_{\widetilde{X}}$-modules. The projection π is affine, therefore $\pi_* : \mathscr{M}(\widetilde{X}) \to \mathscr{M}(\pi_* D_{\widetilde{X}}) = \mathscr{M}(X, \pi_* D_{\widetilde{X}})$ is an equivalence of categories.

Let $\mathfrak{h}^*_{\mathbb{Z}} = \mathrm{Hom}(H, \mathbb{C}^*) \subset \mathfrak{h}^*$ be the lattice of integral weights; for $\chi \in \mathfrak{h}^*$ denote by χ^* the corresponding translation automorphism of $S(\mathfrak{h})$, $\chi^*(h) = h + \chi(h)$, $h \in \mathfrak{h}$. For an ideal $I \subset S(\mathfrak{h})$ and a $\pi_* D_{\widetilde{X}}$-module N we have the $\widetilde{D}$-submodule $N^I \subset N$. Note that $N^{\overline{I}} := \sum_{\chi\in\mathfrak{h}^*_{\mathbb{Z}}} N^{\chi^* I}$ is actually a $\pi_* D_{\widetilde{X}}$-submodule. In particular, for $\overline{\chi} \in \mathfrak{h}^*/\mathfrak{h}^*_{\mathbb{Z}}$ we have the $\pi_* D_{\widetilde{X}}$-submodules $N_{\widetilde{\chi}} := N^{\overline{m}_\chi} \subset N_{\widetilde{\overline{\chi}}} := \bigcup_n N^{\overline{m}_\chi^n}$, where $\chi \in \mathfrak{h}^*$ is any lifting of $\overline{\chi}$, and also the submodule $N_{\text{fin}} := \bigcup N^{\overline{I}}$, $I \subset S(\mathfrak{h})$, has finite codimension.

Clearly $N_{\text{fin}} = \bigoplus_{\overline{\chi}\in\mathfrak{h}^*/\mathfrak{h}^*_{\mathbb{Z}}} N_{\widetilde{\overline{\chi}}}$. Again, we say that N is an $\mathfrak{h}$-*finite* module if $N_{\text{fin}} = N$. We have the corresponding full subcategories $\mathscr{M}(\pi_* D_{\widetilde{X}})_{\overline{\chi}} \subset \mathscr{M}(\pi_* D_{\widetilde{X}})_{\widetilde{\overline{\chi}}} \subset \mathscr{M}(\pi_* D_{\widetilde{X}})_{\text{fin}} \subset \mathscr{M}(\pi_* D_{\widetilde{X}})$ and the decomposition $\mathscr{M}(\pi_* D_{\widetilde{X}})_{\text{fin}} = \prod_{\overline{\chi}\in\mathfrak{h}^*/\mathfrak{h}^*_{\mathbb{Z}}} \mathscr{M}(\pi_* D_{\widetilde{X}})_{\widetilde{\overline{\chi}}}$. The above equivalence $(\pi_*)^{-1}$ sends these subcategories to $\mathscr{M}(\widetilde{X})_{\overline{\chi}} \subset \mathscr{M}(\widetilde{X})_{\widetilde{\overline{\chi}}} \subset \mathscr{M}(\widetilde{X})_{\text{fin}} \subset \mathscr{M}(\widetilde{X})$.

Consider the forgetting of the H-action functor $o : \widetilde{\mathscr{M}(X)} \to \mathscr{M}(\widetilde{X})$. The equivalences $\pi_\cdot$, π_* identify o with the induction functor $\mathscr{M}(\widetilde{D}) \to \mathscr{M}(\pi_* D_{\widetilde{X}})$, $M \mapsto \pi_* D_{\widetilde{X}} \underset{\widetilde{D}}{\otimes} M = \pi_* \mathscr{O}_{\widetilde{X}} \underset{\mathscr{O}_X}{\otimes} M$. Clearly o sends $\mathfrak{h}$-finite modules to $\mathfrak{h}$-finite ones, $\mathscr{M}(X)_\chi$ to $\mathscr{M}(\widetilde{X})_{\overline{\chi}}$, and $\mathscr{M}(X)_{\widetilde{\chi}}$ to $\mathscr{M}(\widetilde{X})_{\widetilde{\overline{\chi}}}$, where $\overline{\chi} = \chi \bmod \mathfrak{h}^*_{\mathbb{Z}}$.

2.5.4. LEMMA. (i) *The functors* $o\colon \mathcal{M}(X)_{\chi} \to \mathcal{M}(\widetilde{X})_{\widetilde{\chi}}$, $\mathcal{M}(X)_{\widetilde{\chi}} \to \mathcal{M}(\widetilde{X})_{\widetilde{\widetilde{\chi}}}$ *are equivalences of categories.*

(ii) *The corresponding functor between the derived categories* $D^b\mathcal{M}(X)_{\widetilde{\chi}} \to D^b\mathcal{M}(\widetilde{X})$ *is fully faithful. It identifies* $D^b\mathcal{M}(X)_{\widetilde{\chi}}$ *with the full subcategory of those complexes that have cohomology in* $\mathcal{M}(\widetilde{X})_{\widetilde{\widetilde{\chi}}}$.

PROOF. (i) The inverse functor to the induction $\mathcal{M}(X)_{\widetilde{\chi}} \to \mathcal{M}(\pi_* D_{\widetilde{X}})_{\widetilde{\widetilde{\chi}}}$ sends a $\pi_* D_{\widetilde{X}}$-module N to the $\widetilde{D}$-module $N_{\widetilde{\chi}} := \bigcup N^{m^n_\chi}$.

(ii) Since our functor is t-exact it suffices to check the first statement (i.e., to show that o induces isomorphism on Ext's). A standard Čech resolvent argument shows that the problem is X-local, hence we can assume that X is affine, $\widetilde{X} = H \times X$. It suffices to verify that Ext's are the same for a family of generators; take one formed by the modules $\pi^{\cdot}(D_X \otimes V)$, V is an $S(\mathfrak{h})$-module killed by some m^n_χ. The Künneth formula reduces the problem to the case X = point, $H = \mathbb{C}^*$, where it is obvious.

2.5.5. REMARKS. (i) For $\varphi \in \mathfrak{h}^*_{\mathbb{Z}} = \mathrm{Hom}(H, \mathbb{C}^*)$ put $\mathcal{O}_\varphi := \{f \in \pi_*\mathcal{O}_{\widetilde{X}} : f(h\widetilde{x}) = \varphi(h)f(\widetilde{x}), h \in H, \widetilde{x} \in \widetilde{X}\}$; this is a line bundle on X which is a $\widetilde{D}$-module in an obvious manner. Note that $\pi_*\mathcal{O}_{\widetilde{X}} = \bigoplus_{\varphi\in\mathfrak{h}^*_{\mathbb{Z}}} \mathcal{O}_\varphi$. We have an autoequivalence $T_\varphi\colon \mathcal{M}(\widetilde{D}) \xrightarrow{\sim} \mathcal{M}(\widetilde{D})$, $T_\varphi(M) = \mathcal{O}_\varphi \underset{\mathcal{O}_X}{\otimes} M$, which preserves $\mathcal{M}(\widetilde{D})_{\mathrm{fin}}$ and sends $\mathcal{M}(X)_{\widetilde{\chi}}$ to $\mathcal{M}(X)_{\widetilde{\chi+\varphi}}$. Clearly $oT_\varphi = o$.

(ii) For a section $\mathfrak{s}\colon \mathfrak{h}^*/\mathfrak{h}^*_{\mathbb{Z}} \to \mathfrak{h}^*$ put $\mathcal{M}(X)_{\widetilde{\mathfrak{s}}} = \pi_{\widetilde{\chi}\in\mathfrak{h}^*/\mathfrak{h}^*_{\mathbb{Z}}}\mathcal{M}(X)_{\widetilde{\mathfrak{s}(\widetilde{\chi})}}$. Then $o\colon \mathcal{M}(X)_{\widetilde{\mathfrak{s}}} \to \mathcal{M}(\widetilde{X})_{\mathrm{fin}}$ is an equivalence of categories. One can say that $\mathcal{M}(\widetilde{X})_{\mathrm{fin}}$ is a quotient of $\mathcal{M}(\widetilde{D})_{\mathrm{fin}}$ with respect to the action of $\mathfrak{h}^*_{\mathbb{Z}}$ by T_φ's.

(iii) $\mathfrak{h}^*/\mathfrak{h}^*_{\mathbb{Z}}$ is the character group of the fundamental group of H; the exponential map identifies $h^*/h^*_{\mathbb{Z}}$ with the dual torus $H^\vee$. The Riemann-Hilbert correspondence identifies tame $D_{\widetilde{X}}$-modules from $\mathcal{M}(\widetilde{X})_{\widetilde{\widetilde{\chi}}}$ with the $\widetilde{\widetilde{\chi}}$-monodromic perverse sheaves on $\widetilde{X}$ (those perverse sheaves that are lissé along the fibers and have $\overline{\chi}$ as eigenvalues of fiberwise monodromy).

(iv) We will use 2.5.4 to transmit the standard results about D-modules to the monodromic $\mathfrak{h}$-finite situation (without repeating the proofs). For example, the derived category $D^b\mathcal{M}(\widetilde{X})^{\mathrm{coh}}_{\mathrm{fin}} \subset D^b\mathcal{M}(\widetilde{X})^{\mathrm{coh}}$, where "coh" means "complexes with coherent cohomology", is stable with respect to the Verdier duality which sends $D^b\mathcal{M}(\widetilde{X})^{\mathrm{coh}}_{\widetilde{\widetilde{\chi}}}$ to $D^b\mathcal{M}(\widetilde{X})^{\mathrm{coh}}_{\widetilde{\widetilde{-\chi}}}$. Therefore, we have a duality on $D^b\mathcal{M}(\widetilde{D})^{\mathrm{coh}}_{\mathrm{fin}}$ which sends $D^b\mathcal{M}(X)^{\mathrm{coh}}_{\widetilde{\chi}}$ to $D^b\mathcal{M}(X)^{\mathrm{coh}}_{\widetilde{-\chi}}$; it induces a duality on the abelian category of holonomic modules $\mathcal{M}(\widetilde{D})^{\mathrm{hol}}_{\mathrm{fin}}$ which sends $\mathcal{M}(X)^{\mathrm{hol}}_{\widetilde{\chi}}$ to $\mathcal{M}(X)^{\mathrm{hol}}_{\widetilde{-\chi}}$. The same thing happens with the $\mathcal{O}$-tensor product.

If $Y \subset X$ is a smooth subvariety with the closure $\overline{Y}$, then Y carries the induced monodromic structure $\widetilde{Y} = \pi^{-1}(Y)$, and we have the Kashiwara equivalence between $\mathcal{M}(Y)_{\widetilde{\chi}}$ and the quotient of the subcategory of $\mathcal{M}(X)_{\widetilde{\chi}}$ which consists of modules supported on $\overline{Y}$ modulo those supported on $\overline{Y}\backslash Y$. We leave to the reader the general functoriality with respect to morphisms between the monodromic varieties.

(v) The monodromic category $\widetilde{\mathcal{M}}(X)$ depends only on H modulo isogeny (if $Q \subset H$ is a finite subgroup, then the $\widetilde{D}$ algebras for the H- and H/Q-monodromic structures $\widetilde{X}$, respectively $Q \setminus \widetilde{X}$, coincide).

2.5.6. Let us discuss briefly the equivariant setting. Let G be an algebraic group, and $\kappa: G \to \operatorname{Aut} H$ be an action of G on H. Since $\operatorname{Aut} H$ is discrete, κ is trivial on the connected component G^o of G and $\kappa(G)$ is finite. An *H-monodromic G-variety* is an H-monodromic variety $(X, \widetilde{X})$ together with a G-action $\mu: G \times \widetilde{X} \to \widetilde{X}$ such that $gh\widetilde{x} = \kappa(g)(h)g\widetilde{x}$ for $g \in G$, $h \in H$, $\widetilde{x} \in \widetilde{X}$; in particular $\widetilde{\mu}$ descends to an action $\mu: G \times X \to X$. Equivalently, this is a variety $\widetilde{X}$ with an action of the κ-semidirect product $G \ltimes H$ such that H acts on $\widetilde{X}$ in a free way. The infinitesimal action $\widetilde{\alpha}: \mathfrak{g} \to \mathcal{T}_{\widetilde{X}}$ sends $\mathfrak{g}$ to $\widetilde{\mathcal{T}} \subset \pi_* \mathcal{T}_{\widetilde{X}}$, therefore G acts on the Lie algebroid $\widetilde{\mathcal{T}}$ and the D-algebra $\widetilde{D}$. Note that the induced action on $\mathfrak{h} \subset \widetilde{\mathcal{T}}$ coincides with κ. A *G-equivariant monodromic module* on X is a $D_{\widetilde{X}}$-module equipped with a weak $(G \ltimes H)$-action that is strong along G.

Such modules form a category $\widetilde{\mathcal{M}}(X, G)$; we have the following equivalence of categories $\pi_\cdot: \widetilde{\mathcal{M}}(X, G) \xrightarrow{\sim} \mathcal{M}(\widetilde{D}, G)$. The $\mathfrak{h}$-finite G-equivariant $D_{\widetilde{X}}$-modules form a full subcategory $\widetilde{\mathcal{M}}(X, G)_{\text{fin}} \subset \widetilde{\mathcal{M}}(X, G)$; for a κ-orbit $\chi^{\cdot} \in \kappa(G) \setminus \mathfrak{h}^*$ we have the corresponding subcategories $\mathcal{M}(X, G)_{\chi^{\cdot}} \subset \mathcal{M}(X, G)_{\widetilde{\chi}^{\cdot}} \subset \widetilde{\mathcal{M}}(X, G)_{\text{fin}}$; this provides a decomposition $\widetilde{\mathcal{M}}(X, G)_{\text{fin}} = \prod \mathcal{M}(X, G)_{\widetilde{\chi}^{\cdot}}$.

The following easy lemma shows that it suffices to consider only one-element orbits.

2.5.7. Lemma. *For $\chi \in \mathfrak{h}^*$ let G_χ be the stabilizer of χ with respect to the κ-action, and let $\chi^{\cdot}$ be the $\kappa(G)$-orbit of χ. Then the functor $\mathcal{M}(X, G)_{\chi^{\cdot}} \to \mathcal{M}(X, G_\chi)_{\widetilde{\chi}}$, $M \mapsto M_{\widetilde{\chi}}$ is an equivalence of categories.*

2.5.8. Consider now the category $\mathcal{M}(\widetilde{X}, G)$ of G-equivariant $D_{\widetilde{X}}$-modules; we have the corresponding full subcategories $\mathcal{M}(\widetilde{X}, G)_{\overline{\chi}^{\cdot}} \subset \mathcal{M}(\widetilde{X}, G)_{\widetilde{\overline{\chi}}^{\cdot}} \subset \mathcal{M}(\widetilde{X}, G)_{\text{fin}} \subset \mathcal{M}(\widetilde{X}, G)$, where $\overline{\chi}^{\cdot} \in \kappa(G)\backslash(\mathfrak{h}^*/\mathfrak{h}^*_{\mathbb{Z}})$, such that $\mathcal{M}(\widetilde{X}, G)_{\text{fin}} = \prod \mathcal{M}(\widetilde{X}, G)_{\widetilde{\overline{\chi}}^{\cdot}}$. The functor $\widetilde{\mathcal{M}}(X, G) \to \mathcal{M}(\widetilde{X}, G)$ of forgetting of the H-action sends $\mathcal{M}(X, G)_{\widetilde{\chi}^{\cdot}}$ to $\mathcal{M}(\widetilde{X}, G)_{\widetilde{\overline{\chi}}^{\cdot}}$. If $\chi \in \mathfrak{h}^*$ is a weight such that the $\kappa(G)$-stabilizers of χ and $\overline{\chi}$ coincide (e.g., if χ is fixed by $\kappa(G)$-action), then this functor is an equivalence of categories.

As in 2.5.5(iv), the elementary functoriality of G-equivariant D-modules translates immediately to the monodromic setting. For example, we have the duality on the subcategory $\widetilde{\mathcal{M}}(X, G)^{\mathrm{hol}}_{\mathrm{fin}} \subset \widetilde{\mathcal{M}}(X, G)_{\mathrm{fin}}$ of holonomic modules that sends $\mathcal{M}(X, G)_{\widetilde{\chi}}$ to $\mathcal{M}(X, G)_{-\widetilde{\chi}}$. For a smooth G-subvariety $Y \subset X$ we have the Kashiwara theorem that identifies the category $\mathcal{M}(Y, G)_{\widetilde{\chi}}$ with the subquotient category of $\mathcal{M}(X, G)_{\widetilde{\chi}}$. If $G' \supset G$ is a larger group and κ extends to $\kappa' \colon G' \to \operatorname{Aut} H$, then we have the induced H-monodromic G'-variety $(X, \widetilde{X}') = (G' \underset{G}{\times} X, G' \underset{G}{\times} \widetilde{X})$ and the pull-back functor for $\widetilde{X} \hookrightarrow \widetilde{X}'$ provides an equivalences of categories $\widetilde{\mathcal{M}}(X', G') \xrightarrow{\sim} \widetilde{\mathcal{M}}(X, G)$, $\mathcal{M}(\widetilde{X}', G') \xrightarrow{\sim} \mathcal{M}(\widetilde{X}, G)$ that identifies the corresponding categories of $\mathfrak{h}$-finite modules.

2.6. Langlands classification. Let $(X, \widetilde{X})$ be an H-monodromic G-variety such that X has only finitely many G-orbits. We present an explicit classification of the irreducible objects in $\widetilde{\mathcal{M}}(X, G)_{\mathrm{fin}} = \mathcal{M}(\widetilde{D}, G)_{\mathrm{fin}}$.

2.6.1. For $x \in X$ consider the action of the stabilizer $G_x \subset G$ on the H-torsor $\widetilde{X}_x = \pi^{-1}(x)$. Since the $(G^\kappa := \operatorname{Ker}\kappa)$- and H-actions commute we see that $G^\kappa_x := G^\kappa \cap G_x$ acts on $\widetilde{X}_x$ via the morphism $\varphi_x \colon G^\kappa_x \to H$. Note that the connected component $(\operatorname{Ker}\varphi_x)^o$ coincides with the connected component $G^o_{\widetilde{x}}$ of the stabilizer $G_{\widetilde{x}}$ of any $\widetilde{x} \in \widetilde{X}_x$. This is a normal subgroup of G_x; put $G_{(x)} := G_x / G^o_{\widetilde{x}}$. Both the G-actions on $\widetilde{X}_x$ and κ factorize through $G_{(x)}$, and φ_x defines the embedding $G^o_{(x)} \hookrightarrow H$. In particular, we have $\mathfrak{g}_{(x)} := \operatorname{Lie} G_{(x)} \hookrightarrow \mathfrak{h}$; together with κ, this defines a Harish-Chandra pair $(G_{(x)}, \mathfrak{h})$.

2.6.2. Assume that our X is a single orbit, so $X = Gx$. Then we can identify $(X, \widetilde{X})$ with an H-monodromic G-variety induced from the H-monodromic G_x-variety $(x, \widetilde{X}_x)$. Hence we have canonical equivalences of categories $\widetilde{\mathcal{M}}(X, G) = \widetilde{\mathcal{M}}(x, G_x) = \mathcal{M}(S(\mathfrak{h}), G_{(x)}) = \mathcal{M}(\mathfrak{h}, G_{(x)})$ (the category of Harish-Chandra modules for the pair $(\mathfrak{h}, G_{(x)})$). We see that $\widetilde{\mathcal{M}}(X, G)_{\mathrm{fin}} = \prod \mathcal{M}(\mathfrak{h}, G_{(x)})_{\widetilde{\chi}}$ has finitely many irreducibles. By 2.5.7 we can identify $\mathcal{M}(\mathfrak{h}, G_{(x)})_{\widetilde{\chi}}$ with $\mathcal{M}(\mathfrak{h}, G_{(x)\chi})_{\widetilde{\chi}}$. The categories $\widetilde{\mathcal{M}}(X, G)_\chi = \mathcal{M}(\mathfrak{h}, G_{(x)})_\chi$ are semisimple. We have a similar description of $\mathcal{M}(\widetilde{X}, G)_{\mathrm{fin}}$ (see 2.5.8). Note that any coherent $M \in \mathcal{M}(\widetilde{X}, G)_{\mathrm{fin}}$ is tame and lissé (i.e., RS holonomic $\mathcal{O}$-coherent in the terminology of [Bo]); any $M \in \mathcal{M}(\widetilde{X}, G)_{\overline{\chi}}$ for $\overline{\chi} \in \mathfrak{h}^*_{\mathbb{Q}}/\mathfrak{h}^*_{\mathbb{Z}} = \mathbb{Q}/\mathbb{Z} \otimes \mathfrak{h}^*_{\mathbb{Z}} = H^\vee_{\mathrm{tors}}$ has finite monodromy and hence is of geometric origin (see [BBD, (6.2.4)]).

2.6.3. Now consider the general case. Let I be the set of G-orbits on X; this is a finite partially ordered set. For $i \in I$ let Q_i be the corresponding orbit; then $i_1 \leqslant i_2$ means that $Q_{i_1} \subset \overline{Q}_{i_2}$. We will say that a subset $J \subset I$

is *closed* if for $j \in J$ any $j' \leqslant j$ lies in J; e.g., for $i \in I$ the set $\bar{\imath} := \{j \in I, j \leqslant i\}$ is closed. The closed subsets of I form a lattice $A(I)$; we can identify $A(I)$ with the lattice of all closed G-invariant subsets of X sending J to $Q_J = \bigcup_{j \in J} Q_j$ (then $Q_{\bar{\imath}} = \overline{Q}_i$). For $J \in A(I)$ denote by $\widetilde{\mathcal{M}}(X, G)_J \subset \widetilde{\mathcal{M}}(X, G)$, $\mathcal{M}(\widetilde{X}, G)_J \subset \mathcal{M}(\widetilde{X}, G)$ the full subcategories of modules supported on $\widetilde{X}_J$.

Let us fix this kind of frame. Below I could be any finite partially ordered set.

2.6.4. Definition. Let C be an abelian category. An *I-stratification on* C is a collection C_J, $J \in A(I)$, of Serre subcategories of C such that for any $J_1, J_2 \subset A(I)$ one has:

(i) $C_{J_1 \cap J_2} = C_{J_1} \cap C_{J_2}$ and $C_{J_1 \cup J_2}$ is the smallest Serre subcategory that contains C_{J_1}, C_{J_2}. In particular if $J_1 \subset J_2$ then $C_{J_1} \subset C_{J_2}$.

(ii) The embeddings

$$\begin{array}{ccccc} & & C_{J_1} & & \\ & \nearrow & & \searrow & \\ C_{J_1 \cap J_2} & & & & C_{J_1 \cup J_2} \\ & \searrow & & \nearrow & \\ & & C_{J_2} & & \end{array}$$

induce equivalences of categories

$$(C_{J_1}/C_{J_1 \cap J_2}) \times (C_{J_2}/C_{J_1 \cap J_2}) \to C_{J_1 \cup J_2}/C_{J_1 \cap J_2}.$$

(iii) For $J_1 \subset J_2$ the projection $C_{J_2} \to C_{J_2}/C_{J_1}$ has left and right adjoints denoted $j_{J_2 \setminus J_1 !}$ and $j_{J_2 \setminus J_1 *}$.

Our categories $\widetilde{\mathcal{M}}(X, G) \supset \widetilde{\mathcal{M}}(X, G)_{\text{fin}} \supset \mathcal{M}(X, G)_{\tilde{\chi}^{\cdot}} \supset \mathcal{M}(X, G)_{\chi^{\cdot}}$, $\mathcal{M}(\widetilde{X}, G) \supset \mathcal{M}(\widetilde{X}, G)_{\text{fin}} \supset \cdots$ are I-stratified.

In any I-stratified category one has the standard devissage pattern. Namely, for $i \in I$ put $C_i := C_{\bar{\imath}}/C_{\bar{\imath} \setminus \{i\}}$; we will call C_i the *i-stratum* of C. We have the functors $j_{i!}, j_{i*}: C_i \to C_{\bar{\imath}}$ left and right adjoint to the projection $j_i^*: C_{\bar{\imath}} \to C_i$. Since $j_i^* j_{i*} = j_i^* j_{i!} = \mathrm{Id}_{C_i}$ we have a natural morphism $j_{i!} \to j_{i*}$. Put $j_{i!*} := \mathrm{Im}(j_{i!} \to j_{i*})$; this functor transforms irreducible objects to irreducible ones.

We say that C is finite if any object of C has finite length. The devissage shows that this is equivalent to the property that objects in C_i have finite length. In this situation any irreducible object of C is isomorphic to some $j_{i!*}(F_i)$, where F_i is an irreducible object in C_i, and the pair (i, F_i) is uniquely determined.

Let us summarize our discussion.

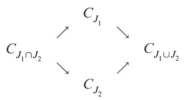

2.6.5. Lemma. (i) *The categories $\mathcal{M}(\widetilde{X}, G)$, $\widetilde{\mathcal{M}}(X, G)$ and their standard subcategories are I-stratified. The i strata coincide with the corresponding*

categories for the orbits X_i. *For* $x \in X_i$ *we have the canonical equivalence* $F_x : \widetilde{\mathcal{M}}(\widetilde{X}, G)_i \xrightarrow{\sim} \mathcal{M}(\mathfrak{h}, G_{(x)})$.

(ii) *The categories* $\widetilde{\mathcal{M}}(X, G)^{\mathrm{coh}}_{\mathrm{fin}}$, $\mathcal{M}(\widetilde{X}, G)^{\mathrm{coh}}_{\mathrm{fin}}$ *are finite. The isomorphism classes of irreducible objects of* $\widetilde{\mathcal{M}}(X, G)_{\mathrm{fin}}$ *are in* 1-1 *correspondence with pairs* (i, V), $i \in I$, V *is an isomorphism class of irreducible* $(\mathfrak{h}, G_{(x)})$-*modules (where* $x \in X_i$*).*

(iii) *Any* $M \in \mathcal{M}(\widetilde{X}, G)^{\mathrm{coh}}_{\mathrm{fin}}$ *is tame. If* $\chi^{\cdot} \in \kappa(G) \setminus H^{\vee}_{\mathrm{tors}} = \kappa(G) \setminus (\mathfrak{h}^*_{\mathbb{Q}}/\mathfrak{h}^*_{\mathbb{Z}})$ *then any irreducible* $M \in \mathcal{M}(\widetilde{X}, G)_{\chi^{\cdot}}$ *is of geometric origin.*

2.6.6. REMARKS. (i) The modules $j_{i!}(V)$, $j_{i*}(V)$, where $V \in \mathcal{M}(\mathfrak{h}, G_{(x)}) = \mathcal{M}(Q_i, G)$, are called !-, $*$-*standard modules* respectively.

(ii) If the embedding $j_i : Q_i \hookrightarrow X$ is affine, then the functors $j_{i!}$, j_{i*} are exact.

(iii) If C is an I-stratified category then the dual category C^o is I-stratified by C^o_J's. The duality interchanges the functors $j_{i!}$ and j_{i*}.

(iv) Let C_a, $a = 1, 2$, be I_a-stratified categories, $\varphi : I_1 \to I_2$ be a morphism of partially ordered sets, and $F : C_1 \to C_2$ be an exact functor. We say that F is a *φ-stratified functor* if $F(C_{1\bar{i}}) \subset C_{2\overline{\varphi(i)}}$ for $i \in I$. Such an F induces the exact functors $F_i : C_{1i} \to C_{2\varphi(i)}$ called the strata of F. Our functor is a *stratified equivalence* if φ is an isomorphism; F is an equivalence of categories and $F(C_{1i}) = C_{2i}$ for any $i \in I_1$. Any stratified equivalence commutes with the $j_{i!}$'s and the j_{i*}'s.

§3. Localization of representations; the structure of K-orbits on the flag variety

3.1. $\mathfrak{g}$-modules. Let $\mathfrak{g}$ be a complex semisimple Lie algebra. Denote by G the algebraic group of automorphisms of $\mathfrak{g}$, so G^o is the adjoint group, and $\mathfrak{g} = \operatorname{Lie} G$; the action of $g \in G$ on $\mathfrak{g}$ will be denoted Ad_g. Let $\mathcal{U}(\mathfrak{g})$ be the universal enveloping algebra and $\mathcal{Z} \subset \mathcal{U}(\mathfrak{g})$ be its center.

Let $\mathfrak{h}$ be the Cartan algebra of $\mathfrak{g}$, $\Delta \subset \mathfrak{h}^*$ the root system, Δ^+ the set of positive roots, $\Sigma \subset \Delta^+$ the set of simple roots, W the Weyl group, and $\rho := 1/2 \sum_{\nu \in \Delta^+} \nu$; for $\alpha \in \Delta$ let h_α be the corresponding co-root and $\sigma_\alpha \in W$ the corresponding reflection. So for any Borel subalgebra $\mathfrak{b} \subset \mathfrak{g}$ and $\mathfrak{n} = \mathfrak{n}_{\mathfrak{b}} := [\mathfrak{b}, \mathfrak{b}]$ we have canonical identification $\mathfrak{h} = \mathfrak{b}/\mathfrak{n}$ invariant under G^o-conjugation, and Δ^+ are weights of $\mathfrak{h}$-action on $\mathfrak{g}/\mathfrak{b} \simeq \mathfrak{n}^*$. ([3])

We will think of W as the group of affine transformations of $\mathfrak{h}^*$ that leave $-\rho$ fixed; this defines an action of W on the algebra $S(\mathfrak{h})$. One has the Harish-Chandra isomorphism $\gamma : \mathcal{Z} \xrightarrow{\sim} S(\mathfrak{h})^W$; let $\gamma^{\cdot} : \mathfrak{h}^* = \operatorname{Spec} S(\mathfrak{h}) \to \operatorname{Spec} \mathcal{Z}$ be the corresponding W-sheeted map of spectra. For $\chi \in \mathfrak{h}^*$ we denote by $m_\chi \subset S(\mathfrak{h})$, $m_{\gamma(\chi)} \subset \mathcal{Z}$ the corresponding maximal ideals.

([3]) People often use the opposite ordering of Δ; we choose the one for which dominant weights correspond to positive line bundles on the flag space.

Denote by $U := \mathscr{U}(\mathfrak{g}) \underset{\mathscr{Z}}{\otimes} S(\mathfrak{h})$ the extended universal enveloping algebra; then $S(\mathfrak{h})$ is the center of U, the group W acts on U (via $S(\mathfrak{h})$), and $\mathscr{U}(\mathfrak{g}) = U^W$. The algebras $\mathscr{U}(\mathfrak{g})$, $S(\mathfrak{h})$, U carry canonical involutions (anti-automorphisms of order 2), denoted by $x \mapsto {}^t x$, compatible with the standard embeddings: ${}^t\xi = -\xi$, ${}^t h = -2\rho(h) - h$ for $\xi \in \mathfrak{g} \subset \mathscr{U}(\mathfrak{g}) \subset U$, $h \in \mathfrak{h} \subset S(\mathfrak{h}) \subset U$; clearly ^{t} commutes with the W-action. Denote by $S(\mathfrak{h})^{\mathrm{reg}}$ the localization of $S(\mathfrak{h})$ off the nonregular hyperplanes for the W-action (so $\mathbb{C}$-points of $S(\mathfrak{h})^{\mathrm{reg}}$ are regular weights); if A is any $S(\mathfrak{h})$-algebra put $A^{\mathrm{reg}} := S(\mathfrak{h})^{\mathrm{reg}} \underset{S(\mathfrak{h})}{\otimes} A$. In particular, we have the algebra U^{reg} and ^{t} extends to U^{reg}. The group G acts on all the above objects in a compatible way; the action on $\mathfrak{h}$, Δ, and W factors through the finite quotient G/G^o. The action of G/G^o on $\mathfrak{h}$ is faithful; we will denote it by κ.

Let $\mathscr{M}(\mathfrak{g})$ and $\mathscr{M}(U)$ be the categories of left $\mathscr{U}(\mathfrak{g})$- and U-modules and let $\mathscr{M}(U)^{\text{f.g.}} \subset \mathscr{M}(U)$ be the subcategory of finitely generated ones. The embedding $\mathscr{U}(\mathfrak{g}) \subset U$ defines an obvious functor $\mathscr{M}(U) \to \mathscr{M}(\mathfrak{g})$. We also consider the categories ${}^r\mathscr{M}$ of the right modules; we will identify ${}^r\mathscr{M}$ with M in a canonical way using ^{t}.

For any ideals $I \subset \mathscr{Z}$, $J \subset S(\mathfrak{h})$ let $\mathscr{M}(\mathfrak{g})^I = \mathscr{M}(\mathscr{U}(\mathfrak{g})/I\mathscr{U}(\mathfrak{g}))$, $\mathscr{M}(U)^J = \mathscr{M}(U/JU)$ be the categories of $\mathfrak{g}$- and U-modules killed by I, J respectively. For $\chi \in \mathfrak{h}^*$ put $U_\chi := U/m_\chi U$; we have the categories $\mathscr{M}(U)_\chi := \mathscr{M}(U_\chi) = \mathscr{M}(U)^{m_\chi}$ and $\mathscr{M}(U)_{\widetilde{\chi}} := \{M \in \mathscr{M}(U)\colon$ any $m \in M$ is killed by some power of $m_\chi\} \subset \mathscr{M}(U)$; we also have the corresponding quotient $\mathscr{U}(\mathfrak{g})_{\gamma(\chi)} = \mathscr{U}(\mathfrak{g})/m_{\gamma(\chi)}\mathscr{U}(\mathfrak{g})$ and the full subcategories $\mathscr{M}(\mathfrak{g})_{\gamma(\chi)} \subset \mathscr{M}(\mathfrak{g})_{\widetilde{\gamma(\chi)}} \subset \mathscr{M}(\mathfrak{g})$. Note that the embedding $\mathscr{U}(\mathfrak{g}) \hookrightarrow U$ induces an isomorphism $\mathscr{U}(\mathfrak{g})_{\gamma(\chi)} \to U_\chi$ for any $\chi \in \mathfrak{h}^*$; if χ is regular then $\mathscr{U}(\mathfrak{g})/m^n_{\gamma(\chi)}\mathscr{U}(\mathfrak{g}) = U/m^n_\chi U$ for any n. The above functor sends $\mathscr{M}(U)_\chi$, $\mathscr{M}(U)_{\widetilde{\chi}}$ to $\mathscr{M}(\mathfrak{g})_{\gamma(\chi)}$, $\mathscr{M}(\mathfrak{g})_{\widetilde{\gamma(\chi)}}$; the functor $\mathscr{M}(U)_\chi \to \mathscr{M}(\mathfrak{g})_{\gamma(\chi)}$ is always an equivalence of categories; the functor $\mathscr{M}(U)_{\widetilde{\chi}} \to \mathscr{M}(\mathfrak{g})_{\widetilde{\gamma(\chi)}}$ is an equivalence if and only if χ is a regular weight.

3.2. The flag variety. Let $X = X_\mathfrak{g}$ be the flag variety of $\mathfrak{g}$; points of X are Borel subalgebras of $\mathfrak{g}$. For $x \in X$ let $\mathfrak{b}_x$ be the corresponding Borel subalgebra, $B_x \subset G^o$ the corresponding Borel subgroup, and $N_x \subset B_x$ the maximal nilpotent subgroup. Then $\operatorname{Lie} B_x = \mathfrak{b}_x$, $\operatorname{Lie} N_x = \mathfrak{n}_x := [\mathfrak{b}_x, \mathfrak{b}_x]$, and $\mathfrak{h} = \mathfrak{b}_x/\mathfrak{n}_x$. Put $H := B_x/N_x$. This torus (the Cartan group of G) does not depend on the choice of x by the same reason as $\mathfrak{h}$ did not; one has $\operatorname{Lie} H = \mathfrak{h}$. The group G acts on X and on H and these actions are compatible with the above actions on Lie algebras. The action of G^o on X is transitive with the stabilizer of $x \in X$ equal to B_x, so $X = G^o/B_x$.

Let $\widetilde{X} = \widetilde{X}_\mathfrak{g}$ be the enhanced flag variety (or "base affine space") of G: its point $\widetilde{x}$ is a pair $(\mathfrak{b}_x, \{a^\alpha_{\widetilde{x}}\})$, where $\mathfrak{b}_x \subset \mathfrak{g}$ is a Borel subalgebra, and $a^\alpha_{\widetilde{x}}$,

$\alpha \in \Sigma$, is a generator for the α-root subspace in $\mathfrak{g}/\mathfrak{b}_x$. The groups G and H act on $\widetilde{X}$ from the left according to formulas $g\widetilde{x} := (\mathrm{Ad}_g(\mathfrak{b}_x), \{\mathrm{Ad}_g(a^\alpha)\})$, $h\widetilde{x} = (\mathfrak{b}_x, \{\exp\alpha(h) \cdot a^\alpha\})$. One has $ghx = \kappa(g)(h)gx$; in particular, G^o commutes with H. The H-action is free, $H \setminus \widetilde{X} = X$, and the G^o-action is transitive. For $\widetilde{x} \in \widetilde{X}$ the stabilizer $G^o_{\widetilde{x}}$ equals N_x; hence we have the isomorphism $G^o/N_x \xrightarrow{\sim} \widetilde{X}$, $gN_x \mapsto g\widetilde{x}$. Note that G^o/N_x carries the H-action $h(gN_x) := gh^{-1}N_x$ (here $H = B_x/N_x$), and the above isomorphism is H-equivariant.

We will consider $\widetilde{X}$ as an H-monodromic G-variety (with the compatibility morphism κ). By 2.5 we get the D-algebra $\widetilde{D} = \mathscr{U}(\widetilde{\mathscr{T}})$ on X equipped with a G-action. The Lie algebra map $\mathfrak{g} \times \mathfrak{h} \to \widetilde{\mathscr{T}}$ defines the morphism of the universal enveloping algebras $\mathscr{U}(\mathfrak{g} \times \mathfrak{h}) = \mathscr{U}(\mathfrak{g}) \otimes S(\mathfrak{h}) \xrightarrow{\widetilde{\delta}} \widetilde{D}$. It is easy to see that $\widetilde{\delta}(z \otimes 1) = \widetilde{\delta}(1 \otimes \gamma(z))$ for $z \in \mathscr{Z}$, hence $\widetilde{\delta}$ factors through a morphism $\widetilde{\delta}\colon U \to \widetilde{D}$ of $S(\mathfrak{h})$-algebras. It induces the morphism $\delta_\chi\colon U_\chi \to D_\chi$ between the m_χ-quotients. Note that $U_0 = \mathscr{U}(\mathfrak{g})_{\gamma(0)}$, $D_0 = D_X$, and δ_0 comes from the infinitesimal $\mathfrak{g}$-action on X.

The above Lie algebra morphism defines a morphism $\widetilde{\mathfrak{g} \times \mathfrak{h}} \to \widetilde{\mathscr{T}}$ of Lie algebroids on X (see 1.2). This morphism is surjective; its kernel is an $\mathscr{O}_X$-Lie algebra $\widetilde{\mathfrak{b}} := \{\gamma \in \mathscr{O}_X \otimes \mathfrak{g}\colon \gamma(x) \in \mathfrak{b}_x \text{ for } x \in X\}$ embedded in $\widetilde{\mathfrak{g} \times \mathfrak{h}}$ by $\gamma \mapsto (\gamma, \gamma \bmod \widetilde{\mathfrak{n}})$. Here $\widetilde{\mathfrak{n}} = [\widetilde{\mathfrak{b}}, \widetilde{\mathfrak{b}}] = \{\gamma \in \mathscr{O}_X \otimes \mathfrak{g}\colon \gamma(x) \in \mathfrak{n}_x \text{ for } x \in X\}$, so $\widetilde{\mathfrak{b}}/\widetilde{\mathfrak{n}} = \widetilde{\mathfrak{h}}$. We see that the induced morphism $\widetilde{\mathfrak{g}} \to \widetilde{\mathscr{T}}$ is also surjective with kernel $\widetilde{\mathfrak{n}}$. Therefore $\widetilde{\mathscr{T}} = \widetilde{\mathfrak{g} \times \mathfrak{h}}/\widetilde{\mathfrak{b}} = \widetilde{\mathfrak{g}}/\widetilde{\mathfrak{n}}$, hence $\widetilde{D} = \mathscr{U}(\widetilde{\mathfrak{g} \times \mathfrak{h}}/\widetilde{\mathfrak{b}}) = \mathscr{U}(\widetilde{\mathfrak{g}}/\widetilde{\mathfrak{n}})$.

REMARK. We see that $\widetilde{\mathfrak{b}}$ and $\widetilde{\mathfrak{n}}$ are normal subalgebras in $\widetilde{\mathfrak{g} \times \mathfrak{h}}$ and $\widetilde{\mathfrak{g}}$.

3.2.1. LEMMA. *The D-algebra $\widetilde{D}$ carries a unique involution ^{t} such that the canonical morphism $\widetilde{\delta}\colon U \to \widetilde{D}$ commutes with ^{t}'s.*

Note that ^{t} induces the duality $D^o_\chi \xrightarrow{\sim} D_{-2\rho-\chi}$. It is easy to check that the morphism $\widetilde{\mathfrak{g}} \to \widetilde{\mathscr{T}_{\Omega_X}}$ (that comes from the action of G on Ω_X) induces an isomorphism of tdo's $D_{-2\rho} \xrightarrow{\sim} D_{\Omega_X}$, and, with respect to this isomorphism, the above duality coincides with the canonical one from 2.4.1.

3.2.2. LEMMA. *The morphism $\widetilde{\delta}\colon U \to \Gamma(X, \widetilde{D})$ is an isomorphism. For all $\chi \in \mathfrak{h}^*$ the corresponding morphisms $\delta_\chi\colon U_\chi \to \Gamma(X, D_\chi)$ are also isomorphisms.*

For a proof see, e.g., [S1].

3.3. Localization. According to 1.6, $\widetilde{\delta}$ defines adjoint functors

$$\mathscr{M}(U) \underset{\Gamma}{\overset{\Delta}{\rightleftarrows}} \mathscr{M}(\widetilde{D}) = \mathscr{M}(X)$$

with $\Gamma(M) := \Gamma(X, M)$, $\Delta(N) := N \underset{U}{\otimes} \widetilde{D}$. These functors are $S(\mathfrak{h})$-linear, hence for any ideal $I \subset S(\mathfrak{h})$ they preserve the subcategories of modules killed by I. Also they commute with (directed) inductive limits. In particular, Γ and Δ induce adjoint functors between the full subcategories

$$\mathscr{M}(U)_\chi \underset{\Gamma_\chi}{\overset{\Delta_\chi}{\rightleftarrows}} \mathscr{M}(X)_\chi, \qquad \mathscr{M}(U)_{\widetilde{\chi}} \underset{\Gamma_{\widetilde{\chi}}}{\overset{\Delta_{\widetilde{\chi}}}{\rightleftarrows}} \mathscr{M}(X)_{\widetilde{\chi}}.$$

Recall that a weight χ is *dominant* if $(\chi + \rho)(h_\gamma) \notin \{-1, -2, \dots\}$ for any positive co-root $h_\gamma \in \mathfrak{h}$. We have the basic

3.3.1. THEOREM. *If χ is a regular dominant weight, then X is D_χ-affine (see 1.6.1), so $(\Gamma_\chi, \Delta_\chi)$ are mutually inverse equivalences of categories. The functors $(\Gamma_{\widetilde{\chi}}, \Delta_{\widetilde{\chi}})$ are also equivalences of categories.*

For a proof of the first statement see [BB1]. An easy devissage then shows that actually X is $\widetilde{D}/m_\chi^n\widetilde{D}$-affine for any $n \geqslant 1$, which implies the second statement.

For the case of nondominant or nonregular χ see [BB2, KL1].

3.3.2. Now assume we have a Harish-Chandra pair $(\mathfrak{g}, K)$, so K is an algebraic group equipped with a morphism $\mathrm{Ad}\colon K \to G$ and a Lie algebra embedding $i_{\mathfrak{k}}\colon \mathfrak{k} := \mathrm{Lie}\, K \hookrightarrow \mathfrak{g}$ which are compatible in an obvious sense. We have the corresponding categories of Harish-Chandra modules $\mathscr{M}(\mathfrak{g}, K) = \mathscr{M}(U(\mathfrak{g}), K)$, $\mathscr{M}(U, K)$, $\mathscr{M}(\widetilde{D}, K)$; for $\chi^\cdot \in \kappa(K) \setminus \mathfrak{h}^*$ we have the corresponding standard subcategories $\mathscr{M}(U, K)_{\chi^\cdot} \subset \mathscr{M}(U, K)_{\widetilde{\chi^\cdot}} \subset \mathscr{M}(U, K)$, etc. (see 2.5.6). As above we have an obvious functor $\mathscr{M}(U, K) \to \mathscr{M}(\mathfrak{g}, K)$ which induces the equivalences $\mathscr{M}(U, K)_{\chi^\cdot} \xrightarrow{\sim} \mathscr{M}(\mathfrak{g}, K)_{\gamma(\chi)^\cdot}$; if χ is regular and the stabilizers of χ and $\gamma(\chi)$ in $\kappa(K)$ coincide (e.g., if $\kappa(K)\chi = \chi$), then $\mathscr{M}(U, K)_{\widetilde{\gamma\chi^\cdot}} \to \mathscr{M}(\mathfrak{g}, K)_{\widetilde{\gamma(\chi)^\cdot}}$ is also an equivalence of categories.

The functors Γ and Δ send, in an obvious way, the K-equivariant modules to K-equivariant ones, therefore we have the adjoint functors

$$\mathscr{M}(U, K) \underset{\Gamma}{\overset{\Delta}{\rightleftarrows}} \mathscr{M}(\widetilde{D}, K) = \widetilde{\mathscr{M}}(X, K)$$

which induce the functors between the full subcategories

$$\mathscr{M}(U, K)_{\chi^\cdot} \underset{\Gamma_\chi}{\overset{\Delta_\chi}{\rightleftarrows}} \mathscr{M}(D_{\chi^\cdot}, K) = \mathscr{M}(X, K)_{\chi^\cdot},$$

$$\mathscr{M}(U, K)_{\widetilde{\chi^\cdot}} \underset{\Gamma_{\widetilde{\chi}}}{\overset{\Delta_{\widetilde{\chi}}}{\rightleftarrows}} \mathscr{M}(\widetilde{D}, K)_{\widetilde{\chi^\cdot}} = \widetilde{\mathscr{M}}(X, K)_{\widetilde{\chi}}.$$

3.3.3. COROLLARY. *If χ is a regular dominant weight, then $(\Gamma_{\chi^\cdot}, \Delta_{\chi^\cdot})$, $(\Gamma_{\widetilde{\chi^\cdot}}, \Delta_{\widetilde{\chi^\cdot}})$ are equivalences of categories.*

According to 2.6.5 these equivalences define an I-stratification on the categories $\mathscr{M}(U, K)_{\chi^\cdot}$, $\mathscr{M}(U, K)_{\widetilde{\chi}}$, where I is the set of K-orbits on X.

3.4. Admissible orbits. In the rest of this section we will collect some geometric information about K-orbits on X that will be used in the construction of the geometric Jantzen filtration. In this section our $(X, \widetilde{X})$ is any H-monodromic K-variety.

For $x \in X$ consider the pair $(\mathfrak{h}, K_{(x)})$ defined in 2.6.1. Put $\mathfrak{h}^*(x) := \{\varphi \in \mathfrak{h}^* : \kappa(K_{(x)})\varphi = \varphi, \ \varphi(i(\mathfrak{k}_{(x)})) = 0\}$ (this is the set of morphisms from $(\mathfrak{h}, K_{(x)})$ to the trivial Harish-Chandra pair $(\mathbb{C}, \{1\})$). Also put $\mathfrak{h}^*_{\mathbb{Z}}(x) := \mathfrak{h}^*_{\mathbb{Z}} \cap \mathfrak{h}^*(x)$. Since $\mathfrak{h}^*(kx) = \kappa(k)\mathfrak{h}^*(x)$, for a fixed K-orbit $Q \subset X$ the spaces $\mathfrak{h}^*(x)$ and $\mathfrak{h}^*_{\mathbb{Z}}(x)$ for $x \in Q$ are canonically identified; we denote them by $\mathfrak{h}^*(Q)$ and $\mathfrak{h}^*_{\mathbb{Z}}(Q)$.

It is easy to see that a weight $\varphi \in \mathfrak{h}^*_{\mathbb{Z}}$ belongs to $\mathfrak{h}^*_{\mathbb{Z}}(Q)$ if and only if there exists a nonzero K-invariant function f_φ on $\widetilde{Q} = \pi^{-1}(Q) \subset \widetilde{X}$ such that $f_\varphi(h\widetilde{x}) = (\exp\varphi)(h) f_\varphi(\widetilde{x})$ for $\widetilde{x} \in \widetilde{Q}$. Such a function f_φ is determined by φ uniquely up to multiplication by a nonzero constant.

Let $\overline{Q}$ be the closure of $\widetilde{Q}$ in $\widetilde{X}$. We say that f_φ is *$\overline{Q}$-regular* if $f_\varphi \in \mathscr{O}(\overline{Q}) \subset \mathscr{O}(\widetilde{Q})$; f_φ is *$\overline{Q}$-invertible* if $f_\varphi \in \mathscr{O}^*(\overline{Q})$; and f_φ is *$\overline{Q}$-positive* if f_φ is $\overline{Q}$-regular and $f_\varphi^{-1}(0) = \overline{Q} \setminus \widetilde{Q}$. Put $\mathfrak{h}^{*o}_{\mathbb{Z}}(Q) = \{\varphi \in \mathfrak{h}^*_{\mathbb{Z}}(Q) : f_\varphi$ is $\overline{Q}$-invertible$\}$, $\mathfrak{h}^{*+}_{\mathbb{Z}}(Q) = \{\varphi \in \mathfrak{h}^*_{\mathbb{Z}}(Q) : f_\varphi$ is $\overline{Q}$-positive$\}$.

3.4.1. DEFINITION. (i) An orbit Q is *admissible* if $\mathfrak{h}^{*+}_{\mathbb{Z}}(Q)$ is not empty.

(ii) The K-action on $(X, \widetilde{X})$ is *admissible* if it has finitely many orbits on X and every orbit is admissible.

3.4.2. LEMMA. (i) *For any admissible orbit Q the embeddings $Q \hookrightarrow X$, $\widetilde{Q} \hookrightarrow \widetilde{X}$ are affine.*

(ii) *$\mathfrak{h}^{*o}_{\mathbb{Z}}(Q)$ is the subgroup of $\mathfrak{h}^*_{\mathbb{Z}}(Q)$, and $\mathfrak{h}^*_{\mathbb{Z}}(Q)/\mathfrak{h}^{*o}_{\mathbb{Z}}(Q)$ has no torsion.*

(iii) *If Q is admissible, then $\mathfrak{h}^{*+}_{\mathbb{Z}}(Q)$ is a subsemigroup of $\mathfrak{h}^*_{\mathbb{Z}}(Q)$ that generates $\mathfrak{h}^*_{\mathbb{Z}}(Q)$ and is invariant under $\mathfrak{h}^{*o}_{\mathbb{Z}}(Q)$-translations. If $\varphi \in \mathfrak{h}^*_{\mathbb{Z}}(Q)$ and $n\varphi \in \mathfrak{h}^{*+}_{\mathbb{Z}}(Q)$ for some $n > 0$, then $\varphi \in \mathfrak{h}^{*+}_{\mathbb{Z}}(Q)$. The quotient $\mathfrak{h}^{*+}_{\mathbb{Z}}(Q)/\mathfrak{h}^{*o}_{\mathbb{Z}}(Q)$ is isomorphic to $\mathbb{Z}^a_+$ for some a.*

(iv) *An orbit is admissible if and only if (some, or any of) its connected components is admissible with respect to the action of the connected component K^o. Hence a K-action is admissible iff its restriction to K^o is admissible.*

(v) *Assume we have a larger group $K' \supset K$ and an extension $\kappa' : K' \to \operatorname{Aut} H$ of κ; let $(X', \widetilde{X}')$ be the induced K'-variety (see 2.5.8). For a K-orbit Q on X let $Q' = K'Q$ be the corresponding K'-orbit on X'. Then $\mathfrak{h}^*_{\mathbb{Z}}(Q) = \mathfrak{h}^*_{\mathbb{Z}}(Q')$, and the same for $\mathfrak{h}^{*o}$, $\mathfrak{h}^{*+}$.*

Hence $(X', \widetilde{X}')$ is an admissible K'-variety if and only if $(X, \widetilde{X})$ is an admissible K-variety.

3.5. Admissible orbits on the flag variety. Let $(\mathfrak{g}, K)$, $(X, \widetilde{X})$ be as in 3.3.2. We will say that our Harish-Chandra pair is *admissible* if the K-action on $(X, \widetilde{X})$ is admissible.

3.5.1. LEMMA. *The pair* $(\mathfrak{g}, N)$, *where* N *is a maximal nilpotent subgroup, is admissible.*

PROOF. Consider a Schubert cell Q_w, where $w \in W$. Let $\mathfrak{h}_{\mathbb{Z}}^{*+}$ be the cone of positive regular integral characters. We will see that for any $w \in W$ one has $\mathfrak{h}_{\mathbb{Z}}^{*+}(Q_w) \supset \rho + \mathfrak{h}_{\mathbb{Z}}^{*+}$, hence Q_w is admissible. For $\chi \in \rho + \mathfrak{h}_{\mathbb{Z}}^{*+}$ take an irreducible G^o-module V with highest weight χ; let $v \in V^N \setminus \{0\}$ be a lowest weight vector. Consider the map $\mathfrak{q}_v : \widetilde{X} = G^o/N \to V\setminus\{0\}$, $\mathfrak{q}_v(g) = gv$. It is clear that if $l \in V^*$ is a linear function on V, then $l\mathfrak{q}_v$ is a χ-homogeneous function on X. Choose $l \in V^*$ such that $l(wv) \neq 0$, $l(\mathfrak{n}wv) = 0$ (here $\mathfrak{n} = \operatorname{Lie} N$). One has $wv \in \mathfrak{q}_v(\widetilde{Q}_w)$, the image $\mathfrak{q}_v(\widetilde{Q}_w)$ lies in the linear N-invariant subspace generated by wv, and $\mathfrak{q}_v(\overline{Q}_w - \widetilde{Q}_w) \subset \mathfrak{n}wv$, where $\overline{Q}_w$ is the closure of $\widetilde{Q}_w$. Hence $l\mathfrak{q}_v$ is the desired χ-homogenous N-invariant function that vanishes on $\overline{Q}_w \setminus \widetilde{Q}_w$.

REMARK. The B-action on X is *not* admissible.

Now assume that $(\mathfrak{g}, K)$ is a symmetric pair, which means that $\mathfrak{k} = \mathfrak{g}^\theta$ for some involution θ of $\mathfrak{g}$. Note that θ is uniquely determined by $\mathfrak{k}$ (its -1 eigenspace coincides with the Killing orthogonal complement to $\mathfrak{k}$); in particular θ commutes with $\operatorname{Ad} K$. For $x \in X$ denote by $\mu_\theta(x) \in W$ the relative position of $(\mathfrak{b}_x, \theta\mathfrak{b}_x)$. Clearly $\mu_\theta(kx) = \kappa(k)(\mu_\theta(x))$ for $k \in K$. In particular μ_θ is constant along the connected components of K-orbits; if Q^o is such a component we will write $\mu_\theta(Q^o) = \mu_\theta(x)$, $x \in Q^o$.

REMARK. The following properties are equivalent:

(i) An orbit Q is closed.

(ii) $\mu(Q) = 1$.

(iii) For $x \in Q$ one has $\dim K \cap N_x = \dim Q$.

(iv) $\dim Q = \dim X_{\mathfrak{k}}$.

3.5.2. LEMMA. *Any symmetric pair is admissible.*

PROOF. According to 3.4.2(iv) we can assume that K is connected.

(i) Let us consider the special case: $\mathfrak{g} = \mathfrak{k} \times \mathfrak{k}$, $i: \mathfrak{k} \to \mathfrak{g}$ is the diagonal embedding. Then θ is the transposition and $\widetilde{X}_{\mathfrak{g}} = \widetilde{X}_{\mathfrak{k}} \times \widetilde{X}_{\mathfrak{k}}$. If $\widetilde{x} \in \widetilde{X}_{\mathfrak{k}}$, then the K-space $\widetilde{X}_{\mathfrak{g}}$ is induced from N_x-space $\widetilde{X}_{\mathfrak{k}} = \widetilde{X}_{\mathfrak{k}} \times \{\widetilde{x}\} \hookrightarrow \widetilde{X}_{\mathfrak{g}}$, and the K-orbits on $X_{\mathfrak{g}}$ are the same as N_x-orbits on $X_{\mathfrak{k}}$: these are $Y_w = K(Q_w \times x)$, $w \in W_{\mathfrak{k}}$. One has $\mathfrak{h}_{\mathfrak{g}\mathbb{Z}}^*(Y_w) = \{(\chi, -w\chi), \chi \in \mathfrak{h}_{\mathfrak{k}\mathbb{Z}}^*\} \subset \mathfrak{h}_{\mathfrak{k}\mathbb{Z}}^* \times \mathfrak{h}_{\mathfrak{k}\mathbb{Z}}^* = \mathfrak{h}_{\mathfrak{g}\mathbb{Z}}^*$, since $K \setminus \widetilde{Y}_w$ is isomorphic to $H_{\mathfrak{k}}$, with the $(H_{\mathfrak{g}} = H_{\mathfrak{k}} \times H_{\mathfrak{k}})$-action given by the formula $(h_1, h_2)h = h_1 w(h_2^{-1})h$. Then clearly $\mathfrak{h}_{\mathfrak{g}\mathbb{Z}}^{*+}(Y_w) = \{(\chi, -w\chi), \chi \in \mathfrak{h}_{\mathfrak{k}\mathbb{Z}}^{*+}(Q_w)\}$, since $f_{(\chi, -w\chi)}|_{\widetilde{X}_w \times \{\widetilde{x}\}} = f_\chi$. Hence we are done by 3.5.1.

(ii) The general case. Consider the embeddings $m_\theta : X \hookrightarrow X \times X$, $\widetilde{m}_\theta : \widetilde{X} \hookrightarrow \widetilde{X} \times \widetilde{X}$, defined by formulas $m_\theta(x) = (x, \theta(x))$ and the same for $\widetilde{m}_\theta$. These maps are equivariant with respect to the K-action on X and the diagonal action of K on $X \times X$ (via $\operatorname{Ad}: K \to G$); one has $m_\theta(x) \in Y_{\mu_\theta(x)}$. For

$w \in W$ consider the locally closed K-invariant subvariety $X_{\theta w} := \mu_\theta^{-1}(w) = m_\theta^{-1}(Y_w) \subset X$. The number of K-orbits on X is finite. This follows from

$$\text{every } K\text{-orbit } Q \subset X_{\theta w} \text{ is open in } X_{\theta w}. \qquad (*)$$

This follows from the corresponding infinitesimal statement for any $x \in X_{\theta w}$:

$$\text{the tangent space } T_{Kx,x} \text{ to the } K\text{-orbit coincides with } dm_\theta^{-1}(T_{Y_w, m_\theta(x)}). \qquad (**)$$

We give a proof: one has $T_{Y_w, m_\theta(x)} = \{(\xi \bmod \mathfrak{b}_x, \xi \bmod \theta\mathfrak{b}_x), \xi \in \mathfrak{g}\} = \{(\xi \bmod \mathfrak{b}_x, \theta(\theta\xi \bmod \mathfrak{b}_x))\}$, $dm_\theta(T_{X,x}) = \{(\xi \bmod \mathfrak{b}_x, \theta(\xi \bmod \mathfrak{b}_x))\}$; hence $\mathscr{T}_{Y_w, m_\theta(x)} \cap dm_\theta(\mathscr{T}_{X,x}) = \{(\xi \bmod \mathfrak{b}_x, \theta(\xi \bmod \mathfrak{b}_x)) : \xi - \theta(\xi) \in \mathfrak{b}_x\} = \{(\eta \bmod \mathfrak{b}_x, \theta(\eta \bmod \mathfrak{b}_x))\}$, where $\eta = (\xi + \theta(\xi))/2 \in \mathfrak{g}^\theta$, which proves $(**)$.

In particular, $(*)$ implies that for any K-orbit Q on X one has $\mu_\theta(Q) \notin \mu_\theta(\overline{Q} \setminus Q)$. Thus for any homogeneous G^o-invariant function f on $\widetilde{Y}_{\mu_\theta(Q)}$ the function $f \circ \widetilde{m}_\theta$ is a homogeneous K-invariant function on $\widetilde{Q}$, and if f is positive, then $f \circ m_\theta$ is also positive. Now the statement (i) above finishes the proof.

3.6. Contravariant duality for standard modules. If $(\mathfrak{g}, K)$ is a finite pair, then we have the Verdier duality on the category of $S(\mathfrak{h})$-finite coherent $(\widetilde{D}, K)$-modules (see 2.5.5(iv)). This duality is local with respect to X and transforms to the Verdier duality on perverse constructible sheaves via the Riemann-Hilbert correspondence. On the other hand, if $(\mathfrak{g}, K)$ is a symmetric pair, or if $K = N$, then one has the usual contravariant duality for $(\mathfrak{g}, K)$-modules. It is an interesting problem to find a geometric (D-modules) description of this duality. At the moment one knows how the contravariant duality acts on the irreducible $(\mathfrak{g}, K)$-modules in terms of their geometric Langlands parameters. We recall this description below.

3.6.1. Consider the involution c on U, $c(u) = w_{\max}{}^t u$, where $w_{\max} \in W$ is the element of maximal length, which acts only on $S(\mathfrak{h})$. It coincides with -1 on $\mathfrak{g}$ and induces on $S(\mathfrak{h})$ the involution $c(\chi) = w_{\max}(-2\rho - \chi)$, $\chi \in \mathfrak{h}^* = \operatorname{Spec} S(\mathfrak{h})$; one has $c(\Delta^+) = \Delta^+$. For a left U-module V let V^o be the dual vector space to V considered as a left U-module via c: for $u \in U$, $v \in V$, $v^* \in V^o$ one has $\langle uv^*, v\rangle = \langle v^*, c(u)v\rangle$. As a $\mathfrak{g}$-module V^o is just the module dual to V; we use c instead of ^{t} since it transforms (regular) dominant weights to (regular) dominant ones, which is handy for localization.

3.6.2. Let us first define the contravariant duality in the case $K = N$. Choose a complementary maximal nilpotent subgroup N', $N' \cap N = \{1\}$. For a dominant regular χ consider the subcategory $\mathscr{M}(U, N)^f_{\widetilde{\chi}} \subset \mathscr{M}(U, N)$

of finitely generated modules. For $V \in \mathcal{M}(U, N)^f_{\widetilde{\chi}}$ denote by V^c the subspace of those vectors in V^o on which $\mathfrak{n}' = \operatorname{Lie} N'$ acts in a locally nilpotent way. Then V^c is a U-submodule of V^o and the action of $\mathfrak{n}'$ on V^c integrates to an algebraic action of N', so $V^c \in \mathcal{M}(U, N')_{c(\chi)}$. One knows that actually V^c is finitely generated and $V^{cc} = V$, so the contravariant duality $c\colon (\mathcal{M}(U, N)^f_\chi)^o \to \mathcal{M}(U, N')^f_{c(\chi)}$ is an equivalence of categories.

The Bruhat decomposition identifies the set of N-orbits on X with the set W equipped with the Bruhat order. By 3.3.3, $\mathcal{M}(U, N)^f_{\widetilde{\chi}}$ is a W-stratified category. For $w \in W$ we have a single irreducible L_w in the corresponding stratum. The corresponding standard modules $j_{w!}(L_w)$ are the Verma modules.

Let c_W be the involution on W, $c_W(w) = w_{\max} w w_{\max}^{-1}$. One has an easy

3.6.3. Lemma. *One has* $L_w^c \approx L_{c_W(w)}$.

Using the fact that the involution c_W preserves the Bruhat order, it is easy to show that the duality c is a c_W-stratified equivalence of categories. In particular it sends !-standard modules to $*$-standard ones (see 2.6.6(iii), (iv)).

3.6.4. Consider now the case of a symmetric subgroup K. Then K is reductive and for a finitely generated $V \in \mathcal{M}(U, K)^f_{\widetilde{\chi}}$ one knows that any irreducible representation of K occurs in V with finite multiplicity. The group K acts on V^o as an abstract group; denote by V^c the maximal subspace on which K acts algebraically. It is easy to see that $V^c = \bigoplus V_\alpha^* \subset \mathcal{P}i V_\alpha^* = V^o$, where $V = \bigoplus V_\alpha$ is K-isotypic decomposition of V. Clearly V^c is a U-submodule of V^o, hence $V^c \in \mathcal{M}(U, K)_{c(\chi)}$. One knows that V^c is also finitely generated and $V^{cc} = V$, so we have the contravariant duality $c\colon (\mathcal{M}(U, K)^f_{\widetilde{\chi}})^o \xrightarrow{\sim} \mathcal{M}(U, K)^f_{\widetilde{c(\chi)}}$.

Let us describe how c acts on the Langlands parameters. Consider, as in 2.6.3, the ordered set I of K-orbits on X. Then $\mathcal{M}(U, K)^f_{\widetilde{\chi}} = \mathcal{M}(X, K)^{\mathrm{coh}}_{\widetilde{\chi}}$ is an I-stratified category. It turns out that c is an I-stratified functor. Let us define explicitly the corresponding involution c_I of I. For $i \in I$ consider the corresponding orbit Q_i. Put

$$\widehat{Q}_i := \{(x_1, x_2) \in X \times X : x_1 \in Q_1, \quad b_{x_1} \cap b_{x_2} \text{ is a } \theta\text{-stable Cartan subalgebra}\}.$$

One knows (see, e.g, [Mil, (A2.3)]) that for $x \in Q_i$ the fiber over x of the first projection $\widehat{Q}_i \to Q_i$, $(x_1, x_2) \mapsto x_1$, is a nonempty $(K \cap N_x)$-torsor. Hence the second projection $\widehat{Q}_i \to X$, $(x_1, x_2) \mapsto x_2$, maps $\widehat{Q}_i$ onto a single K-orbit $Q_{c_I(i)}$. Clearly $c_I\colon I \to I$, $i \mapsto c_I(i)$, is an involution.

3.6.5. Lemma. *The involution c_I preserves the order on the set I and the functor $c\colon \mathcal{M}(U, K)^f_{\widetilde{\chi}} \to \mathcal{M}(U, K)^f_{\widetilde{c(\chi)}}$ is a c_I-stratified involution.*

For a proof see [HMSW2].

Let us describe the action of c on the strata $\mathcal{M}(U, K)^{f}_{\widetilde{\chi} i}$. By 2.6.5(i) for $x \in Q_i$ we have the canonical equivalence $F_x\colon \mathcal{M}(U, K)^{f}_{\widetilde{\chi} i} = \mathcal{M}(X, K)^{\mathrm{coh}}_{\widetilde{\chi} i} \xrightarrow{\sim} \mathcal{M}(\mathfrak{h}, K_{(x)})^{f}_{\widetilde{\chi}}$.

Take $(x, x') \in \widehat{Q}_i$, so $x' \in Q_{c_I(i)}$. It is easy to see that the projections $K_{(x)} \leftarrow K_x \cap K_{x'} \rightarrow K_{(x')}$ are surjective and have the same kernel; therefore, they define an isomorphism $\alpha\colon K_{(x)} \xrightarrow{\sim} K_{(x')}$. This isomorphism extends to an isomorphism of the Harish-Chandra pairs $\alpha\colon (\mathfrak{h}, K_{(x)}) \rightarrow (\mathfrak{h}, K_{(x')})$ that acts on $\mathfrak{h}$ as $w_{\max}$. Denote by $c_{(x,x')}\colon \mathcal{M}(\mathfrak{h}, K_{(x)})^{fo}_{\widetilde{\chi}} \xrightarrow{\sim} \mathcal{M}(\mathfrak{h}, K_{(x')})^{f}_{\widetilde{c(\chi)}}$ the duality functor $c_{(x,x')}(V) := \alpha_*(V^0 \otimes \varphi)$. Here $\alpha_*\colon \mathcal{M}(\mathfrak{h}, K_{(x)}) \rightarrow \mathcal{M}(\mathfrak{h}, K_{(x')})$ is the equivalence defined by α, V^o is the dual module, and φ is the $(\mathfrak{h}, K_{(x)})$-module $\det \mathfrak{n}_x$.

3.6.6. LEMMA. *The equivalences* F_x, $F_{x'}$ *identify the i-stratum of the involution* $c\colon (\mathcal{M}(U, K)^{f}_{\widetilde{\chi} i})^o \xrightarrow{\sim} \mathcal{M}(U, K)^{f}_{\widetilde{c(\chi)} c_I(i)}$ *with* $c_{(x,x')}$.

For a proof see [HMSW2]. This lemma describes how c acts on the irreducible representations in terms of their Langlands parameters. By 3.6.4 and 2.6.6(iii), (iv) c interchanges !- and $*$-standard modules.

§4. The Jantzen filtration

In this section we will define the Jantzen filtration on standard modules; the main point is its relation with the monodromy filtration on nearby cycles.

4.1. The monodromy filtration. We will need a tiny complement to [D2, 1.6]. For an object Q of an abelian category and a nilpotent endomorphism $s \in \operatorname{End} Q$ let $\mu_\cdot = \mu^Q_\cdot$ denote the monodromy filtration on Q (see [D2, (1.6.1)]). Let $P^Q_i = P_i := \operatorname{Ker}(Gr^\mu_i \longrightarrow Gr^\mu_{i-2})$ be the primitive part of $Gr^\mu_\cdot$ [D2, (1.6.3)]; one has the primitive decomposition [D2, (1.6.4)]—a canonical isomorphism of graded $\mathbb{Z}[s]$-modules $Gr^\mu_\cdot \simeq \bigoplus_{j \leqslant 0} P_j \otimes \mathbb{Z}[s]/s^{-j}$, $\deg s = -2$, $\deg P_j = -j$.. Consider the following increasing filtration on $\operatorname{Ker} s$:

$$J_{!i} := \operatorname{Ker} s \cap \operatorname{Im} s^{-i} \quad \text{for } i \leqslant 0, \qquad J_{!i} = \operatorname{Ker} s \quad \text{for } i \geqslant 0.$$

Dually we define an increasing filtration

$$J_{*i} := (\operatorname{Ker} s^i + \operatorname{Im} s)/\operatorname{Im} s$$

on $\operatorname{Coker} s$. We call $J_{!\cdot}$, $J_{*\cdot}$ the *Jantzen filtrations.* Filtrations $J_{!\cdot}$, $J_{*\cdot}$ coincide with the filtrations induced by $\mu_\cdot$ on $\operatorname{Ker} s$, $\operatorname{Coker} s$; one has $Gr^{J_!}_i = P_i$, $Gr^{J*}_i = P_{-i}$ [D2, (1.6.6)]. Consider now $\overline{Q} := Q/\operatorname{Ker} s$ together with the nilpotent endomorphism $\overline{s}$ induced by s, and let $\overline{\mu}_\cdot$ be the corresponding monodromy filtration.

4.1.1. LEMMA. (i) *The exact sequences*

$$0 \to (\operatorname{Ker} s, J_{!\cdot}) \to (Q, \mu_\cdot) \to (\overline{Q}, \overline{\mu}_{\cdot-1}) \to 0,$$
$$0 \to (\overline{Q}, \overline{\mu}_{\cdot+1}) \xrightarrow{s} (Q, \mu_\cdot) \to (\operatorname{Coker} s, J_{*\cdot}) \to 0$$

are strictly compatible with filtrations.

(ii) *Conversely, $\mu_\cdot$ is the unique increasing filtration on Q such that $s\mu_\cdot \subset \mu_{\cdot-2}$ and either one of the above two sequences is strictly compatible with filtrations.*

PROOF. (i) is [D2, 1.6.5].

(ii) Let $\mu'_\cdot$ be another such filtration strictly compatible with, say, the first exact sequence. It suffices to show that $\mu'_i \supset \mu_i$. But $\mu'_i = \mu_i$ for $i \geqslant 0$ (since $J_{!0} = \operatorname{Ker} s$). For $i \leqslant 0$ we have $\mu'_i \supset J_{!i} + s(\mu'_{i-2})$, and $s(\mu'_{i-2}) = s(\mu_{i-2})$, and we are done by downard induction on i.

Assume now that our categories are over a field k of characteristic 0, and let $\boxtimes$ be an exact k-bilinear bifunctor. Let (R, t) be another object with nilpotent endomorphism, and $\mu^R_\cdot$ be its monodromy filtration. Consider the tensor product filtration $\mu^{Q,R}_i := \sum_{a+b=i} \mu^Q_a \boxtimes \mu^R_b$ on $Q \boxtimes R$. We have $Gr^{\mu^{Q,R}}_\cdot = Gr^{\mu^Q}_\cdot \boxtimes Gr^{\mu^R}_\cdot$, and the primitive decomposition together with [D2, (1.6.11, 1.6.12)] implies

4.1.2. LEMMA. (i) *$\mu^{Q,R}_\cdot$ is the monodromy filtration with respect to $s \boxtimes \mathrm{id}_R + \mathrm{id}_Q \boxtimes t$.*

(ii) *One has an "almost canonical" isomorphism $P^{Q\boxtimes R}_{-j} \simeq \bigoplus P^Q_{-j'} \boxtimes P^R_{-j''}$, where (j', j'') run through the set of pairs $\{(j', j''): |j' - j''| \leqslant j \leqslant |j' + j''|, j \equiv j' + j'' \bmod 2\}$.*

4.2. The Jantzen and the monodromy filtration in a geometric setting. Recall the construction of nearby cycles for D-modules [B, K, M, V2]; we follow mainly [B]. Let Y be a smooth variety, $f: Y \to \mathbb{A}^1$ be a function, and $Z := f^{-1}(0) \overset{i}{\hookrightarrow} Y \overset{j}{\hookleftarrow} U := f^{-1}(\mathbb{A}^1 - \{0\})$. For $n > 0$ consider a lissé $D_{\mathbb{A}^1-\{0\}}$-module $I^{(n)}$ with a $\mathbb{C}[s]/s^n$-action, which is a free rank 1 $(\mathcal{O}_{\mathbb{A}^1-\{0\}} \otimes \mathbb{C}[s]/s^n)$-module with generator "$t^s$" such that $t\partial_t(t^s) = st^s$ (here t is the parameter on $\mathbb{A}^1$); we have the obvious projections $I^{(n)} \to I^{(n)}/s^{n-1} = I^{(n-1)}$.

For a D_U-module M_U put $f^s M^{(n)}_U := f^* I^{(n)} \underset{O_U}{\otimes} M_U$: this is a $(D_U \otimes \mathbb{C}[s]/s^n)$-module, $f^s M^{(1)}_U = M_U$, and $f^s M^{(a)}_U = f^s M^{(n)}_U / s^a$ for $a \leqslant n$.

Assume now that M_U is holonomic. Fix some $a \geqslant 0$. Consider the morphism $s^a(n): j_! f^s M^{(n)}_U \to j_* f^s M^{(n)}_U$ of $(D_Y \otimes \mathbb{C}[s]/s^n)$-modules that coincides with s^a on U; one has $s^a(n) \bmod s^{n-1} = s^a(n-1)$. The lemma about b-functions implies that the projective system $\operatorname{Coker} s^a(n)$ stabilizes, so we can put $\pi^a_f(M_U) := \operatorname{Coker} s^a(n)$ for $n \gg 0$. This is a holonomic

D_Y-module with a nilpotent endomorphism s; the restriction to U of $\pi_f^a(M_U)$ is equal to $f^s M_U^{(a)}$.

The most important π's are $\pi_f^0 =: \Psi_f^{\mathrm{un}}$—the part of the nearby cycles functor with unipotent action of monodromy (one has $\Psi_f^{\mathrm{un}}(M_U)|_U = 0$), and $\pi_f^1 =: \Xi_f$—the maximal extension functor (one has $\Xi_f(M_U)|_U = M_U$). We give a list of properties of π_f^a (see [B]):

4.2.1. LEMMA. (i) *$\pi_f^a\colon M(U)_{\mathrm{hol}} \to M(Y)_{\mathrm{hol}}$ is an exact functor.*

(ii) *For $a, b \geqslant 0$ one has canonical exact sequences*

$$0 \to j_!(f^s M_U^{(a)}) \to \pi_f^{a+b}(M_U) \to \pi_f^b(M_U) \to 0,$$

$$0 \to \pi_f^b(M_U) \to \pi^{a+b}(M_U) \to j_*(f^s M_U^{(a)}) \to 0,$$

and

$$\mathrm{Im}(s^a\colon \pi_f^{a+b} \to \pi_f^{a+b}) = \pi_f^b.$$

(ii)′ *In particular one has exact sequences*

$$0 \to j_!(M_U) \to \Xi_f(M_U) \to \Psi_f^{\mathrm{un}}(M_U) \to 0,$$

$$0 \to \Psi_f^{\mathrm{un}}(M_U) \to \Xi_f(M_U) \to j_*(M_U) \to 0$$

with $j_! = \mathrm{Ker}(s\colon \Xi_f \to \Xi_f)$, $j_* = \mathrm{Coker}(s\colon \Xi_f \to \Xi_f)$.

(iii) *π_f^a commutes with the duality.*

Now 4.1 gives us the monodromy filtration $\mu_\cdot^{(a)}$ on π_f^a. On U the term $\mu_i^{(a)}$ coincides with $s^{[(a-i)/2]}\cdot f^s M_U^{(a)}$ (here $[\cdot] :=$ integral part). In particular, we have the monodromy filtrations on Ψ_f^{un} and Ξ_f and the Jantzen filtrations $J_{f!\cdot}$, $J_{f*\cdot}$ on $j_!$, j_* (via Ξ_f and (ii)′ above).

4.2.2. REMARKS. (i) 4.1 implies that, up to a shift, we will get the same Jantzen filtration if we use the isomorphisms 4.2.1(ii), $j_! \simeq \mathrm{Ker}(s\colon \pi^a \to \pi^a)$ for any $a \geqslant 1$; the same holds for j_*.

(ii) One has $J_{f!0} = j_!$, $J_{f!-1} = \mathrm{Ker}(j_! \to j_{!*})$. The embedding $\Psi_f^{\mathrm{un}} \to \Xi_f$ identifies $\mathrm{Ker}(s\colon \Psi_f^{\mathrm{un}} \to \Psi_f^{\mathrm{un}})$ with $J_{f!-1}$; this isomorphism shifts the corresponding Jantzen filtration by one.

Dually, $J_{f*-1} = 0$, $J_{f*0} = j_{!*} \subset j_*$, etc.

(iii) Let $Q_U \subset U$ be a closed subvariety, and Q be the closure of Q_U in Y. Let $\mathscr{M}(Q) \subset \mathscr{M}(Y)$, $\mathscr{M}(Q_U) \subset \mathscr{M}(U)$ be the subcategories of D-modules supported on Q. The above functors π_f^a transform $\mathscr{M}(Q_U)$ to $\mathscr{M}(Q)$, and being restricted to $\mathscr{M}(Q_U)$ they depend on $f|_Q$ only. Since everything is local, we get the functors $\pi_f^a\colon \mathscr{M}(Q_U) \to \mathscr{M}(Q)$ etc., for any regular function f on Q with $Q_U = Q \setminus f^{-1}(0)$.

(iv) The above functors will not change if we multiply f by a nonzero constant $c \in \mathbb{C}$, since one has an isomorphism of $(D_{\mathbb{A}-\{0\}}[s]/s^n)$-modules

$I^n \simeq c^* I^n$, $t^s \mapsto (ct)^s$ (here $c: t \longmapsto ct$ is considered as an automorphism of $\mathbb{A}^1 - \{0\}$).

(v) The above constructions have an obvious counterpart for constructible perverse sheaves compatible with the Riemann-Hilbert correspondence (see [Bo]). One identifies canonically Ψ_f^{un} with the part of the nearby cycles functor $R\Psi_{\overline{\eta}}[-1]$ on which the geometric monodromy acts unipotently, s corresponds to the logarithm of monodromy; here $\overline{\eta}$ is the generic geometric point $\mathrm{Spec}(\bigcup_N \mathbb{C}((t^{1/N})))$ of $\mathbb{C}((t))$.

4.3. The case of standard modules. Assume we are in situation 3.1. Let $Q \subset X$ be an admissible orbit. For $\varphi \in \mathfrak{h}_{\mathbb{Z}}^{*+}(Q)$ consider the corresponding functors $\pi_{f_\varphi}^a : \mathscr{M}(\widetilde{Q}) \to \mathscr{M}(\overline{\widetilde{Q}})$, see 4.2.2 (iii); since, by 4.2.2(iv), they depend on φ only, we will write $\pi_\varphi^a := \pi_{f_\varphi}^a$.

These functors preserve K-equivariance and monodromicity (by construction). Therefore, we have the functors $\pi_\varphi^a : \mathscr{M}(\mathfrak{h}, K_{(x)})_{\widetilde{\chi}}^{f.g.} = \mathscr{M}(\widetilde{Q}, K)_{\widetilde{\chi}}^{\mathrm{hol}} \to \mathscr{M}(\overline{\widetilde{Q}}, K)_{\widetilde{\chi}}^{\mathrm{hol}} \to \mathscr{M}(\widetilde{X}, K)_{\widetilde{\chi}}$ (here $x \in Q$, see 2.6.2) and the Jantzen filtrations $J_{!\cdot}$, $J_{*\cdot}$ on the functors $j_{Q!}, j_{Q*} : \mathscr{M}(\mathfrak{h}, K_{(x)})_{\widetilde{\chi}} \to \mathscr{M}(\widetilde{X}, K)_{\widetilde{\chi}}$. In particular, we have the Jantzen filtrations on standard modules $j_{Q!}(V)$, $j_{Q*}(V)$, where V is an irreducible $(\mathfrak{h}, K_{(x)})_{\widetilde{\chi}}$-module. A priori these filtrations depend on the choice of weight $\varphi \in \mathfrak{h}_{\mathbb{Z}}^{*+}(Q)$.

Note that these constructions can be done directly in terms of the I-stratification pattern (see 2.6.4). Namely, for an orbit Q_α, a point $x \in Q_\alpha$ and a weght $\varphi \in \mathfrak{h}_{\mathbb{Z}}^{*+}(Q_\alpha)$ let $I_\varphi^{(n)}$ be the $(\mathfrak{h}, K_{(x)})$-module $\mathbb{C}[s]/s^n$ such that $h \in \mathfrak{h}$ acts as $\varphi(h)s$ and $K_{(x)}$ acts trivially. The equivalence of categories $F_x : \widetilde{\mathscr{M}}(Q_\alpha, K) \xrightarrow{\sim} \mathscr{M}(\mathfrak{h}, K_{(x)})$ (see 2.6.2) identifies $\mathscr{M} \otimes f_\varphi^*(I^{(n)})$ with $F_x(\mathscr{M}) \otimes I_\varphi^{(n)}$, and we can repeat the constructions of 4.2 using the functors $j_{\alpha!}$, $j_{\alpha*}$, and $\bigotimes I_\varphi^{(n)}$.

If $(\mathfrak{g}, K)$ is an admissible Harish-Chandra pair, we get the Jantzen filtrations on !- and $*$-standard $(U, K)_\chi$-modules ($\chi \in \mathfrak{h}^*$ is a dominant regular weight) using the equivalence 3.3.2. If $K = N$ or K is a symmetric subgroup then $j_!$-extension is contravariant conjugate to j_*-extension (see 3.6); hence the morphism $j_{\alpha!}(V \otimes I_\varphi^{(n)}) \to j_{\alpha*}(V \otimes I_\varphi^{(n)})$ is just the contravariant form. This shows that our definition of $J_{!*}$ coincides with the original Jantzen filtration.

In the Verma modules case one can define the Jantzen filtration by using the deformations of the central character in an arbitrary nondegenerate direction φ, not necessarily in the positive one. According to Barbasch [Ba], the result does not depend on the choice of φ. In the geometric situation we can repeat, in principle, the same constructions and consider for any nonzero meromorprhic function f on X the morphism $j_!(\mathscr{M}_U \otimes f^*(I^{(n)})) \to$

$j_*(\mathscr{M}_U \otimes f^*(I^{(n)}))$, where $U := X \setminus \mathrm{div}(f)$. To define vanishing cycles one needs the stabilization of cokernels when $n \to \infty$. It would be very nice if this fact were true for any f, just as in the case when f (or f^{-1}) is regular on X, but we have no idea how to prove it.

§5. Weight filtrations

5.1. Weights of nearby cycles. Gabber's theorem, which is our main tool, seems not to be published yet. ([4]) Below we reproduce the proof following Gabber's report at IHES in the spring of 1981.

Let us start with the Künneth formula for nearby cycles. Let S be the spectrum of a strictly local Henselian ring; o and η be the closed and the generic points of S; and $\overline{\eta}$ be a geometric point localized at η.

Let $X \to S$ be an S-scheme and $X_0 \overset{i}{\hookrightarrow} X \overset{j}{\hookleftarrow} X_\eta \overset{k}{\longleftarrow} X_{\overline{\eta}}$ be the corresponding fibers. In what follows $D^b(Y)$ will denote either the bounded derived category of étale constructible $\mathbb{Z}/\ell^n$-sheaves on Y (where ℓ is prime to char o) or its $\mathbb{Q}_l$-counterpart [D1].

By [D1, (3.2)], there exist nearby cycles functors $\Psi_{\overline{\eta}} = \Psi_{\overline{\eta}X} : D^b(X_\eta) \to D^b(X_0)$, $\Psi_{\overline{\eta}X} := i^* Rj_* k_* k^*$. Let $Y \to S$ be another S-scheme, and $Z = X \underset{S}{\times} Y \to S$ be the fiber product. Then for $F \in D^b(X_\eta)$, $G \in D^b(Y_\eta)$ one has a canonical morphism in $D^b(Z_0)$:

$$\Psi_{\overline{\eta}X}(F) \boxtimes \Psi_{\overline{\eta}Y}(G) \to \Psi_{\overline{\eta}Z}(F \boxtimes G). \qquad (*)$$

5.1.1. LEMMA. $(*)$ *is an isomorphism.*

REMARK. The transcendental version (hence, the characteristic 0 case) is almost obvious by the ordinary Künneth formula applied to local varieties of vanishing cycles. This, together with the Riemann-Hilbert correspondence, implies a similar fact for tame D-modules. To obtain a similar formula for arbitrary holonomic D-modules one must use the total nearby cycles functor of Deligne [D2].

PROOF. We can assume that the coefficients are $\mathbb{Z}/\ell$ (the $\mathbb{Z}/\ell^n$ and $\mathbb{Q}_\ell$ version follow in a moment). Put $m = \dim X$, $n = \dim Y$. The proof goes by simultaneous induction in m and in n.

Let C be the cone of $(*)$. The induction assumption, together with the trick of Deligne [D1, (3.3)], shows that the cohomology sheaves of C are supported at a finite set of points. So the statement $C = 0$ is equivalent to the statement $R\Gamma(C) = 0$. The problem is local, hence we can assume X, Y to be affine. Then replacing X and Y by their closures we can assume that they are projective S-schemes. In this case $R\Gamma(C) = 0$, since $R\Gamma\Psi_{\overline{\eta}}(F) = R\Gamma(F_{\overline{\eta}})$ in the projective case.

([4]) Added in 1992: a proof appeared recently in a paper of Morihiko Saito.

Now we can pass to Gabber's theorem. Assume we are in a mixed situation, so we consider the schemes over a finite field $\mathbb{F}_q$. Let $\mathcal{M}(X)_{\text{mixed}} \subset D^b(X)_{\text{mixed}}$ be the category of mixed perverse sheaves on X and the corresponding derived category.

Let T be a curve, $o \in T$ be a closed point, and $U := T \setminus \{o\}$. Let S be a strict localization of T at o and $\overline{\eta}$ be the generic geometric point of S. For a T-scheme $f\colon X \to T$ put $X_0 = f^{-1}(o)$, $X_U = f^{-1}(U)$. One has the nearby cycles functor $\Psi_{\overline{\eta}x}\colon D^b(X_U)_{\text{mixed}} \to D^b(X_o)_{\text{mixed}}$ (see [D2]). It is convenient to use the twisted functor $\Psi_f := \Psi_{\overline{\eta}x}[-1]$. This functor is t-exact, i.e., $\Psi_f(\mathcal{M}(X_U)) \subset \mathcal{M}(X_o)$, and commutes with the Verdier duality as follows: $\Psi_f D = D\Psi_f(1)$ (here (1) is the Tate twist). The monodromy group acts on Ψ_f; for a perverse sheaf M let $s \in \operatorname{End}\Psi_f(M_U)$ be the logarithm of the unipotent part of geometric monodromy, and $\mu_.$ be the corresponding monodromy filtration on $\Psi_f(M_U)$.

5.1.2. THEOREM. *If M_U is pure of weight w, then $\mu_{\cdot+w-1}$ coincides with the weight filtration W on $\Psi_f(M_U)$.*

PROOF. The case when f is the identity (or a finite map) is Deligne's theorem [D2, (1.8.4)]. The proof in the general case follows similar lines:

(i) We can assume that M_U is irreducible.

(ii) Replacing T by a finite cover, we can assume that the geometric monodromy is unipotent.

(iii) The weights on $\operatorname{Ker} s$ ($=$ invariants of monodromy action) are $\leqslant w-1$.

PROOF OF (iii). Consider the canonical isomorphism $\operatorname{Ker} s = \operatorname{Ker}(j_! M_U \to j_* M_U)$. Since $j_!$ does not increase weights, the weights of $j_! M_U$ are $\leqslant w$. This implies that the perverse sheaf $j_! M_U / W_{w-1}(j_! M_U)$ is pure of weight w, and hence is semisimple. But the only irreducible quotient of $j_! M_U$ is $j_{!*} M_U$, hence $\operatorname{Ker} s \subset W_{w-1}(j_! M_U)$.

Dually, the weights of $\operatorname{Coker} s$ are $\geqslant w-1$ (since $\operatorname{Coker} s = \operatorname{Coker}(j_! M_U \to j_* M_U)(1)$).

(iv) Since the weight of s is -2, to prove the theorem it suffices to show that the primitive part P_{-i} is pure of weight $w-1-i$. We have $Gr_i^{J_!} = P_i$, $Gr_i^{J_*} = P_{-i}(-i)$ (see 4.1), so (iii) implies the inequalities for weights $\{w_i\}$ of P_i: $w_i \leqslant w-1$, $w_i + 2i \geqslant w-1$, i.e., $w-1-2i \leqslant w_i \leqslant w-1$. In particular, for $i=0$ we are done.

(v) Consider the fiber square $M_U^{\boxtimes 2}[-1]$: this is a perverse sheaf on $X \times_T X$ (at least over the generic point of T—the only thing we need) of weight $2w-1$. Since $\Psi_{f\times f}(M_U^{\boxtimes 2}[-1]) = \Psi_f(M_U)^{\boxtimes 2}$ by 5.1.1, Lemma 4.1.2(ii) implies that $P_{-i} \boxtimes P_{-i}(-i)$ occurs in $P_0(\Psi_{f\times f} M_U^{\boxtimes 2}[-1])$. Hence, by (iv), one has $2w_i + 2i = 2w-2$, or $w_i = w-i-1$.

5.1.3. Assume we have a parameter t at o, $t \in \mathcal{O}_T(T)$. We can define the functors of 4.2 in the mixed situation (see [B]): namely, one has the functors $\pi^a_f\colon M(U)_{\text{mixed}} \to M(X)_{\text{mixed}}$, $(\pi^a_f M_U)|_U$ is a consecutive extension of twists $M_U, M_U(1), \dots, M_U(a-1)$. Now 5.1.2 together with 4.1.1, 4.2.1, and 4.2.2(v) gives

COROLLARY. (i) *If* M_U *is pure of weight* w *then the filtrations* $J_{f!}$ *and* $W_{\cdot+w}$ *on* $j_{U!}(M_U)$ *coincide. The same for the filtrations* J_{f*} *and* $W_{\cdot+w}$ *on* $j_{U*}(M_U)$.

(ii) *The monodromy filtration* $\mu_\cdot$ *on* $\pi^a_f(M_U)$ *coincides with* $W_{\cdot+w+a-1}$. *In particular, for* $\Xi_f(M_U)$ *one has* $\mu_\cdot = W_{\cdot+w}$.

5.2. Pointwise purity and the socle property of the weight filtration. A mixed complex $F^\cdot$ on X is $*$-pointwise pure of weight w if for any closed point $x \in X$ the complex $i^*_x(F^\cdot)$ is pure of weight w (i.e., $H^i i^*_x F^\cdot$ is pure of weight $i+w$; here i_x is the embedding $x \hookrightarrow X$). One defines !-pointwise purity similarly using $i^!_x$ instead of i^*_x; the Verdier duality interchanges $*$- and !-purity. Note that if a pure perverse sheaf is $*$-pointwise pure of weight w, then w coincides with its weight.

Now let $(X, \widetilde{X})$ be a finite H-monodromic K-variety. Recall that any pure monodromic sheaf M has finite geometric monodromy along the fibers of $\widetilde{X} \to X$, hence, if M is geometrically irreducible, it lies in $\mathcal{M}(X, K)_\chi$ for some $\chi \in H^\vee_{\text{tors}}$ (by the local monodromy theorem; note that the restriction of any monodromic sheaf to any fiber of the map $\widetilde{X} \to X$ is tame by [V1]). We will say that $\widetilde{X}$ is $(K, \overline{\chi})$-pointwise pure for $\overline{\chi} \in H^\vee_{\text{tors}}$ if any pure $M \in \mathcal{M}(X, K)_{\overline{\chi}}$ is $*$- and !-pointwise pure, and $\widetilde{X}$ is K-pointwise pure if this holds for any $\overline{\chi}$, i.e., any pure K-equivariant monodromic sheaf is $*$- and !-pure.

5.2.1. EXAMPLES. (i) Here is a simple sufficient condition for $*$-pointwise purity. Let M be a pure perverse sheaf. Assume that for any $x \in X$ there exists an ètale neighborhood U of x such that the canonical map $H^\cdot(U, M) \to H^\cdot i^*_x M$ is surjective. Then M is $*$-pointwise pure (since the weights on $H^\cdot i^*_x M$ are $\leqslant \cdot + w$ by definition). In particular, this implies that "toric" irreducible perverse sheaves on a toric variety are pointwise pure, which leads to an explicit formula for Goresky-MacPherson Betti numbers of toric varieties (J. Bernstein, 1981, unpublished).

(ii) According to Kazhdan-Lusztig [KL2] and Lusztig [L, Chapter 1] the flag variety $\widetilde{X}_{\mathfrak{g}}$ is N-pointwise pure. Lusztig and Vogan [LV] have shown that $X_{\mathfrak{g}}$ is K-pointwise pure if K is a fixed point subgroup of an involution; it seems that their method, together with the decomposition theorem, should prove the K-pointwise purity of $\widetilde{X}_{\mathfrak{g}}$ for any symmetric pair $(\mathfrak{g}, K)$.

Recall that one defines the socle filtration $S_\cdot(M)$ on an object M of an abelian category by induction: $S_{-1} = 0$, $S_0 :=$ maximal semisimple subobject of M, $S_i(M)/S_{i-1}(M) := S_0(M/S_{i-1}(M))$. One defines the cosocle filtration $M = C_0(M) \supset C_{-1}(M) \supset \cdots$ in a dual manner.

If M is a mixed perverse sheaf, then $S_{\cdot}(M)$ and $C_{\cdot}(M)$ will denote the socle and cosocle filtrations on M considered as a geometric sheaf (Frobenius forgotten). Clearly both $S_{\cdot}$ and $C_{\cdot}$ are (being functorial) Frobenius invariant, hence $S_{\cdot}(M)$ and $C_{\cdot}(M)$ are mixed subsheaves of M.

5.2.2. Lemma. *Let $i_Y\colon Y \hookrightarrow X$ be a locally closed subscheme, M_Y a pure perverse sheaf on Y of weight w, and $N \subset {}^pH^0 i_{Y*}M_Y$ a mixed subsheaf such that any irreducible subquotient of N is !-pointwise pure. Then $S_{\cdot}(N) = W_{\cdot+w}(N)$.*

Proof. We have $S_{-1}(N) = 0 = W_{w-1}(N)$ (since i_{Y*} increases weights), $S_0(N) = W_w(N) = i_{Y!*}(i_Y^*N)$ (since, by the adjunction property of i_{Y*}, $S_0({}^pHi_{Y*}M_Y) = i_{!*}M_Y$). Since Gr^w is geometrically semisimple, one has $S_i(N) \supset W_{i+w}(N)$, so it remains to prove that $S_i(N) \subset W_{i+w}(N)$ for $i \geqslant 1$. We will do this by double induction: first in $\dim Y$, then in i. So assume that 5.2.2 is known for any $(Y', M_{Y'}, N')$ with $\dim Y' < \dim Y$, and that $S_j(N) = W_{j+w}(N)$ for $j < i$. Suppose that $S_i(N) \not\subset W_{i+w}(N)$. Then $S_i/S_{i-1} = S_i/W_{w+i-1}$ contains a pure geometrically irreducible subsheaf A of weight $a > i+w$ (possibly, after a finite extension of the finite base field). Note that $\operatorname{Supp} A \subset \overline{Y} \setminus Y$.

(i) Assume that A is supported at a closed point x. Consider the extension $0 \to W_{w+i-1}(N)/W_{w+i-2}(N) \to B \to A \to 0$ defined by N. Since $B \not\subset S_{i-1}(N)$, this extension is geometrically nontrivial, hence it corresponds to a nonzero element in $\operatorname{Hom}_{\mathrm{Frob}}(A, H^1 i_x^! W_{w+i-1}(N)/W_{w+i-2}(N))$. By the !-pointwise purity condition $H^1 i_x^!(W_{w+i-1}(N)/W_{w+i-2}(N))$ has weight $w+i$; but $a > w+i$, hence a contradiction.

(ii) If $\dim \operatorname{supp} A > 0$ we will use induction in $\dim Y$. The conditions of the lemma are local, so we can assume that X is affine. Choose a "generic" hyperplane section $Z \subset X$, namely such that for any irreducible subquotient L of ${}^pH^{\cdot} i_{Y*}M_Y$ a canonical morphism $i_Z^! L(1)[2] \to i_Z^* L$ is an isomorphism. Then $M_{Y\cap Z} := i_{Y\cap Z}^!(M_Y)[1]$ is a pure perverse sheaf of weight $w+1$ on $Y \cap Z$. Consider the complex $i_{Y\cap Z*}(M_{Y\cap Z}) = i_Z^![1] i_{Y*}M_Y$; one has $W_a {}^pH^{\cdot} i_{Y\cap Z*}M_{Y\cap Z} = i_Z^![1] W_{a-1} H^l i_{Y*}M_Y$. A subsheaf $N_Z := i_Z^![1](N)$ of ${}^pH^0 i_{Y\cap Z*}M_{Y\cap Z}$ satifies the conditions of the lemma, hence, by the induction hypothesis, $i_Z^![1](A)$ has weight $i+1$. Since $i_Z^![1](A) \neq 0$ (since $\dim \operatorname{supp} A > 0$) our A has weight i, and we are done.

5.2.3. Corollary. *Let M_1, M_2 be pure perverse sheaves of weights w_1, w_2 that are both $*$- and !-pointwise pure. Suppose that $\operatorname{Ext}^1_{\mathscr{M}_{\mathrm{mixed}}}(M_1, M_2) \neq 0$. Then exactly one of the following conditions holds (here $Y_i := \operatorname{supp} M_i$):*

(i) $Y_1 \subset Y_2$, $Y_1 \neq Y_2$, $w_1 = w_2 + 1$.
(ii) $Y_2 \subset Y_1$, $Y_1 \neq Y_2$, $w_1 = w_2 + 1$.
(iii) $Y_1 = Y_2$.

Proof. Clearly either $Y_1 \subset Y_2$ or $Y_2 \subset Y_1$ (otherwise $\mathrm{Ext}^1 = 0$). Let $0 \to M_2 \to N \to M_1 \to 0$ be a nonsplit mixed extension. If $Y_1 \neq Y_2$ and $Y_1 \subset Y_2$, then a canonical morphism $N \to {}^pH^0 i_{(Y_2\backslash Y_1)*} i^*_{Y_2\backslash Y_1} N = {}^pH^0 i_{(Y_2\backslash Y_1)*}(M_2|_{Y_2\backslash Y_1})$ is injective (since N is nonsplit), so we are in situation (i) by 5.2.2. If $Y_1 \neq Y_2$ and $Y_2 \subset Y_1$, then N is a quotient of $i_{(Y_1\backslash Y_2)!}(M_1|_{Y_1\backslash Y_2})$, and (ii) follows from a statement Verdier dual to 5.2.2.

5.2.4. Let $\widetilde{X}$ be a finite monodromic K-variety. Note that if M_1, M_2 are irreducible objects in $\mathscr{M}(\widetilde{X}, K)_{\widetilde{\chi}}$ such that $\mathrm{Ext}^1_{M(\widetilde{Q}, K)_{\widetilde{\chi}}}(M_1, M_2) \neq 0$, then $\operatorname{supp} M_1 \neq \operatorname{supp} M_2$ (this follows, using the functor $i_{!*}$, from the fact that the category $\mathscr{M}(\widetilde{Q}, K)_{\widetilde{\chi}, 1}$ is semisimple if Q is a single orbit).

Corollary. *Assume that $\widetilde{X}$ is K-pointwise pure. Let M be an object in $\mathscr{M}(\widetilde{X}, K)_{\widetilde{\chi}\,\mathrm{mixed}}$ such that $W_{a-1}(M) = 0$ and $W_a(M) = S_0(M)$. Then $W_{a+i}(M) = S_i(M)$ for any i.*

Proof. This follows by induction in i, using 5.2.2, the previous remark, and also the fact that any subquotient of M is K^0-equivariant.

5.2.5. **Example.** Consider an irreducible $M \in \mathscr{M}(X, K)_{\widetilde{\chi}}$, $\widetilde{\chi} \in H^\vee_{\mathrm{tors}}$. Let $I(M)$ be an injective envelope of M in $\mathscr{M}(X, K)_{\widetilde{\chi}}$. Then $I(M)$ admits a mixed structure (possibly after a finite extension of the base field), and for any such structure the weight filtration coincides with the socle filtration up to a shift.

Proof. The only problem is the existence of a mixed structure. But M clearly has one (being a middle extension of a lissé sheaf with finite monodromy). Any extension of the Frobenius action $M \to \mathrm{Frob}^* M$ to $I(M) \to \mathrm{Frob}^* I(M)$ defines some mixed structure on $I(M)$ (since any irreducible subquotient of $I(M)$ admits a mixed structure, and any Frobenius action on an irreducible perverse sheaf is unique up to a twist).

5.3. Jantzen conjectures. Let us apply the above considerations to $(\mathfrak{g}, K)$-modules. Let $(\mathfrak{g}, K)$ be an admissible Harish-Chandra pair (see 3.2), and $\chi \in \mathfrak{h}^*_{\mathbb{Q}}$ be a fixed rational dominant regular weight. The irreducible objects of $M(\widetilde{X}, K)_{\widetilde{\chi}}$ are of geometric origin (see 2.6.5(iii)), hence the corresponding standard objects each carry a weight filtration defined up to a shift. According to 5.1.3(i) it coincides with the Jantzen filtration. So an array of weight filtration properties also holds for its Jantzen counterpart via the equivalence $\mathscr{M}(U, K)_{\widetilde{\chi}} \xrightarrow{\widetilde{\Delta}_\chi} \mathscr{M}(X, K)_\chi$ (below we use freely the road from $\mathbb{F}$ to $\mathbb{C}$, see [BBD, Section 6]).

5.3.1. **Corollary.** *The Jantzen filtration on standard $(U, K)_\chi$-modules has semisimple consecutive quotients and does not depend on the choice of a positive deformation direction φ (see 4.3).*

5.3.2. COROLLARY. *Assume that* $\widetilde{X}$ *is* $(K, \overline{\chi})$*-pointwise pure (see 5.2).*

(i) *The Jantzen filtration* $J_{*\cdot}$ *on a* $*$*-standard* $(U, K)_\chi$*-module coincides with the socle filtration; the Jantzen filtration* $J_{!\cdot}$ *on a* !*-standard module coincides with the cosocle filtration.*

(ii) *If* $K = N$, *then* $J_{*\cdot}$ *also coincides, up to a shift, with the cosocle filtration, and* $J_{!\cdot}$ *coincides with the socle one.*

PROOF. (i) follows from 5.2.2 plus the Verdier dual statement. (ii) follows from 5.2.4 and the fact that any Verma module contains a unique irreducible submodule.

5.3.3. REMARKS. (i) The statement (ii) above was proved in [Ba] by purely algebraic methods. One can conjecture that it remains valid in the case of an arbitrary symmetric pair.

(ii) In fact, in [Ba] the socle property of $J_{!\cdot}$ for Verma modules was proved for Jantzen filtration defined by means of deformations of the central character in arbitrary nondegenerate directions, and we (in Section 4) used only those deformations in the positive directions. We do not know whether one can use such arbitrary deformations in the definition of $J_{!\cdot}$ for any symmetric pair.

For a regular $\chi \in \mathfrak{h}^*$ put $\Delta^{(\chi)} := \{\alpha \in \Delta \colon \chi(h_\alpha) \in \mathbb{Z}\}$. It is well known that $\Delta^{(\chi)}$ is a root system with the Weyl group $W^{(\chi)} = \{w \in W \colon w\chi - \chi \in \mathfrak{h}^*_{\mathbb{Z}}\}$ (recall that $\mathfrak{h}^*_{\mathbb{Z}} = \mathbb{Z}\Delta$). The orbit $W^{(\chi)}\chi$ contains a unique dominant weight, and for $\chi' \notin W^{(\chi)}\chi$ one has $\mathrm{Hom}(M_{\chi'}, M_\chi) = 0$ and $[M_\chi \colon L_{\chi'}] = 0$ (here $M_\chi \in \mathcal{M}(U, N)$ is the Verma module, L_χ is its irreducible quotient).

Let $\chi_1, \chi_2 \in \mathfrak{h}^*_{\mathbb{Q}}$ be regular weights such that $M_{\chi_1} \subset M_{\chi_2}$. Then for some (unique) dominant weight χ one has $\chi_i = w_i\chi$, where $w_i \in W^{(\chi)}$ and $w_1 \leqslant w_2$ with respect to the usual order on $W^{(\chi)}$.

5.3.4. COROLLARY. *One has* $J_{!i}(M_{\chi_1}) = M_{\chi_1} \cap J_{!i+\ell(w_2)-\ell(w_1)}(M_{\chi_2})$ *(here* ℓ *is the length function on* $W^{(\chi)}$*).*

PROOF. Since $\dim \mathrm{Hom}(M_{\chi_1}, M_{\chi_2}) = 1$, the embedding of the corresponding standard mixed sheaves is pure of certain weight a. Turning back to representations we see that $J_{!\cdot}(M_{\chi_1}) = M_{\chi_1} \cap J_{!\cdot+a}(M_{\chi_2})$. It remains to show that $a = \ell(w_2) - \ell(w_1)$. We can assume that $\ell(w_2) - \ell(w_1) = 1$ (if not, choose a chain $M_{\chi_1} \subset M_{\psi_1} \subset \cdots \subset M_{\psi_{\ell(w_2)-\ell(w_1)-1}} \subset M_{\chi_2}$ of Verma submodules such that each consecutive M has this property, and descend along it). Then Shapovalov's formula for the determinant of contravariant form implies that the vacuum vector of M_{χ_1} lies in $J_{!-1}(M_{\chi_2})/J_{!-2}(M_{\chi_2})$. Hence $a = 1$.

Let $\chi \in \mathfrak{h}^*_{\mathbb{Q}}$ be a dominant regular weight, $w_1, w_2 \in W^{(\chi)}$. Put

$$P_{w_1, w_2} := \sum_i [Gr^{J_!}_{-i}(M_{w_2\chi}) \colon L_{w_1\chi}]t^i.$$

5.3.5. COROLLARY. *This polynomial equals the Kazhdan-Lusztig polynomial for the group* $W^{(\chi)}$.

PROOF. According to [L, Chapter 1], Kazhdan-Lusztig polynomials are the matrix coefficients of the matrix that transforms the basis $j_{w!}(\mathbb{Q}_\ell)$ of the K-group of the category $\mathscr{M}(\widetilde{X}, N)_{\widetilde{\chi}\,\mathrm{mixed}}$ to the basis $j_{w!*}(\mathbb{Q}_\ell)$. Since the Jantzen filtration coincides with the weight filtration, our polynomials correspond to the entries of the inverse matrix. Since these matrices coincide up to standard changes of signs of the coefficients [KL1], we are done.

5.3.6. REMARKS. (i) Corollary 5.3.4 is Jantzen's Conjecture [J, (5.18)], see also [GJ1, (4.2)]. Corollary 5.3.5 was conjectured in [GJ1, GM]; in [GJ1] it was shown that 5.3.4 implies 5.3.5 by purely algebraic arguments.

(ii) It would be nice to get the analogs of 5.3.4 and 5.3.5 for arbitrary symmetric pairs. The only problem is to compute the weights in the space of Hom's between standard modules. Certainly one would like to know the weights in all the Ext's; we are ignorant of this even in the Verma modules case.

(iii) For a regular $\chi \in \mathfrak{h}^*$ let $\mathfrak{g}^{(\chi)}$ be a semisimple Lie algebra with the root system $\Delta^{(\chi)}$, $U^{(\chi)}$ its extended universal enveloping algebra, and $\mathfrak{h}^{(\chi)}$ its Cartan algebra. Then $\mathfrak{h}^{(\chi)}$ is (canonically) a direct summand of $\mathfrak{h}$; let $\chi_{\mathbb{Z}}$ be the $\mathfrak{h}^{(\chi)}$-component of χ, so $\chi_{\mathbb{Z}} \in \mathfrak{h}_{\mathbb{Z}}^{(\chi)*}$. One knows (see [So1] for a stronger statement) that $\mathscr{M}(U, N)_\chi$ is equivalent to a product of several copies of $\mathscr{M}(U^{(\chi)}, N^{(\chi)})_\chi$; this equivalence preserves the Verma modules. This immediately implies that in all the above results about $\mathscr{M}(U, N)_\chi$ we can drop the rationality assumption $\chi \in \mathfrak{h}_{\mathbb{Q}}^*$. Moreover, it sufficies to provide the proof for integral χ's only, which is the same as $\chi = 0$.

(iv) For a treatment of mixed categories of representations and the Koszul and Langlands dualities in this framework see [So1, So2, BGSo].

REFERENCES

[Ba] D. Barbasch, *Filtrations on Verma modules*, Ann. Sci. Ecole Norm. Sup. (4) **16** (1983), 489–494.

[B] A. Beilinson, *How to glue perverse sheaves*, Lecture Notes in Math., vol. 1289, Springer-Verlag, Berlin and New York, 1987, pp. 42–51.

[BB1] A. Beilinson and J. Bernstein, *Localisation of* $\mathfrak{g}$*-modules*, C. R. Acad. Sci. Paris Sér. I Math. **292** (1981), 15–18.

[BB2] ——, *A generalisation of Casselman's submodule theorem*, Representation Theory of Reductive Groups (Utah, 1982), Progr. Math., vol. 40, Birkhäuser, Boston, 1983, pp. 35–52.

[BBD] A. Beilinson, J. Bernstein, and P. Deligne, *Faisceaux pervers*, Astérisque **100** (1982), 5–171.

[BGSo] A. Beilinson, V. Ginsburg, and W. Soergel, *Koszul duality patterns in representation theory* (to appear).

[BGG] J. Bernstein, I. Gelfand, and S. Gelfand, *Category of* $\mathfrak{g}$*-modules*, Funktsional. Anal. i Prilozhen. **10** (1976), no. 2, 1–8; English transl. in Functional Anal. Appl. **10** (1976), no. 2.

[BG] J. Bernstein and S. Gelfand, *Tensor products of finite and infinite dimensional representations of semisimple Lie algebras*, Compositio. Math **41** (1980), 245–285.

[Bo] A. Borel et al., *Algebraic D-modules*, Academic Press, Boston, 1987.

[Br] J. L. Brylinski, *Transformations canoniques, dualité projective, théorie de Lefschetz, transformations de Fourier et sommes trigonometriques*, Astérique **140–141** (1986), 3–134.

[D1] P. Deligne, *Theoremes de finitude en cohomologie* ℓ*-adique*, SGA $4\frac{1}{2}$, Lecture Notes in Math., vol. 569, Springer-Verlag, Berlin and New York, 1977.

[D2] ———, *La conjecture de Weil.* II, Inst. Hautes Études Sci. Publ. Math. **52** (1980), 137–252.

[GJ1] O. Gabber and A. Joseph, *Towards the Kazhdan-Lusztig conjecture*, Ann. Sci. Ecole Norm. Sup. (4) **14** (1981), 261–302.

[GJ2] ———, *The Bernstein-Gelfand-Gelfand resolution and the Duflo sum formula*, Compositio Math. **43** (1981), 107–131.

[GM] S. Gelfand and R. MacPherson, *Verma modules and Schubert cells: a dictionary*, Lecture Notes in Math., vol. 924, Springer-Verlag, Berlin and New York, 1982, pp. 1–50.

[Gi] V. Ginzburg, *Equivariant cohomology and Kahler geometry*, Funktsional. Anal. i Prilozhen. **21** (1987), no. 4, 19–34; English transl. in Functional Anal. Appl. **21** (1987).

[HMSW] H. Hecht, D. Milicic, W. Schmid, and J. A. Wolf, *Localization and standard modules for real semisimple Lie groups* I: *The duality theorem*, Invent. Math. **90** (1987), 297–332.

[HMSW2] ———, *Localization and standard modules for real semisimple Lie groups* II: *Irreducibility, vanishing theorems and classifications* (to appear).

[J] J. C. Jantzen, *Moduln mit einem hohsten Gewicht*, Lecture Notes in Math., vol. 750, Springer-Verlag, Berlin and New York, 1980.

[K] M. Kashiwara, *Vanishing cycle sheaves and holonomic systems of differential equations*, Lecture Notes in Math., vol. 1016, Springer-Verlag, Berlin and New York, 1986, pp. 134–142.

[KL1] D. Kazhdan and G. Lusztig, *Representations of Coxeter groups and Hecke algebras*, Invent. Math **53** (1979), 165–184.

[KL2] ———, *Schubert varieties and Poincaré duality*, Proc. Sympos. Pure Math., vol. 36, Amer. Math. Soc., Providence, RI, 1980, pp. 185–203.

[Lau] G. Laumon, *Champs algebriques*, prepublication, Univ. Paris-Sud.

[L] G. Lusztig, *Characters of reductive groups over a finite field*, Ann. of Math. Studies, vol. 107, Princeton Univ. Press, Princeton, NJ, 1984.

[LV] G. Lusztig and D. Vogan, *Singularities of closures of* K*-orbits on flag manifolds*, Invent. Math. **71** (1983), 365–379.

[M] B. Malgrange, *Polynomes de Bernstein-Sato et évanescence*, Astérisque **101–102** (1983), 243–267.

[Mil] D. Milicic, *Localization and representation theory of reductive Lie groups* (to appear).

[MiVi] I. Mirkovic and K. Villonen, *Characteristic varieties of character sheaves*, Invent. Math. **93** (1988), 405-418.

[Sch] W. Schmid, *Recent developments in representation theory*, Lecture Notes in Math., vol. 1111, Springer-Verlag, Berlin and New York, 1985, pp. 135–153.

[S1] T. Springer, *Quelques applications de la cohomologie d'intersection*, Seminaire Bourbaki, exp. 589, Astérisque **92–93** (1982), 249–273.

[S2] ———, *The orbits in the flag variety under the action of subgroups defined by involutions*, Preprint.

[SGA1] A. Grothendieck (ed.), *Revêments étaes et groupe fondamentale*, Séminaire de Géométrie Algébriques du Bois Marie (SGA 1), 1960–1961, Lecture Notes in Math., vol. 224, Springer-Verlag, Berlin and New York, 1971.

[So1] W. Soergel, *Kategorie $\mathcal{O}$, perverse Garben und Moduln uber Koinvarianten zur Weylgruppe*, J. Amer. Math. Soc. **3** (1990), 421-445.

[So2] ———, *Langland's philosophy and Koszul duality*, Preprint, 1992.

[V1] J.-L. Verdier, *Specialisation de faisceaux et monodromie modere*, Astérisque **101–102** (1983), 332–364.

[V2] ———, *Prolongement des faisceaux pervers monodromiques*, Astérisque **130** (1985), 218–236.

[Vo] D. Vogan, *Representations of real reductive groups*, Progr. Math., vol. 15, Birkhäuser, Boston, 1981.

DEPARTMENT OF MATHEMATICS, MASSACHUSETTS INSTITUTE OF TECHNOLOGY, CAMBRIDGE, MASSACHUSETTS 02139

DEPARTMENT OF MATHEMATICS, HARVARD UNIVERSITY, CAMBRIDGE, MASSACHUSETTS 02138

ADVANCES IN SOVIET MATHEMATICS
Volume 16, Part 1, 1993

String Bases for Quantum Groups of Type A_r

ARKADY BERENSTEIN AND ANDREI ZELEVINSKY

To I. M. Gelfand on his 80th birthday

§0. Introduction

For every positive integer r let $\mathcal{A}_r$ denote the associative algebra with unit over the field of rational functions $\mathbb{Q}(q)$ generated by the elements $x_1, \dots, x_r$ subject to the relations:

$$x_i x_j = x_j x_i \quad \text{for } |i-j| > 1, \tag{0.1}$$

$$x_i^2 x_j - (q + q^{-1}) x_i x_j x_i + x_j x_i^2 = 0 \quad \text{for } |i-j| = 1. \tag{0.2}$$

This is the *quantum deformation* (or q-deformation) of the algebra of polynomial functions on the group N_{r+1} of upper unitriangular $(r+1) \times (r+1)$ matrices. In this paper we introduce and study a class of bases in $\mathcal{A}_r$ which we call *string bases*. The main example of a string basis is given as follows. Let $U_+ = U_{+,r}$ be the quantized universal enveloping algebra of the Lie algebra $\mathfrak{n}_{r+1}$ of N_{r+1} (see, e.g., [10]). Then $\mathcal{A}_r$ is seen to be the graded dual of U_+, and the basis in $\mathcal{A}_r$ dual to Lusztig's canonical basis in U_+ is a string basis. The string bases are defined by means of so called *string axioms*, which we find easier to work with than the axioms imposed by Lusztig or those by Kashiwara. The string axioms seem to be rather strong, and it is even conceivable that the string basis is unique but we do not know this in general. We prove the uniqueness of a string basis for $\mathcal{A}_2$ and $\mathcal{A}_3$.

The main advantage of string bases is that they seem to have nicer multiplicative properties than the canonical basis. We say that $x, y \in \mathcal{A}_r$ *quasicommute* if $xy = q^n yx$ for some integer n. We conjecture that every string basis B has the following property: two elements $b, b' \in B$ quasicommute if and only if $q^N bb' \in B$ for some integer N. We prove this for $\mathcal{A}_2$ and $\mathcal{A}_3$, and provide some supporting evidence for general $\mathcal{A}_r$.

Before giving precise formulations of the results we would like to put this work into historic context. Let $\mathfrak{g}$ be a semisimple complex Lie algebra of rank

1991 *Mathematics Subject Classification.* Primary 17B37; Secondary 20G45.

The second author was partially supported by the NSF (DMS-9104867).

r with fixed Cartan decomposition $\mathfrak{g} = \mathfrak{n}_- \oplus \mathfrak{h} \oplus \mathfrak{n}_+$. Our main motivation was to study "good bases" in irreducible $\mathfrak{g}$-modules. Good bases were introduced independently in [5] and [1]. Let $P \subset \mathfrak{h}^*$ denote the weight lattice of $\mathfrak{g}$, and $P_+ \subset P$ denote the semigroup of dominant integral weights, i.e., weights of the form $n_1\omega_1 + \cdots + n_r\omega_r$, where $\omega_1, \ldots, \omega_r$ are fundamental weights of $\mathfrak{g}$, and $n_1, \ldots, n_r$ are nonnegative integers. For $\lambda \in P_+$ let V_λ denote the irreducible (finite-dimensional) $\mathfrak{g}$-module with the highest weight λ. For $\beta \in P$ we denote by $V_\lambda(\beta)$ the weight subspace of weight β in V_λ. For $\nu = \sum_i n_i\omega_i \in P_+$ we set

$$V_\lambda(\beta; \nu) = \{x \in V_\lambda(\beta) : e_i^{n_i+1}x = 0 \text{ for } i = 1, \ldots, r\},$$

where $e_1, \ldots, e_r$ are standard generators of $\mathfrak{n}_+$ whose weights are simple roots $\alpha_1, \ldots, \alpha_r$ corresponding to fundamental weights $\omega_1, \ldots, \omega_r$. A basis B of V_λ is said to be *good* if every subspace of the form $V_\lambda(\beta; \nu)$ is spanned by a part of B. This definition was motivated by the classical result that for every three irreducible finite-dimensional g-modules V_λ, V_μ, V_ν there is a natural isomorphism

$$\mathrm{Hom}_{\mathfrak{g}}(V_\mu, V_\lambda \otimes V_\nu) \simeq V_\lambda(\mu - \nu; \nu).$$

It follows that a good basis "resolves the multiplicities" in the tensor product $V_\lambda \otimes V_\nu$, i.e., provides its decomposition into irreducible components.

It was conjectured in [5] that good bases always exist (the proof given in [1] turned out to be inadequate). For $\mathfrak{g} = sl_{r+1}$ the existence of good bases follows from [3]: as shown in [5], the special basis constructed in [3] is good. For arbitrary $\mathfrak{g}$ the existence of good bases was proven by Mathieu [14].

Each V_λ has two important realizations dual to each other. First, there is a canonical epimorphism $U(\mathfrak{n}_+) \to V_\lambda$ sending each $u \in U(\mathfrak{n}_+)$ to $u\xi_\lambda$, where ξ_λ is the lowest vector in V_λ. Hence, V_λ can be realized as a quotient of $U(\mathfrak{n}_+)$. For the dual realization we notice that the dual space $U(\mathfrak{n}_+)^*$ can be identified with the algebra $\mathbb{C}[N_+]$ of polynomial functions on the unipotent group N_+ whose Lie algebra is $\mathfrak{n}_+$. Hence, each V_λ can be canonically realized as a subspace of $\mathbb{C}[N_+]$.

It was suggested in [5] and further pursued in [15] that there should exist a basis B in $\mathbb{C}[N_+]$ such that for every $V_\lambda \subset \mathbb{C}[N_+]$ the set $B \cap V_\lambda$ is a good basis for V_λ. Such a basis was constructed for $\mathfrak{g} = sl_3$ in [5] and for $\mathfrak{g} = sp_4$ in [15]. In both cases B consists of some monomials in a finite number of generators. It was conjectured in [15] that there exist some natural conditions which fix B to be *unique* but the problem of finding these conditions remained open.

Two remarkable solutions of this problem were given by Lusztig [10] and Kashiwara [6]. Both solutions provide a system of axioms that determines a basis in $U(\mathfrak{n}_+)$ uniquely. The fundamental idea beyond these axioms is that in order to determine the basis uniquely one has to pass from $U(\mathfrak{n}_+)$ to its q-deformation $U_q(\mathfrak{n}_+)$. Thus, both authors construct a basis in $U_q(\mathfrak{n}_+)$, called *canonical* by Lusztig and (*lower*) *global crystal* by Kashiwara (it was later proven by Lusztig [11] that these bases coincide). One recovers the basis in $U(\mathfrak{n}_+)$ by specializing $q = 1$. As shown in [11, Theorem 4.4(c)], the dual basis B of $\mathbb{C}[N_+]$ has the property that $B \cap V_\lambda$ is a good basis for V_λ for all $V_\lambda \subset \mathbb{C}[N_+]$. This gives another and much more constructive proof of the existence of good bases.

The papers [6, 7, 8, 10, 11, and 12] reveal many important properties and applications of canonical (or crystal) bases. But these exciting developments essentially leave aside the structure of the dual basis.

The algebra $\mathscr{A}_r$ introduced above is the q-deformation of $\mathbb{C}[N_+]$ for $\mathfrak{g} = sl_{r+1}$. Many of the results and arguments below make sense for arbitrary semisimple Lie algebras or even for arbitrary Kac-Moody algebras but for the sake of simplicity we shall treat only this case here.

The material is organized as follows. Main results of the paper are collected in Sections 1 and 2, which can be considered as an expanded introduction. In Section 1 we introduce string axioms and describe the string bases of $\mathscr{A}_r$ for $r = 2, 3$. Our main conjecture on the multiplicative property of string bases is also given here (Conjecture 1.7).

In Section 2 we introduce our main tool for the study of string bases, the notion of the *string* of an element $x \in \mathscr{A}_r$ in a given direction. These strings are certain finite sequences of nonnegative integers, used as combinatorial labels for the elements of a string basis. Precise definitions and our main results on strings are collected in Section 2.

Sections 3 to 10 are devoted to the proofs of all theorems of Sections 1 and 2. More detailed directions to the proof of each theorem can be found after its formulation in Sections 1 and 2.

The article is concluded with the appendix where we discuss basic properties of $\mathscr{A}_r$ and the duality between $\mathscr{A}_r$ and $U_{+,r}$. The results we need (Propositions 1.1 to 1.3) seem to be well known to experts in the field, but we were unable to locate exact references. For the sake of convenience of the reader we sketch the proofs in the appendix.

§1. String bases and quasicommutative monomials

Let $U_+ = U_{+,r}$ be the quantized universal enveloping algebra of the maximal nilpotent subalgebra $\mathfrak{n}_+$ of sl_{r+1}. This is an algebra with unit over the field of rational functions $\mathbb{Q}(q)$ generated by the elements $E_1, \dots, E_r$ subject to the relations:

$$E_iE_j = E_jE_i \quad \text{for } |i-j| > 1, \tag{1.1}$$

$$E_i^2E_j - (q+q^{-1})E_iE_jE_i + E_jE_i^2 = 0 \quad \text{for } |i-j| = 1. \tag{1.2}$$

Let $\alpha_1, \dots, \alpha_r$ be the simple roots of sl_{r+1} in the standard numeration, and Q_+ the semigroup generated by $\alpha_1, \dots, \alpha_r$. The algebra U_+ is Q_+-graded via $\deg(E_i) = \alpha_i$. For $\gamma \in Q_+$ let $U_+(\gamma)$ denote the homogeneous component of degree γ in U_+. Since U_+ is a deformation of $U(\mathfrak{n}_+)$, we have

$$\dim(U_+(\gamma)) = p(\gamma), \tag{1.3}$$

where $p(\gamma)$ is the number of partitions of γ into the sum of positive roots (the Kostant partition function [9]).

Comparing (1.1), (1.2) with (0.1), (0.2) we see that the correspondence $E_i \mapsto x_i$ extends to an algebra isomorphism $U_+ \to \mathscr{A}_r$. We transfer the Q_+-grading from U_+ to $\mathscr{A}_r$ via this isomorphism, so we have $\deg(x_i) = \alpha_i$, and $\dim(\mathscr{A}_r(\gamma)) = p(\gamma)$ for $\gamma \in Q_+$. But it is important for us to keep distinguishing U_+ and $\mathscr{A}_r$. In fact, we wish to identify each graded component $\mathscr{A}_r(\gamma)$ with the dual space $U_+(\gamma)^*$. To do this we introduce an action of U_+ on $\mathscr{A}_r$, which will play the crucial part in the sequel.

PROPOSITION 1.1. *There exists a unique action $(E, x) \mapsto E(x)$ of the algebra U_+ on $\mathscr{A}_r$ satisfying the following properties:*

(a) (*Homogeneity*) *If $E \in U_+(\alpha)$, $x \in \mathscr{A}_r(\gamma)$ then $E(x) \in \mathscr{A}_r(\gamma - \alpha)$.*

(b) (*Leibnitz formula*)

$$E_i(xy) = E_i(x)y + q^{-(\gamma, \alpha_i)} x E_i(y) \quad \text{for } x \in \mathscr{A}_r(\gamma),\ y \in \mathscr{A}_r \tag{1.4}$$

(*here and in the sequel (γ, α) is the usual scalar product on Q_+ defined by means of the Cartan matrix*).

(c) (*Normalization*) *$E_i(x_j) = \delta_{ij}$ for $i, j = 1, \ldots, r$.*

PROPOSITION 1.2. (a) *If $\gamma \in Q_+ \setminus \{0\}$, and x is a nonzero element of $\mathscr{A}_r(\gamma)$ then $E_i(x) \neq 0$ for some $i = 1, \ldots, r$.*

(b) *For every $\gamma \in Q_+$ the mapping $(E, x) \mapsto E(x)$ defines a non-degenerate pairing*

$$U_+(\gamma) \times \mathscr{A}_r(\gamma) \to \mathscr{A}_r(0) = \mathbb{Q}(q).$$

Both propositions will be proven in the appendix.

Now we are in a position to define the string bases in $\mathscr{A}_r$, the main object of study in this paper. We shall use the notation

$$[n] = \frac{q^n - q^{-n}}{q - q^{-1}}, \qquad [n]! = [1][2]\cdots[n],$$

and define the *divided powers* to be

$$E_i^{(n)} = \frac{E_i^n}{[n]!}.$$

For a subset $B \subset \mathscr{A}_r$ we denote by $[B]^+ \subset \mathscr{A}_r$ the set of all linear combinations of elements of B with coefficients from $\mathbb{Z}_+[q, q^{-1}]$.

Let B be a basis in $\mathscr{A}_r$. We say that B is a string basis if it satisfies the following *string axioms*:

(S0) B consists of homogeneous elements and contains 1.

(S1) For every $b, b' \in B$ the product bb' belongs to $[B]^+$.
(S2) $E_i(b) \in [B]^+$ for $b \in B$, $i = 1, \dots, r$.
(S3) If $b \in B$, and l is the maximal integer such that $E_i^l(b) \neq 0$ then $E_i^{(l)}(b) \in B$.

PROPOSITION 1.3. *The basis in $\mathscr{A}_r$ dual to the Lusztig's canonical basis in U_+ is a string basis.*

This will be proven also in the appendix. Note that the axiom (S3) is analogous to [7, Lemma 5.1.1].

Now we construct some elements belonging to every string basis of $\mathscr{A}_r$. We need the q-analogs of the natural coordinates (matrix entries) on the group N_{r+1}. For every two elements $x \in \mathscr{A}_r(\gamma)$, $y \in \mathscr{A}_r(\gamma')$ we define their *q-commutator* as follows:

$$[x, y] = \frac{xy - q^{(\gamma, \gamma')}yx}{q - q^{-1}}. \tag{1.5}$$

Then $[x, y] \in \mathscr{A}_r(\gamma + \gamma')$. Now let $T = T_r = (t_{ij})$ be the $(r+1) \times (r+1)$ matrix with entries in $\mathscr{A}_r$ defined as follows. We set $t_{ij} = 0$ for $i > j$, $t_{ii} = 1$ for all i, $t_{i,i+1} = x_i$ for $i = 1, \dots, r$, and finally for $j > i+1$ we define t_{ij} inductively by $t_{ij} = [t_{i,j-1}, t_{j-1,j}]$. Clearly, t_{ij} is homogeneous of degree $\alpha_i + \alpha_{i+1} + \cdots + \alpha_{j-1}$. In particular,

$$T_2 = \begin{pmatrix} 1 & x_1 & \frac{x_1x_2 - q^{-1}x_2x_1}{q - q^{-1}} \\ 0 & 1 & x_2 \\ 0 & 0 & 1 \end{pmatrix}.$$

For every two increasing sequences of indices $I = (i_1 < i_2 < \cdots < i_s)$, $J = (j_1 < j_2 < \cdots < j_s)$ from $\{1, 2, \dots, r+1\}$ we denote by $\Delta(I; J)$ the *quantum minor* of T with rows I and columns J:

$$\Delta(I; J) = \sum_\sigma (-q)^{l(\sigma)} t_{i_{\sigma(1)}, j_1} \cdots t_{i_{\sigma(s)}, j_s}, \tag{1.6}$$

where the summation is over the symmetric group S_s, and $l(\sigma)$ is the length of a permutation σ.

THEOREM 1.4. *Every string basis of $\mathscr{A}_r$ contains all nonzero minors $\Delta(I; J)$.*

Theorem 1.4 will be proven in Section 7.

Now we are able to describe the string bases for $\mathscr{A}_2$ and $\mathscr{A}_3$. A minor $\Delta(I; J)$ will be called *primitive* if $i_1 < j_1$ and $i_k \leqslant j_{k-1}$ for $k = 2, \dots, s$. Let $\mathscr{P} = \mathscr{P}_r$ denote the set of all primitive minors of T_r. We recall that two elements $x, y \in \mathscr{A}_r$ *quasicommute* if $xy = q^n yx$ for some integer n. We call a product $b_1 b_2 \cdots b_k$ *quasicommutative* if every b_i and b_j quasicommute (some of the elements b_i may coincide).

THEOREM 1.5. *Let $r = 2$ or $r = 3$. Then there is only one string basis B in $\mathcal{A}_r$. Every $b \in B$ has the form $b = q^N b_1 b_2 \cdots b_k$, where $N \in \mathbb{Z}$, and $b_1 \cdots b_k$ is a quasicommutative product of elements of $\mathcal{P}_r$. Conversely, for every quasicommutative product $b_1 \cdots b_k$ of elements of $\mathcal{P}_r$ there is an integer N such that $q^N b_1 b_2 \cdots b_k \in B$.*

The case $r = 2$ will be treated in Section 5, and the case $r = 3$ in Sections 9, 10.

Consider the simplicial complex on $\mathcal{P}_r$ whose simplices are all mutually quasicommutative subsets. Theorem 1.5 says that for $r = 2$ or $r = 3$ the structure of the string basis B in $\mathcal{A}_r$ is controlled by this simplicial complex. The set $\mathcal{P}_2$ consists of four primitive minors:

$$\begin{aligned} \Delta(1;2) = t_{12} = x_1, \qquad & \Delta(2;3) = t_{23} = x_2, \\ \Delta(1;3) = t_{13} = [x_1, x_2], \qquad & \Delta(1,2;2,3) = t_{12}t_{23} - qt_{13} = [x_2, x_1]. \end{aligned} \tag{1.7}$$

In this case Theorem 1.5 can be refined as follows (for the proof see Section 5).

THEOREM 1.6. *The simplicial complex $\mathcal{P}_2$ has two maximal simplices: $\mathcal{P}_2 \setminus \{x_1\}$ and $\mathcal{P}_2 \setminus \{x_2\}$ (i.e., every two elements of $\mathcal{P}_2$ quasicommute with the only exception of x_1 and x_2). The string basis B of $\mathcal{A}_2$ is given by*

$$B = \{q^{\binom{m_1+m_{12}}{2}+\binom{m_2+m_{21}}{2}} x_1^{m_1} x_2^{m_2} [x_1, x_2]^{m_{21}} [x_2, x_1]^{m_{12}}\},$$

where $(m_1, m_2, m_{12}, m_{21})$ runs over all 4-tuples of nonnegative integers such that $\min(m_1, m_2) = 0$.

The set $\mathcal{P}_3$ consists of 12 primitive minors, namely:

$$\begin{gathered} \Delta(1;2), \Delta(2;3), \Delta(3;4), \Delta(1;3), \Delta(2;4), \Delta(1;4), \\ \Delta(1,2;2,3), \Delta(1,2;2,4), \Delta(1,2;3,4), \\ \Delta(1,3;3,4), \Delta(2,3;3,4), \Delta(1,2,3;2,3,4). \end{gathered}$$

The simplicial complex $\mathcal{P}_3$ has 14 maximal simplices; it will be described in Section 9.

Now we return to arbitrary $\mathcal{A}_r$, and state our main conjecture.

CONJECTURE 1.7. *Let B be a string basis in $\mathcal{A}_r$. Two elements $b, b' \in B$ quasicommute if and only if $q^N bb' \in B$ for some integer N.*

Conjecture 1.7 would imply that elements of a string basis B are quasicommutative monomials in some set of generators $\widetilde{\mathcal{P}_r}$. Here $\widetilde{\mathcal{P}_r}$ consists of all elements $b \in B$ that cannot be decomposed into a quasicommutative product of two elements of smaller degree. It follows easily from Theorem 1.5 that Conjecture 1.7 is true for $r = 2, 3$, and in these cases we have $\widetilde{\mathcal{P}_r} = \mathcal{P}_r$. In general, it is not even clear whether $\widetilde{\mathcal{P}_r}$ is finite.

§2. Strings: main results

Here we introduce the main tool for our study of string bases. Let x be a nonzero homogeneous element of $\mathscr{A}_r$. For each $i = 1, \ldots, r$ we set

$$l_i(x) = \max\{l \in \mathbb{Z}_+ : E_i^l(x) \neq 0\}. \tag{2.1}$$

We shall use the following notation:

$$E_i^{(\mathrm{top})}(x) := E_i^{(l_i(x))}(x). \tag{2.2}$$

Now let $\mathbf{i} = (i_1, i_2, \ldots, i_m)$ be a sequence of indices from $\{1, 2, \ldots, r\}$ such that no two consecutive indices are equal to each other. We call such a sequence *admissible*. We associate to x and $\mathbf{i}$ a nonnegative integer vector $a(\mathbf{i}; x) = (a_1, \ldots, a_m)$ defined by

$$a_k = l_{i_k}(E_{i_{k-1}}^{(\mathrm{top})} E_{i_{k-2}}^{(\mathrm{top})} \cdots E_{i_1}^{(\mathrm{top})}(x)). \tag{2.3}$$

We call $a(\mathbf{i}; x)$ the *string* of x in direction $\mathbf{i}$. We abbreviate

$$E_{\mathbf{i}}^{(\mathrm{top})}(x) = E_{i_m}^{(\mathrm{top})} E_{i_{m-1}}^{(\mathrm{top})} \cdots E_{i_1}^{(\mathrm{top})}(x). \tag{2.4}$$

Note that $E_{\mathbf{i}}^{(\mathrm{top})}(x)$ is a nonzero homogeneous element of $\mathscr{A}_r$ of degree $\deg(x) - \sum_k a_k \alpha_{i_k}$.

From now on we fix a string basis B in $\mathscr{A}_r$. By the string axiom (S3), if $x \in B$ then $E_{\mathbf{i}}^{(\mathrm{top})}(x) \in B$ for all $\mathbf{i}$.

THEOREM 2.1. *For every admissible sequence $\mathbf{i}$ an element $b \in B$ is uniquely determined by the string $a(\mathbf{i}; b)$ and the element $E_{\mathbf{i}}^{(\mathrm{top})}(b)$.*

Theorem 2.1 will be proven in Section 4.

Let $W = S_{r+1}$ be the Weyl group of type A_r. For each $w \in W$ we denote by $R(w)$ the set of all reduced decompositions of w, i.e., the set of sequences $\mathbf{i} = (i_1, i_2, \ldots, i_l)$ such that $l = l(w)$, and w is equal to the product of simple reflections $s_{i_1} s_{i_2} \cdots s_{i_l}$.

THEOREM 2.2. *Let $w \in W$, and $\mathbf{i}, \mathbf{i}' \in R(w)$. Then $E_{\mathbf{i}}^{(\mathrm{top})}(b) = E_{\mathbf{i}'}^{(\mathrm{top})}(b)$ for $b \in B$. Furthermore, there is a piecewise-linear automorphism ${}_{\mathbf{i}'}T_{\mathbf{i}} : \mathbb{R}^l \to \mathbb{R}^l$ preserving the lattice $\mathbb{Z}^l$ and such that $a(\mathbf{i}'; b) = {}_{\mathbf{i}'}T_{\mathbf{i}}(a(\mathbf{i}; b))$ for $b \in B$.*

Theorem 2.2 implies that for $b \in B$, $\mathbf{i} \in R(w)$ the element $E_{\mathbf{i}}^{(\mathrm{top})}(b) \in B$ depends only on b and w, so we shall denote it simply by $E_w^{(\mathrm{top})}(b)$. The most important for us will be the case when $w = w_0$, the element of maximal length in W.

THEOREM 2.3. *We have* $E_{w_0}^{(\mathrm{top})}(b) = 1$ *for all* $b \in B$.

THEOREM 2.4. *For every* $\mathbf{i} = (i_1, i_2, \dots, i_m) \in R(w_0)$ *the correspondence* $b \mapsto a(\mathbf{i}; b)$ *is a bijection between* B *and the semigroup* $C_{\mathbb{Z}}(\mathbf{i})$ *of all integral points of some polyhedral convex cone* $C(\mathbf{i}) \subset \mathbb{R}_+^m$.

One easily checks that the number $m = l(w_0)$ is equal to $\binom{r+1}{2}$, and that the sequence $\mathbf{i}(1) = (1, 2, 1, 3, 2, 1, \dots, r, r-1, \dots, 1)$ belongs to $R(w_0)$.

THEOREM 2.5. *The cone* $C(\mathbf{i}(1))$ *is a simplicial cone in* $\mathbb{R}_+^m$ *defined by the inequalities* $a_k \le a_{k-1}$ *whenever* $\mathbf{i}(1)_k < \mathbf{i}(1)_{k-1}$.

The description of the cones $C(\mathbf{i})$ for other reduced decompositions of w_0 is more complicated. They can be computed by means of the maps ${}_{\mathbf{i}'}T_{\mathbf{i}}$ from Theorem 2.2. These maps are closely related to another family of piecewise-linear maps $R_{\mathbf{i}}^{\mathbf{i}'}$ introduced by Lusztig [10, 2.1 and 2.6]. An explicit description of the maps ${}_{\mathbf{i}'}T_{\mathbf{i}}$ is given as follows. First, they are *local* in the following sense.

PROPOSITION 2.6. *Let* $w \in W$, *and* $\mathbf{i} = (i_1, \dots, i_l)$, $\mathbf{i}' = (i'_1, \dots, i'_l) \in R(w)$. *Suppose there are two indices* $s < t$ *such that* $i_k = i'_k$ *for* $k < s$ *or* $k > t$, *and let* $\mathbf{i}_0 = (i_s, \dots, i_t)$, $\mathbf{i}'_0 = (i'_s, \dots, i'_t)$. *Then the map* ${}_{\mathbf{i}'}T_{\mathbf{i}}$ *leaves the components* a_k *with* $k < s$ *or* $k > t$ *unchanged, and transforms the vector* $(a_s, \dots, a_t)$ *according to* ${}_{\mathbf{i}'_0}T_{\mathbf{i}_0}$.

It is well known (cf. [10, Section 2]) that any two reduced decompositions of $w \in W$ can be transformed into each other by a sequence of elementary transformations of two kinds:

$$(\mathbf{i}_1, i, j, \mathbf{i}_2) \mapsto (\mathbf{i}_1, j, i, \mathbf{i}_2) \quad \text{for } |i-j| > 1, \tag{2.5}$$

$$(\mathbf{i}_1, i, j, i, \mathbf{i}_2) \mapsto (\mathbf{i}_1, j, i, j, \mathbf{i}_2) \quad \text{for } |i-j| = 1. \tag{2.6}$$

Taking into account Proposition 2.6, this allows us to reduce the computation of ${}_{\mathbf{i}'}T_{\mathbf{i}}$ to the following two special cases.

THEOREM 2.7. (a) *If* $\mathbf{i} = (i, j)$, $\mathbf{i}' = (j, i)$ *with* $|i-j| > 1$ *then*

$${}_{\mathbf{i}'}T_{\mathbf{i}}(a_1, a_2) = (a_2, a_1). \tag{2.7}$$

(b) *If* $\mathbf{i} = (i, j, i)$, $\mathbf{i}' = (j, i, j)$ *with* $|i-j| = 1$ *then*

$${}_{\mathbf{i}'}T_{\mathbf{i}}(a_1, a_2, a_3) = (\max(a_3, a_2 - a_1), a_1 + a_3, \min(a_1, a_2 - a_3)). \tag{2.8}$$

The proofs of Theorems 2.2 to 2.7 will be completed in Section 6.

Multiplicative properties of string bases are closely related to geometric properties of the maps ${}_{\mathbf{i}'}T_{\mathbf{i}}$. Let $\mathbf{i} = (i_1, \dots, i_m) \in R(w_0)$. By an $\mathbf{i}$-*wall* we mean a hyperplane in $\mathbb{R}^m$ given by the equation $a_k - a_{k+1} + a_{k+2} = 0$ for some index k such that $i_k = i_{k+2} = i_{k+1} \pm 1$. Let $C(\mathbf{i})^0$ be the interior of the cone $C(\mathbf{i})$ from Theorem 2.4. We say that a point $a \in C(\mathbf{i})^0$ is **i**-*regular* if for every $\mathbf{i}' \in R(w_0)$ the point ${}_{\mathbf{i}'}T_{\mathbf{i}}(a)$ does not lie on any $\mathbf{i}'$-wall. We call **i**-*linearity domains* the closures of connected components of the set of **i**-regular points. This term is justified by the following.

PROPOSITION 2.8. *Every* **i**-*linearity domain is a polyhedral convex cone in* $C(\mathbf{i})$. *Two points* $a, a' \in C(\mathbf{i})$ *lie in the same* **i**-*linearity domain if and only if* ${}_{\mathbf{i}'}T_{\mathbf{i}}(a+a') = {}_{\mathbf{i}'}T_{\mathbf{i}}(a) + {}_{\mathbf{i}'}T_{\mathbf{i}}(a')$ *for each* $\mathbf{i}' \in R(w_0)$.

The following theorem adds some support to Conjecture 1.7.

THEOREM 2.9. *Suppose two elements* $b, b' \in B$ *satisfy at least one of the following two conditions*:

(1) b *and* b' *quasicommute*;
(2) $q^N bb' \in B$ *for some integer* N.

Then for every $\mathbf{i} \in R(w_0)$ *the strings* $a(\mathbf{i}; b)$ *and* $a(\mathbf{i}; b')$ *belong to the same* **i**-*linearity domain.*

As a consequence of Theorem 2.9, we obtain the following "first approximation" to Conjecture 1.7.

THEOREM 2.10. *Suppose* $b, b' \in B$ *quasicommute. Then there is exactly one element* $b'' \in B$ *occurring in the* B-*decomposition of* bb' *such that* $a(\mathbf{i}; bb') = a(\mathbf{i}; b'')$ *for every* $\mathbf{i} \in R(w_0)$.

Proposition 2.8 and Theorems 2.9, 2.10 will be proven in Section 8.

We describe the $\mathbf{i}(1)$-linearity domains for $r = 2$ in Section 5, and for $r = 3$ in Section 9. Using this description we can recover the simplicial complex $\mathcal{P}_r$ for $r = 2, 3$ (see Section 1) in the following way.

THEOREM 2.11. *For* $r = 2, 3$ *the edges of* $\mathbf{i}(1)$-*linearity domains are exactly the rays* $\mathbb{R}_+ a(\mathbf{i}(1); b)$ *for* $b \in \mathcal{P}_r$.

THEOREM 2.12. *For* $r = 2, 3$ *there is a decomposition* $C(\mathbf{i}(1)) = \bigcup_{C \in \mathcal{D}} C$ *into the union of simplicial cones, satisfying the following properties*:

(a) *Every* $\mathbf{i}(1)$-*linearity domain is a union of some* $C \in \mathcal{D}$.
(b) *Each edge of a cone* $C \in \mathcal{D}$ *is an edge of some* $\mathbf{i}(1)$-*linearity domain, i.e., has the form* $\mathbb{R}_+ a(\mathbf{i}(1); b)$ *for* $b \in \mathcal{P}_r$.
(c) *Two elements* $b, b' \in \mathcal{P}_r$ *quasicommute if and only if* $\mathbb{R}_+ a(\mathbf{i}(1); b)$ *and* $\mathbb{R}_+ a(\mathbf{i}(1); b')$ *are edges of the same* $C \in \mathcal{D}$.

We expect some analogs of Theorems 2.11, 2.12 to hold for general r.

§3. Bases of PBW type

In this section we collect together some properties of the algebra $\mathcal{A}_r$ which will be used later for the study of string bases. We use freely the notation and terminology introduced above.

We start with a generalization of the Leibnitz formula (1.4). Let $\mathbf{i} = (i_1, \ldots, i_m)$ be a sequence of indices from $\{1, 2, \ldots, r\}$, and $a = (a_1, \ldots, a_m)$ be a nonnegative integral vector of the same length. We define an element $E_{\mathbf{i}}^{(a)} \in U_+$ by the following formula:

$$E_{\mathbf{i}}^{(a)} = E_{i_m}^{(a_m)} E_{i_{m-1}}^{(a_{m-1})} \cdots E_{i_1}^{(a_1)}. \tag{3.1}$$

The element $E_{\mathbf{i}}^{(a)}$ is homogeneous of degree $\sum_{k=1}^m a_k \alpha_{i_k}$.

Proposition 3.1. *Let $x \in \mathcal{A}_r(\gamma)$, $y \in \mathcal{A}_r$. Then*

$$E_{\mathbf{i}}^{(a)}(xy) = \sum_{a', a''} q^{\Phi(a', a'')} E_{\mathbf{i}}^{(a')}(x) E_{\mathbf{i}}^{(a'')}(y), \tag{3.2}$$

where the summation is over all $a', a'' \in \mathbb{Z}_+^m$ with $a' + a'' = a$, and

$$\begin{aligned}\Phi(a', a'') &= \Phi_{\mathbf{i},\gamma}(a', a'') \\ &= \sum_{1 \leqslant k \leqslant m} a'_k a''_k + \sum_{1 \leqslant k < l \leqslant m} a'_k a''_l (\alpha_{i_k}, \alpha_{i_l}) - \Big(\gamma, \sum_{1 \leqslant k \leqslant m} a''_k \alpha_{i_k}\Big).\end{aligned} \tag{3.3}$$

In particular, for $m = 1$, i.e., $a \in \mathbb{Z}_+$ we have

$$E_i^{(a)}(xy) = \sum_{a'=0}^{a} q^{(a-a')(a'-(\gamma, \alpha_i))} E_i^{(a')}(x) E_i^{(a-a')}(y). \tag{3.4}$$

Proof. We first deduce (3.4) from (1.4) using induction on a and representing $E_i^{(a)}$ as $1/[a] E_i^{(a-1)} E_i$. The formula (3.2) follows from (3.4) by induction on m. □

Corollary 3.2. *The algebra $\mathcal{A}_r$ has no zero-divisors.*

Proof. It is enough to show that $xy \neq 0$ for every nonzero homogeneous $x, y \in \mathcal{A}_r$. We prove this by induction on the degree of x (here and in the sequel we use the partial order on the grading semigroup Q_+ given by $\gamma \geqslant \gamma'$ if $\gamma - \gamma' \in Q_+$). There is nothing to prove if $\deg(x) = 0$. Suppose $\deg(x) > 0$. By Proposition 1.2(a), $E_i(x) \neq 0$ for some i. Now we apply (3.4) for $a = l_i(x) + l_i(y)$. Then all the summands on the right-hand side of (3.4) vanish except the one with $a' = l_i(x)$. By the inductive assumption, this remaining summand is nonzero, and we are done. □

The same argument as in the proof of Corollary 3.2 implies the following.

Corollary 3.3. *Let x, y be nonzero homogeneous elements of $\mathcal{A}_r$, and $\deg(x) = \gamma$. Then for every admissible sequence $\mathbf{i} = (i_1, i_2, \ldots, i_m)$ we have*

$$a(\mathbf{i}; xy) = a(\mathbf{i}; x) + a(\mathbf{i}; y), \tag{3.5}$$

$$E_{\mathbf{i}}^{(\mathrm{top})}(xy) = q^{\Phi_{\mathbf{i},\gamma}(a(\mathbf{i}; x), a(\mathbf{i}; y))} E_{\mathbf{i}}^{(\mathrm{top})}(x) E_{\mathbf{i}}^{(\mathrm{top})}(y). \tag{3.6}$$

Our next task is to construct a number of bases in $\mathscr{A}_r$ of the Poincaré-Birkhoff-Witt type. Such a construction was developed by Lusztig (cf. [10, Section 2]). For the convenience of the reader we present an independent and simplified construction in the special cases needed for our purposes.

Fix $m = \binom{r+1}{2}$, and let $\mathscr{R}$ denote the set of all pairs of integers (i, j) such that $1 \leqslant i < j \leqslant r+1$. Let $\mathbb{Z}^{\mathscr{R}}$ denote the integer lattice of rank m with coordinates d_{ij}, $(i, j) \in \mathscr{R}$. We choose the following linear order on $\mathscr{R}$: $(i, j) \prec (i', j')$ if $j < j'$ or $j = j'$, $i > i'$. Let $\varphi: \mathscr{R} \to \{1, 2, \ldots, \binom{r+1}{2}\}$ denote the order-preserving bijection; one checks easily that $\varphi(i, j) = \binom{j}{2} - i + 1$. To illustrate the use of φ we notice that the sequence $\mathbf{i}(1)$ from Theorem 2.5 can be defined by $\mathbf{i}(1)_{\varphi(i,j)} = i$. We define the semigroup Γ by

$$\Gamma = \{(a_1, \ldots, a_m) \in \mathbb{Z}_+^m : a_{\varphi(i,j)} \leqslant a_{\varphi(i',j)} \text{ for } i < i' < j\}. \tag{3.7}$$

(In the notation of Section 2, $\Gamma = C_{\mathbb{Z}}(\mathbf{i}(1))$ is the semigroup of integral points in the cone $C(\mathbf{i}(1))$.)

In Section 1 we associated to each $(i, j) \in \mathscr{R}$ a homogeneous element $t_{ij} \in \mathscr{A}_r$ of degree $\alpha_i + \alpha_{i+1} + \cdots + \alpha_{j-1}$. For every $d = (d_{ij}) \in \mathbb{Z}_+^{\mathscr{R}}$ let t^d denote the monomial $t^d = \prod_{(i,j)\in\mathscr{R}} t_{ij}^{d_{ij}}$, the product taken in the linear order just introduced. We set $\psi(d) := a(\mathbf{i}(1); t^d)$.

Proposition 3.4. (a) *The correspondence $d \mapsto \psi(d)$ is a semigroup isomorphism between $\mathbb{Z}_+^{\mathscr{R}}$ and Γ.*

(b) *For every $d \in \mathbb{Z}_+^{\mathscr{R}}$ the element $E_{\mathbf{i}(1)}^{(\mathrm{top})}(t^d)$ has the form $q^N \in \mathscr{A}_r(0) = \mathbb{Q}(q)$ for some $N \in \mathbb{Z}$.*

Proof. (a) According to (3.5), ψ is a linear map. Therefore, to compute it we have only to compute the string $a(\mathbf{i}(1); t_{ij})$ for every $(i, j) \in \mathscr{R}$. □

Lemma 3.5. *Let $(i, j) \in \mathscr{R}$, and $k = 1, \ldots, r$. Then $E_k(t_{ij}) = 0$ unless $k = j-1$, and $E_{j-1}(t_{ij}) = t_{i,j-1}$.*

Proof of Lemma 3.5. We proceed by induction on $j - i$. If $j - i = 1$ then $t_{ij} = x_{j-1}$, and our statement follows from Proposition 1.1(c). So we can assume that $j - i \geqslant 2$, and that our statement is valid for $t_{i,j-1}$. By definition,

$$t_{ij} = [t_{i,j-1}, x_{j-1}] = \frac{t_{i,j-1}x_{j-1} - q^{-1}x_{j-1}t_{i,j-1}}{q - q^{-1}}.$$

Applying E_k and using the inductive assumption and (1.4), we see that $E_k(t_{ij}) = 0$ unless $k = j-1$ or $k = j-2$, and we have

$$E_{j-1}(t_{ij}) = t_{i,j-1}, \qquad E_{j-2}(t_{ij}) = [t_{i,j-2}, x_{j-1}] = \frac{t_{i,j-2}x_{j-1} - x_{j-1}t_{i,j-2}}{q - q^{-1}}.$$

It remains to show that $[t_{i,j-2}, x_{j-1}] = 0$. Since $t_{i,j-2}$ is a polynomial in variables $x_i, \ldots, x_{j-3}$, it commutes with x_{j-1} in view of (0.1).

Using Lemma 3.5 and the definition (3.3), we conclude that

$$a(\mathbf{i}(1); t_{ij}) = \sum_{i'=i}^{j-1} e_{\varphi(i', j)}, \tag{3.8}$$

where e_k stands for the vector $(a_1, \ldots, a_m)$ with $a_i = \delta_{ki}$. Clearly, the vectors of the form (3.8) for all $(i, j) \in \mathscr{R}$ form a $\mathbb{Z}$-basis of the semigroup Γ. This completes the proof of Proposition 3.4(a).

(b) According to (3.6), it is enough to prove our statement for $t^d = t_{ij}$. But in this case it follows from (3.8) and Lemma 3.5. □

For every $\gamma \in Q_+$ we set

$$\mathbb{Z}_+^{\mathscr{R}}(\gamma) = \Big\{(d_{ij}) \in \mathbb{Z}_+^{\mathscr{R}} : \sum_{(i,j)\in\mathscr{R}} d_{ij}(\alpha_i + \alpha_{i+1} + \cdots + \alpha_{j-1}) = \gamma\Big\}, \tag{3.9}$$

$$\Gamma(\gamma) = \Big\{(a_1, \ldots, a_m) \in \Gamma : \sum_{(i,j)\in\mathscr{R}} a_{\varphi(i,j)}\alpha_i = \gamma\Big\}. \tag{3.10}$$

COROLLARY 3.6. *Let* $\gamma \in Q_+$. *Then*

(a) *The elements* t^d *for* $d \in \mathbb{Z}_+^{\mathscr{R}}(\gamma)$ *form a basis of* $\mathscr{A}_r(\gamma)$.

(b) *The elements* $E_{\mathbf{i}(1)}^{(a)}$ *for* $a \in \Gamma(\gamma)$ *form a basis of* $U_+(\gamma)$.

PROOF. We prove both statements at the same time. Clearly, $t^d \in \mathscr{A}_r(\gamma)$ for $d \in \mathbb{Z}_+^{\mathscr{R}}(\gamma)$, and $E_{\mathbf{i}(1)}^{(a)} \in U_+(\gamma)$ for $a \in \Gamma(\gamma)$. By definition,

$$\operatorname{card}(\mathbb{Z}_+^{\mathscr{R}}(\gamma)) = p(\gamma),$$

the Kostant partition function. By Proposition 3.4(a) and (3.8), ψ is a bijection between $\mathbb{Z}_+^{\mathscr{R}}(\gamma)$ and $\Gamma(\gamma)$. It follows that

$$\operatorname{card}(\Gamma(\gamma)) = p(\gamma).$$

Taking into account (1.3), it remains to show that each of the families $\{t^d : d \in \mathbb{Z}_+^{\mathscr{R}}(\gamma)\}$ and $\{E_{\mathbf{i}(1)}^{(a)} : a \in \Gamma(\gamma)\}$ is linearly independent. It is enough to show that the pairing matrix $(c_{\mathrm{ad}} = E_{\mathbf{i}(1)}^{(a)}(t^d))$ is nondegenerate.

Consider the *lexicographic linear ordering* of $\mathbb{Z}^m$ defined as follows: $(a'_1, \ldots, a'_m) \prec (a_1, \ldots, a_m)$ if the first nonzero difference among $a_1 - a'_1, \ldots, a'_m - a_m$ is positive. By definition of the string, we have

$$E_{\mathbf{i}(1)}^{(a)}(t^d) = 0 \quad \text{for } \psi(d) \prec a. \tag{3.11}$$

Furthermore, if $a = \psi(d)$ then $E_{\mathbf{i}(1)}^{(a)}(t^d) \neq 0$ in view of Proposition 3.4(b). It follows that if we identify $\mathbb{Z}_+^{\mathscr{R}}(\gamma)$ and $\Gamma(\gamma)$ by means of ψ, and order both sets lexicographically then the pairing matrix (c_{ad}) becomes triangular with nonzero diagonal entries, hence nondegenerate. □

REMARK. The basis in Corollary 3.6(b) is analogous to the Verma bases constructed in [13].

Now we fix a number $s \in \{1, 2, \dots, r\}$, and consider the sequence

$$\mathbf{i}(s) = (s, \overline{s-1, s}, \overline{s-2, s}, \dots, \overline{1, s}, \overline{s+1, 1}, \overline{s+2, 1}, \dots, \overline{r, 1}), \tag{3.12}$$

where for $i < j$ the symbol $\overline{i, j}$ stands for the sequence $i, i+1, \dots, j$, and $\overline{j, i}$ stands for the sequence $j, j-1, \dots, i$. The sequence $\mathbf{i}(1)$ has been introduced earlier. It is easy to see that $\mathbf{i}(s) \in R(w_0)$ for all s. We shall extend Proposition 3.4 and Corollary 3.6 to $\mathbf{i}(s)$.

Let $x \mapsto x^*$ be the $\mathbb{Q}(q)$-linear antiautomorphism of $\mathscr{A}_r$ such that $x_i^* = x_i$ for $i = 1, \dots, r$ (it is well defined in view of the defining relations (0.1), (0.2)). Clearly, $x \mapsto x^*$ is a degree-preserving involution of $\mathscr{A}_r$. In particular, for every $(i, j) \in \mathscr{R}$ the element t_{ij}^* is homogeneous of the same degree $\alpha_i + \alpha_{i+1} + \dots + \alpha_{j-1}$ as t_{ij}.

Now we define $t(s)_{ij}$ to be t_{ij} for $j > s+1$, and to be t_{ij}^* for $j \leqslant s+1$. For every $d = (d_{ij}) \in \mathbb{Z}_+^{\mathscr{R}}$ let $t(s)^d$ denote the monomial $t(s)^d = \prod_{(i,j)\in\mathscr{R}} t(s)_{ij}^{d_{ij}}$, the product in the same linear order as above. Finally, let σ_s denote the automorphism of the semigroup $\mathbb{Z}_+^{\mathscr{R}}$ given by $\sigma_s(d)_{ij} = d_{ij}$ for $j > s+1$, and $\sigma_s(d)_{ij} = d_{s+2-j, s+2-i}$ for $j \leqslant s+1$.

PROPOSITION 3.7. (a) *For every $d \in \mathbb{Z}_+^{\mathscr{R}}$ we have $a(\mathbf{i}(s); t(s)^d) = \psi(\sigma_s(d))$. Hence the correspondence $d \mapsto a(\mathbf{i}(s); t(s)^d)$ is a semigroup isomorphism between $\mathbb{Z}_+^{\mathscr{R}}$ and Γ.*

(b) *For every $d \in \mathbb{Z}_+^{\mathscr{R}}$ the element $E_{\mathbf{i}(s)}^{(\mathrm{top})}(t(s)^d)$ has the form $q^N \in \mathscr{A}_r(0) = \mathbb{Q}(q)$ for some $N \in \mathbb{Z}$.*

The proof of Proposition 3.7 is totally analogous to that of Proposition 3.4. We have only to replace Lemma 3.5 by the following lemma.

LEMMA 3.8. *Let $(i, j) \in \mathscr{R}$, and $k = 1, \dots, r$. Then $E_k(t_{ij}^*) = 0$ unless $k = i$, and $E_i(t_{ij}^*) = t_{i+1,j}^*$.*

COROLLARY 3.9. *Let $\gamma \in Q_+$. Then*

(a) *The elements $t(s)^d$ for $d \in \mathbb{Z}_+^{\mathscr{R}}(\gamma)$ form a basis of $\mathscr{A}_r(\gamma)$.*

(b) *The elements $E_{\mathbf{i}(s)}^{(a)}$ for $a \in \psi\sigma_s\psi^{-1}(\Gamma(\gamma))$ form a basis of $U_+(\gamma)$.*

This follows from Proposition 3.7 in exactly the same way as Corollary 3.6 from Proposition 3.4.

Let $E \mapsto E^*$ be the $\mathbb{Q}(q)$-linear antiautomorphism of $U_{+,r}$ such that $E_i^* = E_i$ for $i = 1, \dots, r$ (it is well defined in view of the defining relations (1.1), (1.2)). This is a degree-preserving involution of $U_{+,r}$.

Proposition 3.10. *Suppose $E \in U_{+,r}$ and $z \in \mathscr{A}_r$ are homogeneous elements of the same degree. Then $E(z) = E^*(z^*)$.*

Proof. It is enough to consider the case when E is the monomial $E_{\mathbf{i}}^{(a)}$ from (3.1), and $z = xy$ is a product of two elements of smaller degree. By definition, $(E_{\mathbf{i}}^{(a)})^* = E_{\mathbf{i}_{\mathrm{opp}}}^{(a_{\mathrm{opp}})}$, where $\mathbf{i}_{\mathrm{opp}}$ and a_{opp} stand for the sequences $(i_m, \dots, i_1)$ and $(a_m, \dots, a_1)$ respectively. Now we expand $E(z) = E_{\mathbf{i}}^{(a)}(xy)$ with the help of (3.2). Since $\deg(E) = \deg(xy)$, for each nonzero summand in (3.2) we have

$$\sum_{1\leqslant k\leqslant m} a'_k \alpha_{i_k} = \deg(x), \qquad \sum_{1\leqslant k\leqslant m} a''_k \alpha_{i_k} = \deg(y). \tag{3.13}$$

Using induction on degree of E we can assume that

$$E_{\mathbf{i}}^{(a')}(x) = E_{\mathbf{i}_{\mathrm{opp}}}^{(a'_{\mathrm{opp}})}(x^*), \qquad E_{\mathbf{i}}^{(a'')}(y) = E_{\mathbf{i}_{\mathrm{opp}}}^{(a''_{\mathrm{opp}})}(y^*). \tag{3.14}$$

Using (3.3) and (3.13), we check directly that

$$\Phi_{\mathbf{i}, \deg(x)}(a', a'') = \Phi_{\mathbf{i}_{\mathrm{opp}}, \deg(y)}(a''_{\mathrm{opp}}, a'_{\mathrm{opp}}). \tag{3.15}$$

Substituting the expressions (3.14), (3.15) into (3.2) we conclude that $E_{\mathbf{i}}^{(a)}(xy) = E_{\mathbf{i}_{\mathrm{opp}}}^{(a_{\mathrm{opp}})}(y^*x^*)$, which proves our statement. □

We conclude this section with some commutation relations for the elements t_{ij}.

Proposition 3.11. (a) *If $i \leqslant j < i' \leqslant j'$ then $[t_{ij}, t_{i',j'}] = 0$.*

(b) *If $i < i' \leqslant j' \leqslant j$ then $[t_{ij}, t_{i',j'}] = 0$.*

(c) *If $i < i' \leqslant j < j'$ then $[t_{ij}, t_{i',j'}] = t_{i,j'} t_{i',j}$.*

Proof. Part (a) is clear since t_{ij} is a (noncommutative) polynomial in variables $x_i, x_{i+1}, \dots, x_{j-1}$, $t_{i',j'}$ is a (noncommutative) polynomial in variables $x_{i'}, x_{i'+1}, \dots, x_{j'-1}$, and each of $x_i, x_{i+1}, \dots, x_{j-1}$ commutes with each $x_{i'}, x_{i'+1}, \dots, x_{j'-1}$. We prove (b) and (c) simultaneously by induction on $(j - i) + (j' - i')$. Part (b) is evident if $i' = j'$, so we can assume that $i' < j'$.

First consider the case $j' < j$ in (b). By Proposition 1.2(a), it is enough to show that $E_k([t_{ij}, t_{i',j'}]) = 0$ for all k. According to (1.4) and Lemma 3.5, we have only to consider $k = j' - 1$, $j - 1$, and in these cases we have

$$E_{j'-1}([t_{ij}, t_{i',j'}]) = [t_{ij}, t_{i',j'-1}], \qquad E_{j-1}([t_{ij}, t_{i',j'}]) = [t_{i,j-1}, t_{i',j'}],$$

both expressions being 0 by the inductive assumption.

In the remaining case $i < i' < j' = j$ in (b) we proceed in the same way. Now the only nontrivial thing is to show that $E_{j-1}([t_{ij}, t_{i',j}]) = 0$. By a straightforward calculation using (1.4), Lemma 3.5, and the inductive assumption that $t_{ij} t_{i',j-1} = t_{i',j-1} t_{ij}$, we obtain

$$E_{j-1}([t_{ij}, t_{i',j}]) = [t_{i,j-1}, t_{i',j}] - t_{ij} t_{i',j-1}.$$

But the last expression is 0 by the inductive assumption in part (c).

To prove (c) we start with the case $i < i' = j < j'$, when it becomes the identity

$$[t_{ij}, t_{j,j'}] = t_{i,j'} \quad \text{for } i < j < j'. \tag{3.16}$$

If $j = j' - 1$ then (3.16) is the definition of $t_{i,j'}$, so we can assume that $j < j' - 1$. Again we apply all E_k to both sides of (3.16). It suffices to consider $k = j' - 1, j - 1$. We have

$$E_{j'-1}([t_{ij}, t_{j,j'}]) = [t_{ij}, t_{j,j'-1}] = t_{i,j'-1} = E_{j'-1}(t_{i,j'}),$$
$$E_{j-1}([t_{ij}, t_{j,j'}]) = [t_{i,j-1}, t_{j,j'}] = 0 = E_{j-1}(t_{i,j'})$$

(in the first case we use the inductive assumption, and in the second the part (a) which is already proven).

It remains to consider the case $i < i' < j < j'$. As before, it is enough to check that both parts in (c) give the same result under the action of E_{j-1} and $E_{j'-1}$. Applying E_{j-1} we get

$$E_{j-1}([t_{ij}, t_{i',j'}]) = [t_{i,j-1}, t_{i',j'}] = t_{i,j'}t_{i',j-1} = E_{j-1}(t_{i,j'}t_{i',j}),$$

as required. Applying $E_{j'-1}$ we have to distinguish two cases: $j < j' - 1$, and $j = j' - 1$. If $j < j' - 1$ then we get

$$E_{j'-1}([t_{ij}, t_{i',j'}]) = [t_{i,j}, t_{i',j'-1}] = t_{i,j'-1}t_{i',j} = E_{j'-1}(t_{i,j'}t_{i',j}),$$

as required. Finally, for $j = j' - 1$ we use the following identity (the proof is straightforward):

$$E_j([t_{ij}, t_{i',j+1}]) = t_{ij}t_{i',j} + q^{-1}[t_{ij}, t_{i',j}]. \tag{3.17}$$

By the inductive assumption in (b), the second summand in (3.17) is 0, and we are done. □

§4. String parametrizations of string bases

From now on we fix a string basis B in $\mathcal{A}_r$. In this section we show that for every $s = 1, \ldots, r$ the strings in direction $\mathbf{i}(s)$ provide a parametrization of B. This allows us to prove Theorem 2.1 and Theorem 2.4 for $\mathbf{i} = \mathbf{i}(s)$.

We shall use the following notation: for $x \in \mathcal{A}_r$, $b \in B$ let $[x : b]$ denote the coefficient of b in the B-expansion of x. We say that b is a *constituent* of x if $[x : b] \neq 0$. Recall that $[B]^+$ stands for the set of all $x \in \mathcal{A}_r$ such that $[x : b] \in \mathbb{Z}_+[q, q^{-1}]$ for all $b \in B$. The following statement is an immediate consequence of the definition of strings and string axioms (S2) and (S3).

PROPOSITION 4.1. *Suppose $x \in [B]^+$ is nonzero, and $\mathbf{i}$ is an admissible sequence. Then for every constituent b of x the string $a(\mathbf{i}; b)$ either is equal to $a(\mathbf{i}; x)$ or precedes $a(\mathbf{i}; x)$ in the lexicographic order. There is at least one* *constituent b of x such that $a(\mathbf{i}; b) = a(\mathbf{i}; x)$; for every such b the element $E_{\mathbf{i}}^{(\mathrm{top})}(b)$ is a constituent of $E_{\mathbf{i}}^{(\mathrm{top})}(x)$.*

Recall from Section 3 that to each pair of indices $(i, j) \in \mathcal{R}$ (i.e., such that $1 \leqslant i < j \leqslant r + 1$) there are associated two homogeneous elements $t_{ij}, t_{ij}^* \in \mathcal{A}_r$ of the same degree $\alpha_i + \alpha_{i+1} + \cdots + \alpha_{j-1}$.

PROPOSITION 4.2. *The elements t_{ij}, t_{ij}^* belong to B for all $(i, j) \in \mathcal{R}$.*

PROOF. We deal only with the elements t_{ij}, the case of t_{ij}^* being totally similar. We proceed by induction on $j-i$. In view of (S0) and (S3), $t_{i,i+1} = x_i \in B$ because $E_i^2(x_i) = 0$ and $E_i(x_i) = 1 \in B$. Now let $j \geqslant i+1$ and assume that $t_{i,j'} \in B$ for all $j' = i+1, \ldots, j$. We have to show that $t_{i,j+1} \in B$. Consider the product $x_j t_{ij}$. By (S1) and the inductive assumption, $x_j t_{ij} \in [B]^+$. Using (1.4) and Lemma 3.5, we see that $E_k(x_j t_{ij}) = 0$ unless $k = j-1, j$, and

$$E_{j-1}(x_j t_{ij}) = q x_j t_{i,j-1}, \qquad E_j(x_j t_{ij}) = t_{ij}.$$

Furthermore, $E_j^2(x_j t_{ij}) = 0$, hence $E_j^{(\mathrm{top})}(x_j t_{ij}) = t_{ij}$. By Proposition 4.1, $x_j t_{ij}$ has a constituent $b \in B$ such that $E_j(b) = t_{ij}$. Clearly, $[x_j t_{ij} : b] = 1$. It remains to show that $b = t_{i,j+1}$.

By Proposition 1.2(a), it is enough to prove that $E_k(b) = E_k(t_{i,j+1})$ for all k. By Lemma 3.5 and our choice of b, we have $E_j(b) = E_j(t_{i,j+1}) = t_{ij}$, and $E_k(b) = E_k(t_{i,j+1}) = 0$ for $k \neq j-1, j$. It remains to show that $E_{j-1}(b) = 0$. Suppose this is not so, i.e., $E_{j-1}(b) = b' \neq 0$. Since $E_{j-1}^2(x_j t_{ij}) = 0$, it follows that $E_{j-1}^2(b) = 0$ hence $b' = E_{j-1}^{(\mathrm{top})}(b) \in B$. Therefore, b' is a constituent of $E_{j-1}(x_j t_{ij}) = q x_j t_{i,j-1}$, and $[q x_j t_{i,j-1} : b'] = 1$. We get a contradiction by showing that $x_j t_{i,j-1} \in B$.

Let $\gamma = \alpha_i + \alpha_{i+1} + \cdots + \alpha_{j-2} + \alpha_j$, so that $x_j t_{i,j-1} \in \mathcal{A}_r(\gamma)$. Since x_j commutes with each of $x_i, \ldots, x_{j-2}$ it follows that every element of $\mathcal{A}_r(\gamma)$ has the form $x_j y$ for some $y \in \mathcal{A}_r(\gamma - \alpha_j)$. By (S1) and the inductive assumption, $x_j t_{i,j-1} \in [B]^+$. Let $x_j y \in B$ be a constituent of $x_j t_{i,j-1}$. We have $E_j(x_j y) = y$, $E_j^2(x_j y) = 0$, hence $y \in B$ in view of (S3). On the other hand, by Proposition 4.1, y is a constituent of $t_{i,j-1}$. Hence, $y = t_{i,j-1}$, and we are done. □

For $s = 1, \ldots, r$ let $\mathbf{i}(s)$ be the reduced decomposition of w_0 defined by (3.12). Our next result is a special case of Theorems 2.3 and 2.4.

PROPOSITION 4.3. (a) *The correspondence $b \mapsto a(\mathbf{i}(s); b)$ is a bijection between B and Γ.*

(b) *We have $E_{\mathbf{i}(s)}^{(\mathrm{top})}(b) = 1$ for all $b \in B$.*

PROOF. Consider the basis in $\mathcal{A}_r$ formed by monomials $t(s)^d$ for $d \in \mathbb{Z}_+^{\mathcal{R}}$ (see Corollary 3.9 (a)). By Proposition 4.2 and the string axiom (S1), all these monomials belong to $[B]^+$. By Proposition 4.1, every monomial $t(s)^d$ has a constituent $b = b(s, d) \in B$ such that $a(\mathbf{i}(s); b) = a(\mathbf{i}; t(s)^d)$. In view of Proposition 3.7(a), all elements $b(s, d)$ are distinct, hence they exhaust B. Now both statements of Proposition 4.3 follow from the corresponding statements of Proposition 3.7. □

PROOF OF THEOREM 2.1. Using induction on the length of an admissible sequence $\mathbf{i}$ we can assume that $\mathbf{i}$ has length 1, i.e., consists of one index s. Suppose $b, b' \in B$ are such that $l_s(b) = l_s(b')$, and $E_s^{(\mathrm{top})}(b) = E_s^{(\mathrm{top})}(b')$. Since the sequence $\mathbf{i}(s)$ has the first term s, it follows that $a(\mathbf{i}(s); b) = a(\mathbf{i}(s); b')$. Hence by Proposition 4.3 (a), $b = b'$, as required. □

The above proof actually establishes the following.

PROPOSITION 4.4. *If $b, b' \in B$ and $E_s^n(b) = E_s^n(b') \neq 0$ for some $s = 1, \dots, r$, $n \geqslant 0$ then $b = b'$.*

We conclude this section with two helpful corollaries of Theorem 2.1.

PROPOSITION 4.5. *Let x be a nonzero homogeneous element of $\mathscr{A}_r$. Then $l_s(b) \leqslant l_s(x)$ for each constituent b of x and each $s = 1, \dots, r$.*

PROOF. Fix s and let l be the maximal possible value of $l_s(b)$ for a constituent b of x. Let $b_1, \dots, b_k$ be all the constituents of x with $l_s(b) = l$. By Theorem 2.1, $E_s^{(l)}(b_1), \dots, E_s^{(l)}(b_k)$ are distinct elements of B, hence are linearly independent. But $E_s^{(l)}(x)$ is their nontrivial linear combination, hence $E_s^{(l)}(x) \neq 0$, and so $l \leqslant l_s(x)$, as required. □

For every $\gamma \in Q_+$, $\nu = (n_1, \dots, n_r) \in \mathbb{Z}_+^r$ we denote

$$\mathscr{A}_r(\gamma; \nu) = \{x \in \mathscr{A}_r(\gamma) : l_s(x) \leqslant n_s \text{ for } s = 1, \dots, r\}. \tag{4.1}$$

Proposition 4.5 can be reformulated as follows.

PROPOSITION 4.6. *Every subspace of the type $\mathscr{A}_r(\gamma; \nu)$ is spanned by a subset of B.*

REMARKS. (a) Using Proposition 4.5, one can show that Proposition 4.1 remains true even without the assumption that $x \in [B]^+$.

(b) Let $\mathscr{A}_r^\nu = \bigoplus_\gamma \mathscr{A}_r(\gamma; \nu)$. One can show that for every ν the subspace $\mathscr{A}_r^\nu \subset \mathscr{A}_r$ is invariant under the action of all E_i. Moreover, the action of U_+ on $\mathscr{A}_r^\nu$ extends to an action of the whole algebra $U_q(sl_{r+1})$ so that $\mathscr{A}_r^\nu$ becomes an irreducible $U_q(sl_{r+1})$-module with lowest weight $-\nu$ (cf. [10, Section 8]). We see that each string basis B gives rise to a basis in every irreducible finite-dimensional $U_q(sl_{r+1})$-module. Specializing $q = 1$ we obtain a basis in every irreducible finite-dimensional sl_{r+1}-module. If B is dual to the Lusztig's canonical basis then the bases in irreducible finite-dimensional sl_{r+1}-modules constructed in this way are *good* (see Introduction).

§5. The string basis for sl_3

In this section we prove Theorems 1.5, 1.6, and all theorems in Section 2 in the case $r = 2$, i.e., for the algebra $\mathcal{A}_2$ having only two generators x_1 and x_2.

PROOF OF THEOREMS 1.5, 1.6. The general definition of the elements t_{ij}, t_{ij}^* given above takes the form

$$t_{13} = [x_1, x_2], \quad t_{13}^* = [x_2, x_1] \tag{5.1}$$

(recall that $[x, y]$ stands for the q-commutator (1.5)). Comparing (5.1) with (1.7), we see that $\mathcal{P}_2 = \{x_1, x_2, t_{13}, t_{13}^*\}$. One easily checks that the defining relations (0.2) can be rewritten as

$$[x_1, t_{13}] = [t_{13}, x_2] = 0, \tag{5.2}$$

or, equivalently as

$$[x_2, t_{13}^*] = [t_{13}^*, x_1] = 0. \tag{5.3}$$

This implies the first statement in Theorem 1.6: every two elements of $\mathcal{P}_2$ quasicommute with the only exception of x_1 and x_2.

As in Section 4, assume that we are given a string basis B in $\mathcal{A}_2$. Let M denote the set of all 4-tuples $m = (m_1, m_2, m_{12}, m_{21})$ of nonnegative integers such that $\min(m_1, m_2) = 0$. For every $m \in M$ we set

$$b(m) = q^{\binom{m_1+m_{12}}{2}+\binom{m_2+m_{21}}{2}} x_1^{m_1} x_2^{m_2} t_{13}^{m_{21}} (t_{13}^*)^{m_{12}}, \tag{5.4}$$

and temporarily denote by B_0 the set of all monomials $b(m)$ for $m \in M$. To complete the proofs of Theorems 1.5 and 1.6 we have to show that $B = B_0$.

First of all, $\mathcal{P}_2 \subset B$ by Proposition 4.2. By (S1), we have $B_0 \subset [B]^+$.

Now let $\mathbf{i} = (1, 2, 1)$, $\mathbf{i}' = (2, 1, 2)$ be two reduced decompositions of $w_0 \in W = S_3$ (in the notation of previous sections, $\mathbf{i} = \mathbf{i}(1)$, $\mathbf{i}' = \mathbf{i}(2)$). The following formulas are proven by a straightforward computation using Proposition 3.1, (3.5), and Lemmas 3.5 and 3.8. For $m = (m_1, m_2, m_{21}, m_{12}) \in M$ we have

$$\begin{aligned} a(\mathbf{i}; b(m)) &= (m_1 + m_{12}, m_2 + m_{12} + m_{21}, m_{21}), \\ a(\mathbf{i}'; b(m)) &= (m_2 + m_{21}, m_1 + m_{12} + m_{21}, m_{12}), \end{aligned} \tag{5.5}$$

$$E_{\mathbf{i}}^{(\mathrm{top})}(b(m)) = E_{\mathbf{i}'}^{(\mathrm{top})}(b(m)) = 1. \tag{5.6}$$

In particular,

$$l_1(b(m)) = m_1 + m_{12}, \quad l_2(b(m)) = m_2 + m_{21}. \tag{5.7}$$

We abbreviate $\eta(m) = a(\mathbf{i}; b(m))$, $\eta'(m) = a(\mathbf{i}'; b(m))$. Recall that $\Gamma = \{(a_1, a_2, a_3) \in \mathbb{Z}_+^3 : a_2 \geqslant a_3\}$. □

PROPOSITION 5.1. *Each of the maps η and η' is a bijection between M and Γ.*

The proof follows directly from (5.5): the inverse bijections are given by

$$\begin{aligned}\eta^{-1}(a_1, a_2, a_3) &= (\max(0, a_1 + a_3 - a_2), \\ &\qquad \max(0, a_2 - a_1 - a_3), \min(a_1, a_2 - a_3), a_3), \\ \eta'^{-1}(a_1, a_2, a_3) &= (\max(0, a_2 - a_1 - a_3), \\ &\qquad \max(0, a_1 + a_3 - a_2), a_3, \min(a_1, a_2 - a_3)).\end{aligned} \tag{5.8}$$

For $a \in \Gamma$ we denote by $b(1; a)$ and $b(2; a)$ the elements from B such that $a(\mathbf{i}; b(1; a)) = a(\mathbf{i}'; b(2; a)) = a$ (see Proposition 4.3(a)). Taking into account Proposition 4.1, we obtain the following.

COROLLARY 5.2. *The monomials $b(m)$ for $m \in M$ form a basis in $\mathcal{A}_2$, and we have*

$$[b(m) : b(1; \eta(m))] = [b(m) : b(2; \eta'(m))] = 1 \tag{5.9}$$

for all $m \in M$.

To show that $B_0 = B$ it remains to prove that all monomials $b(m)$ belong to B. In view of Proposition 4.6 and (5.6), this is a consequence of the following.

PROPOSITION 5.3. *Let $m \in M$, and let $\gamma \in Q_+$ be the degree of $b(m)$, and $\nu = (l_1(b(m)), l_2(b(m)))$. Then $\mathcal{A}_2(\gamma; \nu)$ is a one-dimensional space spanned by $b(m)$.*

PROOF. Fix $\gamma = g_1\alpha_1 + g_2\alpha_2$. Without loss of generality we can assume that $g_1 \leqslant g_2$. Then the condition $b(m) \in \mathcal{A}_2(\gamma)$ means that $m_1 = 0$ and

$$m_{12} + m_{21} = g_1, \qquad m_2 + m_{12} + m_{21} = g_2. \tag{5.10}$$

It follows that the elements m such that $b(m) \in \mathcal{A}_2(\gamma)$ are exactly those of the form $m = (0, g_2 - g_1, n, g_1 - n)$ for $n = 0, 1, \ldots, g_1$; we abbreviate $b(0, g_2 - g_1, n, g_1 - n)$ as b_n. By (5.7),

$$l_1(b_n) = n, \qquad l_2(b_n) = g_2 - n. \tag{5.11}$$

Now choose $\nu = (n, g_2 - n)$ and suppose that $x \in \mathcal{A}_2(\gamma; \nu)$. Then the condition $l_1(x) \leqslant n$ implies that x is a linear combination of elements $b_k \in \mathcal{A}_2(\gamma)$ with $k \leqslant n$. Likewise, the condition $l_2(x) \leqslant g_2 - n$ implies that x is a linear combination of elements $b_l \in \mathcal{A}_2(\gamma)$ with $l \geqslant n$. It follows that x is proportional to b_n, and we are done. □

Theorems 1.5 and 1.6 are proven.

PROPOSITION 5.4. *Theorems 2.2 to 2.12 are valid for $\mathcal{A}_2$.*

PROOF. The first statement in Theorem 2.2, and Theorem 2.3 follow from (5.6). Theorems 2.4 and 2.5 follow from Proposition 5.1. To establish existence of the map ${}_{\mathbf{i}'}T_{\mathbf{i}}: \mathbb{R}^3 \to \mathbb{R}^3$ in Theorem 2.2, we notice that according to (5.9), $b(m) = b(1; \eta(m)) = b(2; \eta'(m))$ for all $m \in M$. It follows that the map ${}_{\mathbf{i}'}T_{\mathbf{i}}$ exists and is equal to $\eta'\eta^{-1}$. Computing this composition with the help of (5.5) and (5.8), we obtain (2.8). As for Proposition 2.6 and (2.7), there is nothing to prove in our case.

In our situation the cone $C = C(\mathbf{i}(1))$ has the form

$$C = \{(a_1, a_2, a_3) \in \mathbb{R}^3 : a_1 \geqslant 0, a_2 \geqslant a_3 \geqslant 0\}. \tag{5.12}$$

By definition (see Section 2), there are two $\mathbf{i}(1)$-linearity domains given by

$$\begin{aligned} C_+ &= \{(a_1, a_2, a_3) \in C : a_2 \geqslant a_1 + a_3\}, \\ C_- &= \{(a_1, a_2, a_3) \in C : a_2 \leqslant a_1 + a_3\}. \end{aligned} \tag{5.13}$$

Since the only nontrivial transformation ${}_{\mathbf{i}'}T_{\mathbf{i}}$ is given by (2.8), Proposition 2.8 becomes clear (we have only to check that if $a = (a_1, a_2, a_3) \in C_+^0$, $a' = (a_1', a_2', a_3') \in C_-^0$ then ${}_{\mathbf{i}'}T_{\mathbf{i}}(a + a') \neq {}_{\mathbf{i}'}T_{\mathbf{i}}(a) + {}_{\mathbf{i}'}T_{\mathbf{i}}(a')$).

Each of the cones C, C_+, C_- is simplicial: C is generated by e_1, e_2, $e_2 + e_3$, C_+ is generated by e_2, $e_1 + e_2$, $e_2 + e_3$, and C_- is generated by e_1, $e_1 + e_2$, $e_2 + e_3$, where e_1, e_2, e_3 is the standard basis in $\mathbb{R}^3$. On the other hand, by (5.5) we have

$$a(\mathbf{i}; x_1) = e_1, \quad a(\mathbf{i}; x_2) = e_2, \quad a(\mathbf{i}; t_{13}) = e_2 + e_3, \quad a(\mathbf{i}; t_{13}^*) = e_1 + e_2.$$

Theorems 2.9, 2.11, and 2.12 follow by inspection (the decomposition in Theorem 2.12 is simply $C = C_+ \cup C_-$). Finally, we do not have to bother about Theorem 2.10, because the above results establish the stronger Conjecture 1.7 (for $r = 2$). □

We conclude this section with some corollaries on the structure of the dual algebra U_+. We retain the notation of the proof of Proposition 5.3, i.e., fix $\gamma = g_1\alpha_1 + g_2\alpha_2$ with $g_1 \leqslant g_2$, and abbreviate $b_n = b(0, g_2 - g_1, n, g_1 - n)$. For $n = 0, \ldots, g_1$ we set

$$u_n = E_1^{(g_1 - n)} E_2^{(g_2)} E_1^{(n)}, \qquad u_n' = E_2^{(n)} E_1^{(g_1)} E_2^{(g_2 - n)}. \tag{5.14}$$

PROPOSITION 5.5. (a) *The elements* u_n *form a basis in* $U_+(\gamma)$ *dual to the basis* $\{b_n : 0 \leqslant n \leqslant g_1\}$, *i.e.,* $u_n(b_k) = \delta_{kn}$.

(b) *The transition matrix between* (u_n) *and* (u_n') *is unitriangular, i.e.,*

$$\begin{aligned} u_n &= u_n' + \textit{linear combination of } u_0', \ldots, u_{n-1}', \\ u_n' &= u_n + \textit{linear combination of } u_0, \ldots, u_{n-1}. \end{aligned}$$

PROOF. Part (a) is proven by a straightforward computation. In view of part (a), to prove part (b) it is enough to show that $u_n'(b_k) = 0$ for $k > n$, and $u_n'(b_n) = 1$, which is also quite straightforward (we can actually spare some of the calculations by observing that ${}_{\mathbf{i}'}T_{\mathbf{i}}(n, g_2, g_1 - n) = (g_2 - n, g_1, n)$). □

Recalling Proposition 1.3, we see that Proposition 5.5(a) recovers the Lusztig's result that the canonical basis in $U_+(\gamma)$ consists of the elements u_n (cf. [10, 3.4]).

§6. Proofs of Theorems 2.2 to 2.7: general case

PROOFS OF THEOREM 2.2, PROPOSITION 2.6, AND THEOREM 2.7. We prove all these statements simultaneously. The argument is divided into several steps.

STEP 1. PROOF OF THEOREM 2.2 IN THE CASE OF THEOREM 2.7(a). Let $\mathbf{i} = (i, j)$, $\mathbf{i}' = (j, i)$ with $|i - j| > 1$. Let $b \in B$, and suppose $a(\mathbf{i}; b) = (a_1, a_2)$, $E_{\mathbf{i}}^{(\mathrm{top})}(b) = b_0$. We must show that $a(\mathbf{i}'; b) = (a_2, a_1)$, $E_{\mathbf{i}'}^{(\mathrm{top})}(b) = b_0$. Since E_i and E_j commute, it suffices to show that

$$a_1 = l_i(b), \qquad a_2 = l_j(b). \tag{6.1}$$

The equality $a_1 = l_i(b)$ is just the definition (2.3), and the inequality $a_2 \leqslant l_j(b)$ is clear since

$$b_0 = E_j^{(a_2)} E_i^{(a_1)}(b) = E_i^{(a_1)} E_j^{(a_2)}(b) \neq 0.$$

To prove the reverse inequality $a_2 \geqslant l_j(b)$ we first show that

$$E_i(b_0) = E_j(b_0) = 0. \tag{6.2}$$

Indeed, $E_j E_j^{(a_2)} E_i^{(a_1)}$ is proportional to $E_j^{(a_2+1)} E_i^{(a_1)}$, and $E_i E_j^{(a_2)} E_i^{(a_1)}$ is proportional to $E_j^{(a_2)} E_i^{(a_1+1)}$ since E_i and E_j commute. Hence (6.2) follows from the definition (2.4). Now consider the element $x = x_i^{a_1} x_j^{a_2} b_0$. By (S1) and (S3), $x \in [B]^+$. Using (1.4) and (6.2), we see that $a(\mathbf{i}; x) = (a_1, a_2)$, and $E_{\mathbf{i}}^{(\mathrm{top})}(x) = b_0$. Using Proposition 4.1 and Theorem 2.1, we conclude that b is a constituent of x. Hence $l_j(b) \leqslant l_j(x) = a_2$, which completes the proof of (6.1). Theorem 2.2 for $w = s_i s_j$, and the formula (2.7) are proven.

STEP 2. PROOF OF THEOREM 2.2 IN THE CASE OF THEOREM 2.7(b). Let $\mathbf{i} = (i, j, i)$, $\mathbf{i}' = (j, i, j)$ with $|i - j| = 1$. Let $b \in B$, and suppose $a(\mathbf{i}; b) = (a_1, a_2, a_3)$. We have to show that $a(\mathbf{i}'; b)$ is given by (2.8), and $E_{\mathbf{i}'}^{(\mathrm{top})}(b) = E_{\mathbf{i}}^{(\mathrm{top})}(b)$.

Without loss of generality we can assume that $i = s$, $j = s - 1$, i.e., $\mathbf{i}$ is the initial segment of the sequence $\mathbf{i}(s)$ (see (3.12)). It follows that the string $a := a(\mathbf{i}(s); b)$ begins with a_1, a_2, a_3. In particular, by Proposition 4.3, $a_2 \geqslant a_3$. Consider the string $a^{(0)}$ obtained from a by replacing the initial segment (a_1, a_2, a_3) with $(0, 0, 0)$. Clearly, $a^{(0)} \in \Gamma$, hence by Proposition 4.3, there is a unique $b_0 \in B$ such that $a(\mathbf{i}(s); b_0) = a^{(0)}$. It follows that $E_{s-1}(b_0) = E_s(b_0) = 0$. We claim that

$$E_{\mathbf{i}}^{(\mathrm{top})}(b) = b_0. \tag{6.3}$$

Let $b_1 \in B$ be the element with $a(\mathbf{i}(s); b_1) = a - a^{(0)} = (a_1, a_2, a_3, 0, \ldots, 0)$. By (S3), $b_1 b_0 \in [B]^+$. Using (3.5) and (3.6), we see that $a(\mathbf{i}(s); b_1 b_0) = a$, and $E_{\mathbf{i}}^{(\mathrm{top})}(b_1 b_0) = b_0$. By Propositions 4.1 and 4.3, b is a constituent of $b_1 b_0$. Hence $E_{\mathbf{i}}^{(\mathrm{top})}(b)$ is a constituent of $E_{\mathbf{i}}^{(\mathrm{top})}(b_1 b_0) = b_0$. But $E_{\mathbf{i}}^{(\mathrm{top})}(b) \in B$ according to (S3). This proves (6.3).

The degree of b_1 is equal to $\gamma = a_2 \alpha_{s-1} + (a_1 + a_3)\alpha_s$. Hence b_1 belongs to the subalgebra of $\mathcal{A}_r$ generated by x_{s-1}, x_s, and we can apply to it the results of Section 5. In particular, we see that $a(\mathbf{i}'; b_1) = (a_1', a_2', a_3')$ is given by (2.8), and $E_{\mathbf{i}'}^{(\mathrm{top})}(b_1) = 1$. Since $E_{s-1}(b_0) = E_s(b_0) = 0$, it follows that $a(\mathbf{i}'; b_1 b_0) = (a_1', a_2', a_3')$ and $E_{\mathbf{i}'}^{(\mathrm{top})}(b_1 b_0) = b_0$. Since b is a constituent of $b_1 b_0$, we have $l_{s-1}(b) \leqslant l_{s-1}(b_1 b_0) = a_1'$. By Proposition 4.1, the string $a(\mathbf{i}'; b)$ is either equal to (a_1', a_2', a_3') (and in this case $E_{\mathbf{i}'}^{(\mathrm{top})}(b) = b_0$) or precedes (a_1', a_2', a_3') in the lexicographic order. It remains only to exclude the latter possibility.

By Proposition 5.5, the element $E_{\mathbf{i}}^{(a_1, a_2, a_3)} \in U_+(\gamma)$ has the form $E_{\mathbf{i}'}^{(a_1', a_2', a_3')}$ + (linear combination of other elements $E_{\mathbf{i}'}^{(a_1'', a_2'', a_3'')}$ of the same degree). If $a_2 \geqslant a_1 + a_3$ then by Proposition 5.5(b), all the terms $E_{\mathbf{i}'}^{(a_1'', a_2'', a_3'')}$ appearing in this decomposition have $a_1'' > a_1'$, hence vanish at b. If $a_2 \leqslant a_1 + a_3$ then by Proposition 5.5(a), all the terms $E_{\mathbf{i}'}^{(a_1'', a_2'', a_3'')}$ vanish at b_1, hence vanish at $b_1 b_0$, hence vanish at b too. We conclude that in both cases

$$E_{\mathbf{i}}^{(a_1, a_2, a_3)}(b) = E_{\mathbf{i}'}^{(a_1', a_2', a_3')}(b).$$

Theorem 2.2 for $w = s_i s_j s_i$, and the formula (2.8) are proven.

STEP 3. END OF THE PROOF. The results of the two previous steps imply Theorem 2.2 and Proposition 2.6 in the case when the transformation $\mathbf{i} \mapsto \mathbf{i}'$ is given by (2.5) or (2.6). Since any two reduced decompositions of the same element $w \in W$ can be transformed into each other by a sequence of these elementary transformations, we conclude that Theorem 2.2 is true in general. Once we know that the maps ${}_{\mathbf{i}'}T_{\mathbf{i}}$ exist, Proposition 2.6 becomes an immediate consequence of the definitions (2.3), (2.4), and the first statement in Theorem 2.2. □

PROOF OF THEOREM 2.3. Taking into account Proposition 1.2(a), we see that Theorem 2.3 is a consequence of the following.

PROPOSITION 6.1. *Let $w \in W$, $b \in B$. Then $E_i(E_w^{(\mathrm{top})}(b)) = 0$ for all $i = 1, \ldots, r$ such that $l(ws_i) < l(w)$.*

PROOF. Clearly, the condition $l(ws_i) < l(w)$ means that there is a reduced decomposition $\mathbf{i} = (i_1, \ldots, i_l) \in R(w)$ with $i_l = i$. Hence $E_i(E_w^{(\mathrm{top})}(b)) = E_i(E_{\mathbf{i}}^{(\mathrm{top})}(b))$ vanishes by definition of $E_{\mathbf{i}}^{(\mathrm{top})}(b)$. □

PROOFS OF THEOREMS 2.4 AND 2.5. Theorem 2.4 for $\mathbf{i} = \mathbf{i}(1)$ and Theorem 2.5 are established by Proposition 4.3(a) because the semigroup Γ coincides with $C_{\mathbb{Z}}(\mathbf{i}(1))$. Now let $\mathbf{i}$ be an arbitrary reduced decomposition of w_0. By Theorem 2.2, the correspondence $b \mapsto a(\mathbf{i}; b)$ is a bijection between B and ${}_{\mathbf{i}}T_{\mathbf{i}(1)}(\Gamma)$. Let $C = {}_{\mathbf{i}}T_{\mathbf{i}(1)}(C(\mathbf{i}(1)))$. Since ${}_{\mathbf{i}}T_{\mathbf{i}(1)}$ is a piecewise-linear automorphism preserving $\mathbb{Z}^m$, it follows that C is a polyhedral cone, and ${}_{\mathbf{i}}T_{\mathbf{i}(1)}(\Gamma)$ coincides with the set $C_{\mathbb{Z}} = C \cap \mathbb{Z}^m$ of integral points in C. It remains to show that C is convex. Clearly, $C_{\mathbb{Q}} = C \cap \mathbb{Q}^m$ is dense in C. So it is enough to show that $C_{\mathbb{Q}}$ is closed under taking rational convex combinations. This in turn follows from the fact that $C_{\mathbb{Z}}$ is a semigroup. But the last statement follows from (3.5) and Proposition 4.1: for every $b', b'' \in B$ there is a constituent b of $b'b''$ such that $a(\mathbf{i}; b) = a(\mathbf{i}; b') + a(\mathbf{i}; b'')$. □

§7. Proof of Theorem 1.4

We recall that the quantum minors $\Delta(I; J)$ are defined by (1.6). Here $I = (i_1 < i_2 < \cdots < i_s)$, $J = (j_1 < j_2 < \cdots < j_s)$ are two increasing sequences of indices from $\{1, 2, \ldots, r+1\}$.

PROPOSITION 7.1. *We have* $E_j(\Delta(I; J)) = 0$ *unless* $j \notin J$, $(j+1) \in J$. *If* $j \notin J$, $(j+1) \in J$ *then*

$$E_j(\Delta(I; J)) = \Delta(I; J \cup \{j\} \setminus \{j+1\}). \tag{7.1}$$

PROOF. Applying E_j term by term to the expression (1.6), and using (1.4) and Lemma 3.5, we obtain (7.1) and the fact that $E_j(\Delta(I; J)) = 0$ whenever $(j+1) \notin J$. It remains to check that $E_j(\Delta(I; J)) = 0$ if $j, j+1 \in J$. Concentrating on the contribution to $\Delta(I; J)$ of the jth and $(j+1)$th column, we see that it suffices to show that $E_j(\Delta(i, i'; j, j+1)) = 0$ for all $i < i'$. The only nontrivial case is $i < i' < j$, when $E_j(\Delta(i, i'; j, j+1))$ is easily seen to be proportional to $[t_{ij}, t_{i',j}]$ (see (1.5)). But $[t_{ij}, t_{i',j}] = 0$ by Proposition 3.11(b). □

We say that $\Delta(I; J)$ is *nontrivial* if $i_k \leqslant j_k$ for $k = 1, \ldots, s$.

COROLLARY 7.2. *We have* $\Delta(I; J) \neq 0$ *if and only if* $\Delta(I; J)$ *is nontrivial.*

PROOF. Since the matrix $T = (t_{ij})$ is unitriangular, $\Delta(I; J) = 0$ unless $\Delta(I; J)$ is nontrivial. Conversely, suppose $\Delta(I; J)$ is nontrivial. If $I = J$ then $\Delta(I; J)$ is unitriangular, and hence equal to 1. So we can assume that $I \neq J$. Let k be the minimal index such that $i_k < j_k$. Then $j_{k-1} = i_{k-1} < i_k \leqslant j_k - 1$, hence $(j_k - 1) \notin J$. Applying (7.1) for $j = j_k - 1$ and using induction on $(j_1 - i_1) + \cdots + (j_s - i_s)$, we conclude that $E_j(\Delta(I; J)) \neq 0$ hence $\Delta(I; J) \neq 0$. □

According to Corollary 7.2, we can state Theorem 1.4 as follows.

THEOREM 7.3. *Every string basis* B *in* $\mathscr{A}_r$ *contains all nontrivial minors* $\Delta(I; J)$.

PROOF. As in the proof of Corollary 7.2, we use induction on $(j_1 - i_1) + \cdots + (j_s - i_s)$. If $I = J$ then $\Delta(I; J) = 1$ belongs to B in view of the string axiom (S0). Hence we assume that $I \neq J$. As above, choose $j = j_k - 1$, where k is the minimal index such that $i_k < j_k$. Let $J' = J \cup \{j\} \setminus \{j+1\}$. By Proposition 7.1, $E_j(\Delta(I; J)) = \Delta(I; J')$, and $E_i(\Delta(I; J)) = 0$ for $i < j$.

Let $x = x_j\Delta(I; J')$. By induction, we can assume that $\Delta(I; J') \in B$, hence by (S1), $x \in [B]^+$. Clearly, $E_j(x) = \Delta(I; J')$, and $E_j^2(x) = 0$. By Proposition 4.1, there is a constituent b of x such that $E_j(b) = \Delta(I; J')$. It remains to show that $\Delta(I; J) = b$. By Proposition 1.2(a), it suffices to prove that $E_i(\Delta(I; J)) = E_i(b)$ for all i.

If $i < j$ then $E_i(\Delta(I; J)) = E_i(x) = 0$ by Proposition 7.1 and our choice of j. By (S2), $E_i(b) = 0$, as required. If $i = j$ we have nothing to prove. So it remains to treat $i > j$.

First consider the case $i > j + 1$. Consider two admissible sequences $\mathbf{i} = (j, i)$, $\mathbf{i}' = (i, j)$. Clearly,

$$a(\mathbf{i}; b) = a(\mathbf{i}; \Delta(I; J)), \qquad E_{\mathbf{i}}^{(\mathrm{top})}(b) = E_{\mathbf{i}}^{(\mathrm{top})}(\Delta(I; J)).$$

Moreover, by (7.1) we have only two possibilities: the string $a(\mathbf{i}; \Delta(I; J))$ can be equal either to $(1, 0)$ or to $(1, 1)$. Furthermore, (7.1) implies that if $a(\mathbf{i}; \Delta(I; J)) = (1, 0)$ (resp. $(1, 1)$) then $a(\mathbf{i}'; \Delta(I; J)) = (0, 1)$ (resp. $(1, 1)$); we have also $E_{\mathbf{i}'}^{(\mathrm{top})}(\Delta(I; J)) = E_{\mathbf{i}}^{(\mathrm{top})}(\Delta(I; J))$. On the other hand, by Theorem 2.7(a) the element b satisfies the same properties. Hence $a(\mathbf{i}'; b) = a(\mathbf{i}'; \Delta(I; J))$ and $E_{\mathbf{i}'}^{(\mathrm{top})}(b) = E_{\mathbf{i}'}^{(\mathrm{top})}(\Delta(I; J))$. If $a(\mathbf{i}'; b) = a(\mathbf{i}'; \Delta(I; J)) = (0, 1)$ then $E_i(\Delta(I; J)) = E_i(b)$, as required. If $a(\mathbf{i}'; b) = a(\mathbf{i}'; \Delta(I; J)) = (1, 1)$ then $E_i(b) \in B$ by (S3), and $E_i(\Delta(I; J)) \in B$ by the inductive assumption. Since two basis elements $E_i(b)$ and $E_i(\Delta(I; J))$ have the same string in direction (j), and $E_j(b) = E_j(\Delta(I; J))$ they must coincide in view of Proposition 4.4.

It remains to treat the case $i = j + 1$. The proof in this case is parallel to the previous one. Consider two admissible sequences $\mathbf{i} = (j, i, j)$, $\mathbf{i}' = (i, j, i)$. Clearly,

$$a(\mathbf{i}; b) = a(\mathbf{i}; \Delta(I; J)), \qquad E_{\mathbf{i}}^{(\mathrm{top})}(b) = E_{\mathbf{i}}^{(\mathrm{top})}(\Delta(I; J)).$$

By (7.1), there are only two possibilities for the string $a(\mathbf{i}; \Delta(I; J))$: it can be equal to either $(1, 0, 0)$ or $(1, 1, 0)$. Furthermore, $E_i(\Delta(I; J)) = 0$. On the other hand, the string $a(\mathbf{i}'; b)$ can be computed by (2.8). Applying (2.8) to each of $(1, 0, 0)$ and $(1, 1, 0)$ we obtain the string with $a_1 = 0$. Hence, $E_i(b) = 0$, which completes the proof of Theorem 7.3. □

REMARKS. (a) It follows easily from Proposition 7.1 that for every nontrivial minor $\Delta(I; J)$ and every admissible sequence $\mathbf{i}$ the string $a(\mathbf{i}; \Delta(I; J))$ consists of zeros and ones. The arguments similar to those in the proof of Theorem 7.3 imply the following converse statement: if an element $b \in B$

has the property that the string $a(\mathbf{i}; b)$ in every admissible direction consists of zeros and ones then b coincides with some nontrivial minor $\Delta(I; J)$.

(b) One can show that the set of all nontrivial minors $\Delta(I; J)$ is invariant under the antiautomorphism $x \mapsto x^*$ (see Section 3). In particular, we have $t_{ij}^* = \Delta(i, i+1, \ldots, j-1; i+1, i+2, \ldots, j)$.

§8. Linearity domains: proofs of Proposition 2.8 and Theorems 2.9, 2.10

PROOF OF PROPOSITION 2.8. Fix $\mathbf{i} \in R(w_0)$. We call **i**-*chambers* the connected components of the set of **i**-regular points (so **i**-linearity domains are the closures of **i**-chambers). Clearly, two **i**-regular points a, a' lie in the same **i**-chamber if and only if for every $\mathbf{i}' \in R(w_0)$ the points ${}_{\mathbf{i}'}T_{\mathbf{i}}(a)$ and ${}_{\mathbf{i}'}T_{\mathbf{i}}(a')$ lie on the same side of each $\mathbf{i}'$-wall. Since all maps ${}_{\mathbf{i}'}T_{\mathbf{i}}$ are homogeneous of degree 1, it follows that every **i**-chamber is a cone. To complete the proof of Proposition 2.8 it remains only to show the following. □

LEMMA 8.1. *Let* $a, a' \in C(\mathbf{i})$.

(a) *If* a, a' *belong to the same* **i**-*chamber* C^0 *then* $a + a'$ *also belongs to* C^0, *and*

$$ {}_{\mathbf{i}'}T_{\mathbf{i}}(a+a') = {}_{\mathbf{i}'}T_{\mathbf{i}}(a) + {}_{\mathbf{i}'}T_{\mathbf{i}}(a') \tag{8.1}$$

for each $\mathbf{i}' \in R(w_0)$.

(b) *If* a, a' *do not belong to the same* **i**-*linearity domain then* ${}_{\mathbf{i}'}T_{\mathbf{i}}(a+a') \neq {}_{\mathbf{i}'}T_{\mathbf{i}}(a) + {}_{\mathbf{i}'}T_{\mathbf{i}}(a')$ *for some* $\mathbf{i}' \in R(w_0)$.

PROOF OF LEMMA 8.1. We know that each $\mathbf{i}' \in R(w_0)$ can be reached by **i** by a number of moves of type (2.5) and (2.6). Let $d(\mathbf{i}, \mathbf{i}')$ denote the minimal number of moves of type (2.6) needed for the transition from **i** to $\mathbf{i}'$. We prove the equality (8.1) in (a) by induction on $d(\mathbf{i}, \mathbf{i}')$. If $d(\mathbf{i}, \mathbf{i}') = 0$ then by Proposition 2.6 and Theorem 2.7(a), ${}_{\mathbf{i}'}T_{\mathbf{i}}$ is some permutation of coordinates $(a_1, \ldots, a_m)$ hence is a linear map, and we are done. If $d(\mathbf{i}, \mathbf{i}') = d > 0$ then we can decompose ${}_{\mathbf{i}'}T_{\mathbf{i}}$ into a product ${}_{\mathbf{i}'}T_{\mathbf{i}_2} \circ {}_{\mathbf{i}_2}T_{\mathbf{i}_1} \circ {}_{\mathbf{i}_1}T_{\mathbf{i}}$, where $d(\mathbf{i}, \mathbf{i}_1) = 0$, $d(\mathbf{i}_2, \mathbf{i}') = d-1$, and $\mathbf{i}_2$ is obtained from $\mathbf{i}_1$ by a move of type (2.6). Clearly, the points $a_1 = {}_{\mathbf{i}_1}T_{\mathbf{i}}(a)$ and $a_1' = {}_{\mathbf{i}_1}T_{\mathbf{i}}(a')$ lie in the same $\mathbf{i}_1$-chamber, in particular, are on the same side of each $\mathbf{i}_1$-wall. It follows from Proposition 2.6 and Theorem 2.7(b) that

$$ {}_{\mathbf{i}_2}T_{\mathbf{i}_1}(a_1 + a_1') = {}_{\mathbf{i}_2}T_{\mathbf{i}_1}(a_1) + {}_{\mathbf{i}_2}T_{\mathbf{i}_1}(a_1'). \tag{8.2}$$

Applying ${}_{\mathbf{i}'}T_{\mathbf{i}_2}$ to both sides of (8.2) and using induction, we obtain (8.1).

To complete the proof of (a) we assume that $a, a' \in C^0$ but $(a+a') \notin C^0$. By definition, this means that there exist $\mathbf{i}' \in R(w_0)$ and an $\mathbf{i}'$-wall U such that ${}_{\mathbf{i}'}T_{\mathbf{i}}(a)$ and ${}_{\mathbf{i}'}T_{\mathbf{i}}(a')$ are on one side of U, but ${}_{\mathbf{i}'}T_{\mathbf{i}}(a+a')$ is on the other side. But this contradicts (8.1).

To prove (b) suppose that a, a' do not belong to the same $\mathbf{i}$-linearity domain. This means that there exist $\mathbf{i}' \in R(w_0)$ and an $\mathbf{i}'$-wall U separating ${}_{\mathbf{i}'}T_{\mathbf{i}}(a)$ from ${}_{\mathbf{i}'}T_{\mathbf{i}}(a')$. Let $\mathbf{i}' \mapsto \mathbf{i}''$ be the move of type (2.6) corresponding to the wall U. It follows easily from (2.8) that

$${}_{\mathbf{i}''}T_{\mathbf{i}'}({}_{\mathbf{i}'}T_{\mathbf{i}}(a) + {}_{\mathbf{i}'}T_{\mathbf{i}}(a')) \neq {}_{\mathbf{i}''}T_{\mathbf{i}'}({}_{\mathbf{i}'}T_{\mathbf{i}}(a)) + {}_{\mathbf{i}''}T_{\mathbf{i}'}({}_{\mathbf{i}'}T_{\mathbf{i}}(a')).$$

This implies (b). Lemma 8.1 and hence Proposition 2.8 are proven. □

PROOF OF THEOREM 2.9. Let $b, b' \in B$. First suppose that b, b' satisfy (2), i.e., $b'' = q^N bb' \in B$ for some $N \in \mathbb{Z}$. In view of (3.5), $a(\mathbf{i}; b'') = a(\mathbf{i}; b) + a(\mathbf{i}; b')$ for all $\mathbf{i} \in R(w_0)$. It follows that

$${}_{\mathbf{i}'}T_{\mathbf{i}}(a(\mathbf{i}; b) + a(\mathbf{i}; b')) = {}_{\mathbf{i}'}T_{\mathbf{i}}(a(\mathbf{i}; b)) + {}_{\mathbf{i}'}T_{\mathbf{i}}(a(\mathbf{i}; b'))$$

for each $\mathbf{i}' \in R(w_0)$. By Proposition 2.8, $a(\mathbf{i}; b)$ and $a(\mathbf{i}; b')$ belong to the same $\mathbf{i}$-linearity domain, as claimed.

Now suppose that b and b' quasicommute. To show that the strings $a(\mathbf{i}; b)$ and $a(\mathbf{i}; b')$ belong to the same $\mathbf{i}$-linearity domain, it is enough to show that they are not separated by any $\mathbf{i}$-wall. Suppose an $\mathbf{i}$-wall U corresponds to the move

$$\mathbf{i} = (\mathbf{i}_1, \mathbf{i}_0, \mathbf{i}_2) \mapsto (\mathbf{i}_1, \mathbf{i}_0', \mathbf{i}_2) = \mathbf{i}'$$

of type (2.6), where $\mathbf{i}_0 = (i, j, i)$, $\mathbf{i}_0' = (j, i, j)$, $|i-j| = 1$. By (3.6), the elements $E_{\mathbf{i}_1}^{(\mathrm{top})}(b)$ and $E_{\mathbf{i}_1}^{(\mathrm{top})}(b')$ quasicommute. Replacing if needed b and b' by $E_{\mathbf{i}_1}^{(\mathrm{top})}(b)$ and $E_{\mathbf{i}_1}^{(\mathrm{top})}(b')$ we can assume that $\mathbf{i} = (\mathbf{i}_0, \mathbf{i}_2)$, $\mathbf{i}' = (\mathbf{i}_0', \mathbf{i}_2)$. Let $a(\mathbf{i}_0; b) = a = (a_1, a_2, a_3)$, $a(\mathbf{i}_0'; b) = a' = (a_1', a_2', a_3')$. We have to show that the numbers $a_2 - a_1 - a_3$ and $a_2' - a_1' - a_3'$ are of the same sign.

Let $b_0 = E_{\mathbf{i}_0}^{(\mathrm{top})}(b)$, $b_0' = E_{\mathbf{i}_0}^{(\mathrm{top})}(b')$. Then we have $b'b = q^n bb'$, $b_0'b_0 = q^{n_0} b_0 b_0'$ for some integers n, n_0. Applying (3.6), we obtain

$$E_{\mathbf{i}_0}^{(\mathrm{top})}(bb') = q^{\Phi_{\mathbf{i}_0, \gamma}(a, a')} b_0 b_0', \qquad E_{\mathbf{i}_0}^{(\mathrm{top})}(b'b) = q^{\Phi_{\mathbf{i}_0, \gamma'}(a', a)} b_0' b_0,$$

where $\gamma = \deg(b)$, $\gamma' = \deg(b')$. It follows that

$$\Phi_{\mathbf{i}_0, \gamma}(a, a') - \Phi_{\mathbf{i}_0, \gamma'}(a', a) = n_0 - n. \tag{8.3}$$

By Theorem 2.2, we have $b_0 = E_{\mathbf{i}_0'}^{(\mathrm{top})}(b)$, $b_0' = E_{\mathbf{i}_0'}^{(\mathrm{top})}(b')$. Using the same argument and applying Theorem 2.7(b), we obtain

$$\Phi_{\mathbf{i}_0', \gamma}(T(a), T(a')) - \Phi_{\mathbf{i}_0', \gamma'}(T(a'), T(a)) = n_0 - n, \tag{8.4}$$

where T is the operator given by (2.8).

Combining (8.3) and (8.4), we see that

$$\Phi_{\mathbf{i}_0, \gamma}(a, a') - \Phi_{\mathbf{i}_0, \gamma'}(a', a) = \Phi_{\mathbf{i}_0', \gamma}(T(a), T(a')) - \Phi_{\mathbf{i}_0', \gamma'}(T(a'), T(a)). \tag{8.5}$$

Using (3.3), we can rewrite (8.5) in the form

$$\Psi(a, a') = \Psi(T(a), T(a')), \tag{8.6}$$

where $\Psi(a, a')$ is a skew-symmetric bilinear form on $\mathbb{R}^3$ given by

$$\Psi(a, a') = 2(a_1 a_3' - a_3 a_1') + a_2(a_1' - a_3') - (a_1 - a_3)a_2'. \tag{8.7}$$

To complete the proof it remains to show that (8.6) implies that $a_2 - a_1 - a_3$ and $a_2' - a_1' - a_3'$ are of the same sign. Suppose this is not so, i.e., say $a_2 - a_1 - a_3 > 0$, $a_2' - a_1' - a_3' < 0$. Then $T(a) = (a_2 - a_1, a_1 + a_3, a_1)$, $T(a') = (a_3', a_1' + a_3', a_2' - a_3')$. Substituting these vectors into (8.7), we obtain after a straightforward calculation that

$$\Psi(T(a), T(a')) - \Psi(a, a') = 2(a_2 - a_1 - a_3)(a_2' - a_1' - a_3') < 0,$$

which contradicts (8.6). Theorem 2.9 is proven. □

The argument in the proof of Theorem 2.9 implies the following.

PROPOSITION 8.2. *Let $w \in W$, and $\mathbf{i}, \mathbf{i}' \in R(w)$. If $b, b' \in B$ quasicommute then*

$$_{\mathbf{i}'}T_{\mathbf{i}}(a(\mathbf{i}; b) + a(\mathbf{i}; b')) = {_{\mathbf{i}'}T_{\mathbf{i}}}(a(\mathbf{i}; b)) + {_{\mathbf{i}'}T_{\mathbf{i}}}(a(\mathbf{i}; b')).$$

PROOF OF THEOREM 2.10. Choose an arbitrary decomposition from $R(w_0)$, say $\mathbf{i}(1)$. By Proposition 4.1 and (3.5), there is exactly one constituent b'' of bb' such that $a(\mathbf{i}(1); b'') = a(\mathbf{i}(1); b) + a(\mathbf{i}(1); b')$. By Proposition 8.2,

$$a(\mathbf{i}; b'') = a(\mathbf{i}; b) + a(\mathbf{i}; b') \quad \text{for every } \mathbf{i} \in R(w_0),$$

and we are done. □

§9. The string basis for sl_4

In this section we prove Theorems 1.5, 2.11, and 2.12 in the case $r = 3$. Recall that the reduced expression $\mathbf{i}(1)$ of $w_0 \in W$ is

$$\mathbf{i}(1) = (1, 2, 1, 3, 2, 1),$$

and the cone $C = C(\mathbf{i}(1))$ is given by

$$C = \{a \in \mathbb{R}^6 : a_1 \geqslant 0,\ a_2 \geqslant a_3 \geqslant 0,\ a_4 \geqslant a_5 \geqslant a_6 \geqslant 0\}.$$

It will be convenient for us to rename the elements $b \in \mathscr{P}_3$ according to the first column of Table 1 on the next page.

The strings $a(\mathbf{i}(1); b)$ and the values of $l_1(b)$, $l_2(b)$, $l_3(b)$ displayed in Table 1 are obtained by using repeatedly Proposition 7.1 and Corollary 7.2. The functions $n_1(b)$, $n_2(b)$, and $n_3(b)$ will be defined and used in Section 10.

TABLE 1. Generators of the string basis in $\mathcal{A}_3$

$b \in \mathcal{P}_3$	$\deg(b)$	$a(\mathbf{i}(1); b)$	l_1	l_2	l_3	n_1	n_2	n_3
$x_1 = t_{12}$	α_1	$(1, 0, 0, 0, 0, 0)$	1	0	0	1	0	0
$x_2 = t_{23}$	α_2	$(0, 1, 0, 0, 0, 0)$	0	1	0	0	1	0
$x_{21} = t_{13}$	$\alpha_1 + \alpha_2$	$(0, 1, 1, 0, 0, 0)$	0	1	0	1	0	0
$x_{12} = t_{13}^* = \Delta(1, 2; 2, 3)$	$\alpha_1 + \alpha_2$	$(1, 1, 0, 0, 0, 0)$	1	0	0	0	1	0
$x_3 = t_{34}$	α_3	$(0, 0, 0, 1, 0, 0)$	0	0	1	0	0	1
$x_{32} = t_{24}$	$\alpha_2 + \alpha_3$	$(0, 0, 0, 1, 1, 0)$	0	0	1	0	1	0
$x_{23} = t_{24}^* = \Delta(2, 3; 3, 4)$	$\alpha_2 + \alpha_3$	$(0, 1, 0, 1, 0, 0)$	0	1	0	0	0	1
$x_{132} = \Delta(1, 2; 2, 4)$	$\alpha_1 + \alpha_2 + \alpha_3$	$(1, 0, 0, 1, 1, 0)$	1	0	1	0	1	0
$x_{213} = \Delta(1, 3; 3, 4)$	$\alpha_1 + \alpha_2 + \alpha_3$	$(0, 1, 1, 1, 0, 0)$	0	1	0	1	0	1
$x_{321} = t_{14}$	$\alpha_1 + \alpha_2 + \alpha_3$	$(0, 0, 0, 1, 1, 1)$	0	0	1	1	0	0
$x_{123} = t_{14}^* = \Delta(1, 2, 3; 2, 3, 4)$	$\alpha_1 + \alpha_2 + \alpha_3$	$(1, 1, 0, 1, 0, 0)$	1	0	0	0	0	1
$x_{2132} = \Delta(1, 2; 3, 4)$	$\alpha_1 + 2\alpha_2 + \alpha_3$	$(0, 1, 1, 1, 1, 0)$	0	1	0	0	1	0

Let $M = M(3)$ denote the set of all 12-tuples $m = (m_1, m_2, m_{21}, m_{12}, m_3, m_{32}, m_{23}, m_{132}, m_{213}, m_{321}, m_{123}, m_{2132})$ of nonnegative integers satisfying at least one of the following 14 conditions:

$$m_1 = m_2 = m_{21} = m_{12} = m_{132} = m_{213} = 0, \tag{C1}$$
$$m_1 = m_{21} = m_{12} = m_3 = m_{132} = m_{213} = 0, \tag{C2}$$
$$m_1 = m_2 = m_{21} = m_{12} = m_{32} = m_{132} = 0, \tag{C3}$$
$$m_1 = m_{21} = m_3 = m_{23} = m_{132} = m_{213} = 0, \tag{C4}$$
$$m_1 = m_3 = m_{32} = m_{23} = m_{132} = m_{213} = 0, \tag{C5}$$
$$m_1 = m_2 = m_{21} = m_{12} = m_{23} = m_{213} = 0, \tag{C6}$$
$$m_2 = m_{21} = m_{12} = m_{32} = m_{23} = m_{213} = 0, \tag{C7}$$
$$m_2 = m_{21} = m_{12} = m_{32} = m_{23} = m_{132} = 0, \tag{C8}$$
$$m_2 = m_{12} = m_3 = m_{32} = m_{23} = m_{132} = 0, \tag{C9}$$
$$m_1 = m_2 = m_{21} = m_3 = m_{23} = m_{213} = 0, \tag{C10}$$
$$m_2 = m_{21} = m_3 = m_{32} = m_{23} = m_{213} = 0, \tag{C11}$$
$$m_2 = m_3 = m_{32} = m_{23} = m_{132} = m_{213} = 0, \tag{C12}$$
$$m_1 = m_{12} = m_3 = m_{32} = m_{132} = m_{213} = 0, \tag{C13}$$
$$m_1 = m_2 = m_{12} = m_3 = m_{32} = m_{132} = 0. \tag{C14}$$

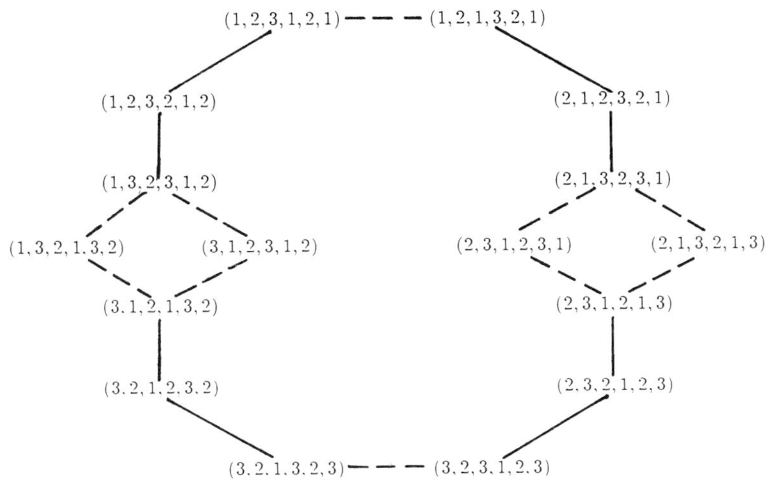

FIGURE 1. Reduced decompositions of w_0 for S_4.

For each $j = 1, \ldots, 14$ let $M_j \subset M$ denote the semigroup of points satisfying (Cj). Let C_j denote the cone generated by strings $a(\mathbf{i}(1); x_\rho)$ for all indices $\rho = 1, 2, \ldots, 2132$ not appearing in the condition (Cj).

For each $m \in M$ we abbreviate $x^m = x_1^{m_1} x_2^{m_2} \cdots x_{2132}^{m_{2132}}$. Iterating (3.6), we see that $E^{(\mathrm{top})}_{\mathbf{i}(1)}(x^m)$ is an integer power of q. We set

$$b(m) = (E^{(\mathrm{top})}_{\mathbf{i}(1)}(x^m))^{-1} x^m, \tag{9.1}$$

so that $E^{(\mathrm{top})}_{\mathbf{i}(1)}(b(m)) = 1$. The following theorem refines Theorems 1.5 and 2.12.

THEOREM 9.1. (a) *There is a unique string basis in $\mathscr{A}_3$. It is formed by the elements $b(m)$ for $m \in M$.*

(b) *Two elements $b(m)$ and $b(m')$ for $m, m' \in M$ quasicommute if and only if m and m' belong to the same semigroup M_j for some $j = 1, \ldots, 14$.*

(c) *The collection $\mathscr{D}$ of cones $C_1, \ldots, C_{14}$ satisfies all conditions of Theorem* 2.12.

PROOF. We start with the description of the $\mathbf{i}(1)$-linearity domains.

PROPOSITION 9.2. *There are thirteen $\mathbf{i}(1)$-linearity domains in C: the simplicial cones $C_1, \ldots, C_{12}$, and the cone $\widetilde{C}_{13} = C_{13} \cup C_{14}$.*

PROOF OF PROPOSITION 9.2. The set $R(w_0)$ is displayed in Figure 1. Here dotted lines correspond to moves of type (2.5), and solid lines correspond to moves of type (2.6).

Now we can find the $\mathbf{i}(1)$-linearity domains in C by a tedious but completely straightforward calculation using Proposition 2.6 and Theorem 2.7. We get thirteen domains defined by linear constraints. This is again a straightforward task to identify these domains with the cones $C_1, \dots, C_{12}$ and $\widetilde{C}_{13}$. The description in terms of linear constraints is as follows:

$$\begin{aligned}
C_1 &= \{a_2 \geqslant a_1 + a_3,\ a_5 \geqslant a_3 + a_6,\\
&\qquad a_3 + a_4 \geqslant a_2 + a_5,\ a_1 \geqslant 0,\ a_3 \geqslant 0,\ a_6 \geqslant 0\},\\
C_2 &= \{a_4 \geqslant a_1 + a_5,\ a_5 \geqslant a_3 + a_6,\\
&\qquad a_2 + a_5 \geqslant a_3 + a_4,\ a_1 \geqslant 0,\ a_3 \geqslant 0,\ a_6 \geqslant 0\},\\
C_3 &= \{a_2 \geqslant a_1 + a_3,\ a_3 + a_6 \geqslant a_5 \geqslant a_6,\ a_4 \geqslant a_2 + a_6,\ a_1 \geqslant 0,\ a_6 \geqslant 0\},\\
C_4 &= \{a_2 \geqslant a_1 + a_3,\ a_1 + a_5 \geqslant a_4 \geqslant a_5 \geqslant a_3 + a_6,\ a_3 \geqslant 0,\ a_6 \geqslant 0\},\\
C_5 &= \{a_2 \geqslant a_1 + a_3,\ a_1 + a_5 \geqslant a_4 \geqslant a_5 \geqslant a_6 \geqslant 0,\ a_3 + a_6 \geqslant a_5\},\\
C_6 &= \{a_1 + a_3 \geqslant a_2 \geqslant a_3,\\
&\qquad a_3 + a_4 \geqslant a_2 + a_5 \geqslant a_1 + 2a_3 + a_6,\ a_3 \geqslant 0,\ a_6 \geqslant 0\},\\
C_7 &= \{a_2 \geqslant a_3,\ a_3 + a_4 \geqslant a_2 + a_5,\ a_5 \geqslant a_3 + a_6,\\
&\qquad a_1 + 2a_3 + a_6 \geqslant a_2 + a_5,\ a_3 \geqslant 0,\ a_6 \geqslant 0\},\\
C_8 &= \{a_1 + a_3 \geqslant a_2 \geqslant a_3,\ a_3 + a_6 \geqslant a_5 \geqslant a_6 \geqslant 0,\ a_4 \geqslant a_2 + a_6\},\\
C_9 &= \{a_1 + a_3 \geqslant a_2 \geqslant a_3,\ a_3 + a_4 \geqslant a_2 + a_5,\ a_5 \geqslant a_6 \geqslant 0,\ a_2 + a_6 \geqslant a_4\},\\
C_{10} &= \{a_1 + a_3 \geqslant a_2,\ a_2 + a_5 \geqslant a_3 + a_4,\\
&\qquad a_4 \geqslant a_5,\ a_2 + a_5 \geqslant a_1 + 2a_3 + a_6,\ a_3 \geqslant 0,\ a_6 \geqslant 0\},\\
C_{11} &= \{a_1 + 2a_3 + a_6 \geqslant a_2 + a_5 \geqslant a_3 + a_4,\\
&\qquad a_4 \geqslant a_5 \geqslant a_3 + a_6,\ a_3 \geqslant 0,\ a_6 \geqslant 0\},\\
C_{12} &= \{a_1 + a_3 \geqslant a_2,\ a_2 + a_5 \geqslant a_3 + a_4,\ a_4 \geqslant a_5,\ a_3 + a_6 \geqslant a_5 \geqslant a_6 \geqslant 0\},\\
\widetilde{C}_{13} &= \{a_2 \geqslant a_1 + a_3,\ a_4 \geqslant a_1 + a_5,\\
&\qquad a_3 + a_6 \geqslant a_5 \geqslant a_6 \geqslant 0,\ a_2 + a_6 \geqslant a_4,\ a_1 \geqslant 0\}.
\end{aligned}$$

It remains to show that $\widetilde{C}_{13} = C_{13} \cup C_{14}$. This follows easily once we represent C_{13} and C_{14} in terms of linear constraints:

$$\begin{aligned}
C_{13} &= \{a_4 \geqslant a_1 + a_5,\ a_3 + a_6 \geqslant a_5 \geqslant a_6 \geqslant 0,\ a_2 + a_5 \geqslant a_3 + a_4,\ a_1 \geqslant 0\},\\
C_{14} &= \{a_2 \geqslant a_1 + a_3,\ a_3 + a_4 \geqslant a_2 + a_5,\ a_2 + a_6 \geqslant a_4,\ a_5 \geqslant a_6 \geqslant 0,\ a_1 \geqslant 0\}
\end{aligned}$$

(the wall between C_{13} and C_{14} in $\widetilde{C}_{13}$ is given by the equation $a_2 + a_5 = a_3 + a_4$). □

In particular, looking at the edges of all cones $C_1, \dots, C_{12}$ and $\widetilde{C}_{13}$, we obtain Theorem 2.11.

We define $\eta: M \to \mathbb{Z}_+^6$ by $\eta(m) = a(\mathbf{i}(1); b(m))$. In view of (3.5), η is a restriction to M of a linear map $\mathbb{R}^{12} \to \mathbb{R}^6$. Therefore, it is totally determined by the column $a(\mathbf{i}(1); b)$ of Table 1. By definition, for each $j = 1, \dots, 14$ the map η is a semigroup isomorphism between M_j and the semigroup of integral points in C_j. Since $C = \bigcup_j C_j$, we obtain the following proposition.

PROPOSITION 9.3. *The map η is a bijection between M and the semigroup Γ of integral points in C.*

As before, we fix a string basis B of $\mathcal{A}_3$. Let B_0 denote the set of all elements $b(m)$ for $m \in M$. By Theorem 1.4 and the string axiom (S3), all $b(m)$ belong to $[B]^+$. Hence, Propositions 9.3, 4.1, and 4.3 imply that B_0 is a basis of $\mathcal{A}_3$.

Table 2 (see p. 82) exhibits the expansion of the product of any two elements of $\mathcal{P}_3$ in the basis B_0. Here the entry in the row x_ρ and the column $x_{\rho'}$ is equal to the product $x_\rho x_{\rho'}$. All the entries can be checked by induction on degrees with the help of Proposition 1.2(a). We exclude from the table the last three elements of $\mathcal{P}_3$, namely x_{321}, x_{123}, and x_{2132}, because it is easy to check that each of them quasicommutes with all elemens of $\mathcal{P}_3$.

Table 2 allows us to check by inspection that if $m, m' \in M$ belong to the same M_j then $b(m)$ and $b(m')$ quasicommute. This proves a half of Theorem 9.1(b). To prove the converse statement we notice that according to Theorem 2.9, if $b(m)$ and $b(m')$ quasicommute then $\eta(m)$ and $\eta(m')$ belong to the same $\mathbf{i}(1)$-linearity domain in C. By Proposition 9.2, we have only to check that if $m \in C_{13} \setminus C_{14}$, $m' \in C_{14} \setminus C_{13}$ then $b(m)$ and $b(m')$ do not quasicommute. This also follows from Table 2. Theorem 9.1(b) is proven. Combining Theorem 9.1(b) and Proposition 9.2, we obtain Theorem 9.1(c).

To complete the proof of Theorem 9.1 it remains to show the uniqueness of a string basis in $\mathcal{A}_3$. It is enough to show that $B_0 \subset B$. We deduce it from the following result whose proof will be given in Section 10.

PROPOSITION 9.4. *Let $m, m' \in M$ be two distinct elements such that $b(m)$ and $b(m')$ have the same degree. Then there exists a product E of divided powers $E_1^{(n_1)}$, $E_2^{(n_2)}$, $E_3^{(n_3)}$ such that $E(b(m)) = 0$, $E(b(m')) \neq 0$.*

Using Proposition 9.4, we complete the proof of Theorem 9.1 as follows. Suppose $b(m) \notin B$ for some $m \in M$. Choose $m \in M$ so that $b(m) \notin B$, and the string $\eta(m)$ is lexicographically smallest possible. We have mentioned already that $b(m) \in [B]^+$. By our choice, $b(m)$ has a constituent $b' \in B$ which has the form $b' = b(m')$ for some $m' \in M$, $m' \neq m$. But then applying to $b(m)$ and $b(m')$ a monomial E from Proposition 9.4, we obtain a contradiction with the string axiom (S2).

REMARKS. (a) Table 2 provides us with a straightening type algorithm for expanding every monomial in the elements of $\mathcal{P}_3$ in the basis B_0. More precisely, we assign weights to the elements of $\mathcal{P}_3$ as follows:

$$\omega(x_1) = \omega(x_{21}) = \omega(x_{12}) = \omega(x_3) = \omega(x_{32}) = \omega(x_{23}) = 3,$$
$$\omega(x_2) = \omega(x_{132}) = \omega(x_{213}) = 4, \qquad \omega(x_{123}) = \omega(x_{321}) = \omega(x_{2132}) = 0.$$

(We use the following rule: the weight of a generator x_ρ is defined as the number of other generators $x_{\rho'}$ such that x_ρ and $x_{\rho'}$ do not quasicommute.)

TABLE 2. Products of pairs of elements from $\mathscr{P}_3$

	x_1	x_2	x_{21}	x_{12}	x_3	x_{32}	x_{23}	x_{132}	x_{213}
x_1	x_1^2	$qx_{21} + x_{12}$	x_1x_{21}	x_1x_{12}	x_1x_3	$qx_{321} + x_{132}$	$qx_{213} + x_{123}$	x_1x_{132}	x_1x_{213}
x_2	$qx_{12} + x_{21}$	x_2^2	x_2x_{21}	x_2x_{12}	$qx_{32} + x_{23}$	x_2x_{32}	x_2x_{23}	$qx_{12}x_{32} + x_{2132}$	$q^{-1}x_{21}x_{23} + x_{2132}$
x_{21}	$q^{-1}x_1x_{21}$	qx_2x_{21}	x_{21}^2	$x_{21}x_{12}$	$qx_{321} + x_{213}$	$qx_2x_{321} + x_{2132}$	$x_{21}x_{23}$	$q^{-1}x_1x_{2132} + x_{12}x_{321}$	$x_{21}x_{213}$
x_{12}	qx_1x_{12}	$q^{-1}x_2x_{12}$	$x_{21}x_{12}$	x_{12}^2	$qx_{132} + x_{123}$	$x_{12}x_{32}$	$q^{-1}x_2x_{123} + x_{2132}$	$x_{12}x_{132}$	$x_{21}x_{123} + x_1x_{2132}$
x_3	x_1x_3	$qx_{23} + x_{32}$	$qx_{213} + x_{321}$	$qx_{123} + x_{132}$	x_3^2	x_3x_{32}	x_3x_{23}	x_3x_{132}	x_3x_{213}
x_{32}	$qx_{132} + x_{321}$	$q^{-1}x_2x_{32}$	$q^{-1}x_2x_{321} + x_{2132}$	$x_{12}x_{32}$	qx_3x_{32}	x_{32}^2	$x_{32}x_{23}$	$x_{32}x_{132}$	$x_{23}x_{321} + x_3x_{2132}$
x_{23}	$qx_{123} + x_{213}$	qx_2x_{23}	$x_{21}x_{23}$	$qx_2x_{123} + x_{2132}$	$q^{-1}x_3x_{23}$	$x_{32}x_{23}$	x_{23}^2	$q^{-1}x_3x_{2132} + x_{32}x_{123}$	$x_{23}x_{213}$
x_{132}	qx_1x_{132}	$q^{-1}x_{12}x_{32} + x_{2132}$	$q^{-1}x_{12}x_{321} + x_1x_{2132}$	$q^{-1}x_{12}x_{132}$	qx_3x_{132}	$q^{-1}x_{32}x_{132}$	$q^{-1}x_{32}x_{123} + x_3x_{2132}$	x_{132}^2	$x_1x_3x_{2132} + x_{321}x_{123}$
x_{213}	$q^{-1}x_1x_{213}$	$qx_{21}x_{23} + x_{2132}$	$qx_{21}x_{213}$	$qx_{21}x_{123} + q^{-1}x_1x_{2132}$	$q^{-1}x_3x_{213}$	$qx_{23}x_{321} + q^{-1}x_3x_{2132}$	$qx_{23}x_{213}$	$q^{-2}x_1x_3x_{2132} + x_{321}x_{123}$	x_{213}^2

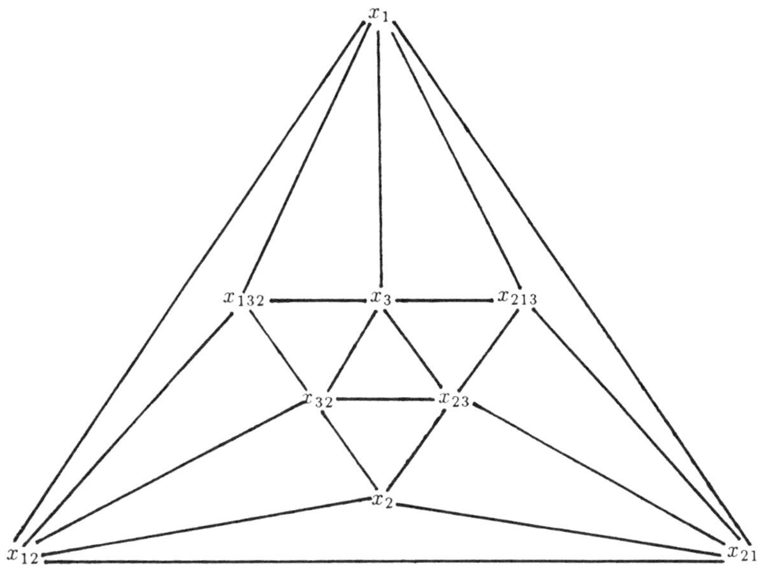

FIGURE 2. The simplicial complex on $\mathscr{P}_3 \setminus \{x_{321}, x_{123}, x_{2132}\}$.

Inspecting Table 2, one observes the following: for each product of the generators $x = x_\rho x_{\rho'}$ that is not proportional to an element from B_0 the weight $\omega(x)$ is greater than the weight of every monomial occurring in the decomposition of x given by Table 2. Now the standard argument shows that every monomial in $\mathscr{P}_3$ can be transformed into a linear combination of elements of B_0 by a sequence of operations consisting of rearrangements of terms and replacements of each monomial $x = x_\rho x_{\rho'}$ as above by its expression from Table 2. Using this algorithm, one can verify in a straightforward way that B_0 is a string basis.

(b) Consider $\mathscr{P}_3$ as a simplicial complex whose simplices are subsets of mutually quasicommuting elements (see Section 1). By Theorem 9.1(b), there are fourteen maximal simplices: to each of the conditions (Cj) ($j = 1, \ldots, 14$) is associated a simplex $\{x_{\rho_1}, \ldots, x_{\rho_6}\}$, where $m_{\rho_1}, \ldots, m_{\rho_6}$ are the exponents not occurring in (Cj). A more transparent description of the simplices can be given as follows. First, each of the elements x_{321}, x_{123}, and x_{2132} has the property that adding it to each simplex of $\mathscr{P}_3$ gives us again a simplex. Hence, it is enough to describe simplices in $\mathscr{P}_3 \setminus \{x_{321}, x_{123}, x_{2132}\}$. One checks readily that these simplices correspond to complete subgraphs of the graph displayed in Figure 2.

§10. Invariants $n_i(b)$ and the proof of Proposition 9.4

To prove Proposition 9.4 we need some notation which makes sense for general $\mathscr{A}_r$. For $i = 1, \ldots, r$ and $x \in \mathscr{A}_r$ we define $n_i(x) = l_i(x^*)$, where the antiautomorphism $x \mapsto x^*$ was defined in Section 3. It follows readily from Proposition 3.10 that

$$n_i(x) = \max\{n \in \mathbb{Z}_+ : E_i^{(n)}(E(x)) \text{ is a nonzero scalar for some monomial } E\}. \quad (10.1)$$

Here by a monomial E we mean any product of divided powers $E_j^{(n_j)}$. In view of (3.5) we have also

$$n_i(xy) = n_i(x) + n_i(y) \quad (10.2)$$

for $i = 1, \dots, r$.

Now we return to the case $r = 3$. We retain the notation of Section 9. A direct calculation shows that the map $x \mapsto x^*$ leaves $\mathscr{P}_3$ invariant (cf. Remark (b) in the end of Section 7). More precisely, we have $x_\rho^* = x_{\rho_{\text{opp}}}$, where the index sequence ρ_{opp} is obtained from ρ by reversing the order of indices (we use the convention that two consecutive indices 1 and 3 can be interchanged without changing an element x_ρ, so $x_{132}^* = x_{213}$, $x_{213}^* = x_{132}$, $x_{2132}^* = x_{2132}$). The values of $n_1(b)$, $n_2(b)$, $n_3(b)$ for $b \in \mathscr{P}_3$ are given in the last three columns of Table 1. By (3.5) and (10.2), each of the functions $m \mapsto l_i(b(m))$ and $m \mapsto n_i(b(m))$ for $i = 1, 2, 3$ is a restriction to M of some linear form on $\mathbb{R}^{12}$. By slight abuse of notation, we denote these linear forms by the same symbols l_i or n_i. They can be read off the last six columns of Table 1, e.g., we have $l_1(m) = m_1 + m_{12} + m_{132} + m_{123}$.

We also define the linear forms $g_1(m)$, $g_2(m)$, $g_3(m)$ as the coefficients in the formula

$$\deg(b(m)) = g_1(m)\alpha_1 + g_2(m)\alpha_2 + g_3(m)\alpha_3. \quad (10.3)$$

These forms can be read off from the second column of Table 1, e.g., we have

$$g_1(m) = m_1 + m_{21} + m_{12} + m_{132} + m_{213} + m_{321} + m_{123} + m_{2132}.$$

We shall deduce Proposition 9.4 from the following lemma.

LEMMA 10.1. (a) *Every $m \in M$ satisfies the following linear constraints:*

$$l_1 + l_2 + l_3 \geqslant g_1, \quad (10.4)$$

$$l_1 + l_2 + l_3 \geqslant g_3, \quad (10.5)$$

$$l_1 + l_2 + l_3 + n_2 \geqslant g_2, \quad (10.6)$$

$$l_1 + l_2 + n_1 \geqslant g_1, \quad (10.7)$$

$$l_1 + n_1 \geqslant 2g_1 - g_2, \quad (10.8)$$

$$l_2 + n_2 \geqslant 2g_2 - g_1 - g_3, \quad (10.9)$$

$$l_3 + n_3 \geqslant 2g_3 - g_2. \quad (10.10)$$

(b) *The inequality* (10.4) *becomes an equality on* $M_9 \cup M_{12}$, (10.5) *becomes an equality on* $M_1 \cup M_3$, (10.6) *becomes an equality on* M_{14}, (10.7) *becomes an equality on* M_{10}, (10.8) *becomes an equality on* $M_7 \cup M_8 \cup M_{11}$, (10.9) *becomes an equality on* $M_2 \cup M_4 \cup M_5 \cup M_{13}$, *and* (10.10) *becomes an equality on* M_6.

Proof. All the inequalities (10.4) to (10.10) are treated in the same way. For instance, consider (10.9). It follows from Table 1 that

$$l_2 + n_2 - (2g_2 - g_1 - g_3) = m_1 + m_3 + m_{132} + m_{213}. \tag{10.11}$$

This implies (10.9). Inspecting conditions (C1) to (C14), we see that the right-hand side of (10.11) vanishes on $M_2 \cup M_4 \cup M_5 \cup M_{13}$, which proves the equality condition in (b). □

Proof of Proposition 9.4. Suppose the statement is not true, and let $m, m' \in M$ provide a counter-example of minimal degree. Using (2.1) and (10.1), we see that

$$l_i(m') \leqslant l_i(m), \qquad n_i(m') \leqslant n_i(m)$$

for $i = 1, 2, 3$. Furthermore, suppose $l_i(m') = l_i(m) = l$ for some i. By Proposition 9.3, we have

$$E_i^{(l)}(b(m)) = b(m_0), \qquad E_i^{(l)}(b(m')) = b(m_0')$$

for some distinct $m_0, m_0' \in M$. Then m_0 and m_0' also provide a counter-example to our statement. Since we have chosen m, m' as a counter-example of minimal degree, we conclude that $l = 0$. Summarizing, we have reduced the proof of Proposition 9.4 to the following statement.

Lemma 10.2. *Suppose $m, m' \in M$ satisfy the following properties:*
(1) $g_i(m) = g_i(m')$ *for* $i = 1, 2, 3$.
(2) $l_i(m') \leqslant l_i(m)$, $n_i(m') \leqslant n_i(m)$ *for* $i = 1, 2, 3$.
(3) *If* $l_i(m') = l_i(m) = l$ *for some* i *then* $l = 0$.
Then $m' = m = 0$.

Proof of Lemma 10.2. We shall repeatedly use the following statement: if one of the inequalities (10.4) to (10.10), say (10.?), becomes an equality at m then we have $l_i(m) = 0$ for all the forms l_i appearing in the left-hand side of (10.?). This follows at once by combining (10.?) with the conditions (1) to (3). Now we proceed in several steps.

Step 1. Suppose $m \in M_1 \cup M_3 \cup M_9 \cup M_{12} \cup M_{14}$. By Lemma 10.1(b), at least one of (10.4), (10.5), and (10.6) becomes an equality at m, hence $l_1(m) = l_2(m) = l_3(m) = 0$. This obviously implies that $m' = m = 0$.

Step 2. Suppose $m \in M_{10}$. By Lemma 10.1(b), (10.7) becomes an equality at m, hence $l_1(m) = l_2(m) = 0$. Combining this condition with (C10), we conclude that all the components of m are equal to 0 except maybe m_{32} and m_{321}. But then $m \in M_1$, and we find ourselves in the situation of Step 1.

The same argument works for m in each of the classes M_7, M_8, or M_{11}, which make (10.8) an equality. Now we obtain $l_1(m) = 0$. Combining this with (C8), we conclude that all the components of m are equal to 0 except maybe m_3, m_{213}, m_{321}, and m_{2132}. But then $m \in M_3$, which again brings us to the situation of Step 1.

For $m \in M_6$ we use (10.10), and the same routine implies that m must belong to M_1, so we are again in the situation of Step 1.

STEP 3. Finally, for $m \in M_2 \cup M_4 \cup M_5 \cup M_{13}$ we use (10.9) and conclude that all the components of m are equal to 0 except maybe m_{12}, m_{32}, m_{321}, and m_{123}. But then $m \in M_{10}$, which brings us to the situation of Step 2.

Since we have covered all fourteen classes M_j, the proof of Lemma 10.2, and hence that of Proposition 9.4, is completed. □

REMARKS. (a) It is well known that the canonical basis in $U_{+,r}$ is invariant under the antiautomorphism $E \mapsto E^*$ (this was shown for arbitrary symmetrizable Kac-Moody algebras in [8, Section 2]). Using Proposition 3.10, we conclude that the dual basis in $\mathcal{A}_r$ is invariant under $x \mapsto x^*$. We do not know whether this is true for arbitrary string bases.

(b) The results in Sections 9 and 10 allow us to construct explicitly good bases of irreducible finite-dimensional sl_4-modules. This construction implies all the conjectures on good bases made in [2] for the case of sl_4. These questions will be treated in detail in a separate publication.

Appendix: Proofs of Propositions 1.1 to 1.3

PROOFS OF PROPOSITIONS 1.1 AND 1.2. We recall that Q_+ is the semigroup generated by simple roots of the root system of type A_r, and that $U_+ = U_{+,r}$ is the Q_+-graded algebra over $\mathbb{Q}(q)$ generated by the elements $E_1, \ldots, E_r$ subject to the relations (1.1), (1.2). For $\gamma \in Q_+$ let $U_+^*(\gamma) = \mathrm{Hom}_{\mathbb{Q}(q)}(U_+(\gamma), \mathbb{Q}(q))$ be the dual of the homogeneous component in U_+, and let $U_+^* = \bigoplus_{\gamma \in Q_+} U_+^*(\gamma)$. To prove Propositions 1.1 and 1.2 we turn things around: we make U_+^* an associative algebra and check that it satisfies all the desired properties, and then show that it can be identified with our algebra $\mathcal{A}_r$.

Let $\mathcal{K} = \mathbb{Q}(q)[K_1, K_1^{-1}, \ldots, K_r, K_r^{-1}]$ be the (commutative) algebra of Laurent polynomials in variables $K_1, \ldots, K_r$. Let $\mathcal{K}U_+$ be an associative algebra over $\mathbb{Q}(q)$ generated by $\mathcal{K}$ and U_+ subject to the relations

$$K_iE_j = E_jK_i \quad \text{for } |i-j| > 1, \tag{A.1}$$

$$K_iE_j = q^{-1}E_jK_i \quad \text{for } |i-j| = 1. \tag{A.2}$$

The algebra $\mathcal{K}U_+$ is a subalgebra of $U_q(sl_{r+1})$ (see, e.g., [10, Section 1]). The following proposition is well known (cf. [11, 4.1]).

PROPOSITION A.1. *The algebra $\mathcal{K}U_+$ is a Hopf algebra with the comultiplication $\Delta: \mathcal{K}U_+ \to \mathcal{K}U_+ \otimes \mathcal{K}U_+$ given by*

$$\Delta(E_i) = E_i \otimes 1 + K_i \otimes E_i, \qquad \Delta(K_i) = K_i \otimes K_i. \tag{A.3}$$

We extend the Q_+-grading from U_+ to $\mathcal{K}U_+$ by setting $(\mathcal{K}U_+)(\gamma) =$

$\mathscr{K}(U_+(\gamma))$. We define

$$(\mathscr{K}U_+)^* = \bigoplus_{\gamma\in Q_+} (\mathscr{K}U_+)^*(\gamma),$$

where

$$(\mathscr{K}U_+)^*(\gamma) = \mathrm{Hom}_{\mathbb{Q}(q)}((\mathscr{K}U_+)(\gamma), \mathbb{Q}(q)).$$

The multiplication map $(\mathscr{K}U_+)^* \otimes (\mathscr{K}U_+)^* \to (\mathscr{K}U_+)^*$ adjoint to the comultiplication $\Delta: \mathscr{K}U_+ \to \mathscr{K}U_+ \otimes \mathscr{K}U_+$ makes $(\mathscr{K}U_+)^*$ a Q_+-graded associative algebra over $\mathbb{Q}(q)$.

Let $\varepsilon: \mathscr{K} \to \mathbb{Q}(q)$ be an algebra homomorphism specializing all K_i to 1. For $x \in U_+^*$ let $\widetilde{x} \in (\mathscr{K}U_+)^*$ denote the extension of a linear form x given by $\widetilde{x}(KE) = \varepsilon(K)x(E)$ for $K \in \mathscr{K}$, $E \in U_+$. Clearly, the correspondence $x \mapsto \widetilde{x}$ is an embedding of graded spaces $U_+^* \to (\mathscr{K}U_+)^*$, and its image $\widetilde{U}_+^*$ consists of all forms $y \in (\mathscr{K}U_+)^*$ such that $y(KH) = \varepsilon(K)y(H)$ for $K \in \mathscr{K}$, $H \in \mathscr{K}U_+$.

Proposition A.2. *The image* $\widetilde{U}_+^*$ *of the map* $x \mapsto \widetilde{x}$ *is a subalgebra of* $(\mathscr{K}U_+)^*$.

Proof. Let $y, z \in \widetilde{U}_+^*$. It is enough to show that $(yz)(K_iE) = (yz)(E)$ for $E \in U_+$. Using definitions and (A.3), we see that

$$\begin{aligned}(yz)(K_iE) &= (y \otimes z)(\Delta(K_iE)) = (y \otimes z)(\Delta(K_i)\Delta(E)) \\ &= (y \otimes z)((K_i \otimes K_i)\Delta(E)) = (y \otimes z)(\Delta(E)) = (yz)(E),\end{aligned}$$

as desired. □

Using Proposition A.2, we transfer the multiplicative structure from $\widetilde{U}_+^*$ to U_+^* via the bijection $x \mapsto \widetilde{x}$. Thus, U_+^* becomes a graded associative algebra.

Let $(E, x) \mapsto E(x)$ be the action of U_+ on U_+^* adjoint to the action of U_+ on itself by right multiplication, i.e., given by $(E(x), E') = (x, E'E)$. For $i = 1, \dots, r$ we define an element $x_i \in U_+^*(\alpha_i)$ by the normalization condition $(x_i, E_i) = 1$. Then all the statements in Propositions 1.1, 1.2 are valid if we replace $\mathscr{A}_r$ by U_+^*. They follow directly from definitions (for Proposition 1.2(a) we use also the fact that U_+ is generated by $E_1, \dots, E_r$). It remains to show that U_+^* can be identified with $\mathscr{A}_r$. More precisely, it suffices to prove the following.

Proposition A.3. *The elements* $x_1, \dots, x_r \in U_+^*$ *generate the algebra* U_+^*. *They satisfy relations* (0.1) *and* (0.2) *from the introduction, and these relations form a defining set of relations for* U_+^*.

Proof. First we show that relations (0.1) and (0.2) are formal consequences of Propositions 1.1, 1.2, hence must hold in U_+^*. By Proposition 1.2(a), it is enough to show that (0.1) and (0.2) remain true after applying every E_i to both sides. If $|i-j| > 1$ then by Proposition 1.1(b), (c) we have $E_k(x_ix_j) = E_k(x_jx_i) = 0$ for $k \neq i, j$, $E_i(x_ix_j) = E_i(x_jx_i) = x_j$, and $E_j(x_ix_j) = E_j(x_jx_i) = x_i$. This implies (0.1). If $|i-j| = 1$, and x is the left-hand side of (0.2) then using again Proposition 1.1(b), (c) we obtain $E_k(x) = 0$ for $k \neq i, j$, and

$$E_i(x) = (1+q^{-2})x_ix_j - (q+q^{-1})(x_jx_i + q^{-1}x_ix_j) + (q+q^{-1})x_jx_i = 0,$$
$$E_j(x) = q^2x_i^2 - (q+q^{-1})qx_i^2 + x_i^2 = 0.$$

This implies (0.2).

It remains to prove two statements:

(1) $x_1, \dots, x_r$ generate U_+^*.
(2) The relations (0.1) and (0.2) form a system of defining relations for U_+^*.

Looking through the arguments in Section 3, we observe that both statements actually were not used there. We needed only the relations (0.1) and (0.2), Propositions 1.1 and 1.2, and the fact that $\mathscr{A}_r$ and U_+ have the same dimensions of homogeneous components. If we replace $\mathscr{A}_r$ by U_+^* the latter statement becomes obvious since $U_+^*(\gamma) = U_+(\gamma)^*$. We conclude that all the results in Section 3 remain true for U_+^*. In particular, the analog of Corollary 3.6(b) implies the statement (1) above. The statement (2) follows from the fact that $\dim(\mathscr{A}_r(\gamma)) = \dim(U_+^*(\gamma))$ for all γ. This completes the proof of Proposition A.3, and hence the proofs of Propositions 1.1, 1.2. □

REMARK. We see that the q-deformations of the (commutative) algebra of functions $\mathbb{C}[N_+]$ and of the universal enveloping algebra $U(\mathfrak{n}_+)$ are isomorphic to each other. This remarkable property was first observed by Drinfeld [4].

PROOF OF PROPOSITION 1.3. Let B^* be the Lusztig's canonical basis in U_+, and B be the basis in $\mathscr{A}_r$ dual to B^*. In the course of the above proof of Propositions 1.1, 1.2 we have interpreted the multiplication in $\mathscr{A}_r$ and the action of U_+ on $\mathscr{A}_r$ in terms of the Hopf algebra structure on $\mathscr{H} U_+$. This allows us to reformulate each of the properties (S0) to (S3) in terms of B^*. Now the property (S0) is clear, (S1) is dual to [12, Theorem 11.5(b)], (S2) is dual to [11, Proposition 7.2(a)], and (S3) is dual to [11, Theorem 7.5]. □

Acknowledgments

We are happy and proud to dedicate this work to our teacher Israel M. Gelfand on the occasion of his eightieth anniversary.

Our work greatly benefitted from the conversations with M. Kashiwara, D. Kazhdan, G. Lusztig, and Y. Soibelman. We are grateful to all of them.

This work was partly done during the stay of Arkady Berenstein at the RIMS, Kyoto University in August-September 1992. He is grateful to RIMS and, in particular, to Professor M. Kashiwara for the kind hospitality and support.

REFERENCES

1. K. Baclawski, *A new rule for computing Clebsch-Gordan series*, Adv. in Appl. Math. **5** (1984), 416–432.
2. A. Berenstein and A. Zelevinsky, *Tensor product multiplicities and convex polytopes in partition space*, J. Geom. Phys. **5** (1988), 453–472.
3. C. De Concini and D. Kazhdan, *Special bases for S_n and GL_n*, Israel J. Math. **40** (1981), 275–290.
4. V. Drinfeld, *Quantum groups*, Proceedings Internat. Congr. Math. (Berkeley, 1986), vol. 1, Amer. Math. Soc., Providence, RI, 1986, pp. 798–820.
5. I. M. Gelfand and A. V. Zelevinsky, *Polytopes in the pattern space and canonical bases for irreducible representations of gl_3*, Funktsional. Anal. i Prilozhen. **19** (1985), no. 2, 72–75; English transl. in Functional Anal. Appl. **19** (1985), no. 2.
6. M. Kashiwara, *On crystal bases of the q-analogue of universal enveloping algebra*, Duke Math. J. **63** (1991).
7. ——, *Global crystal bases of quantum groups*, Preprint RIMS–756, 1991.
8. ——, *Crystal base and Littelman's refined Demazure character formula*, Preprint RIMS–880, 1991.
9. B. Kostant, *A formula for the multiplicity of a weight*, Trans. Amer. Math. Soc. **93** (1959), 53–73.
10. G. Lusztig, *Canonical basis arising from quantized enveloping algebras*, J. Amer. Math. Soc. **3** (1990), 447–498.
11. ——, *Canonical basis arising from quantized enveloping algebras*. II, Progr. Theoret. Phys. **102** (1990).
12. ——, *Quivers, perverse sheaves, and quantized enveloping algebras*, J. Amer. Math. Soc. **4** (1991), 365–421.
13. S.-P. Li, R. Moody, M. Nicolescu, and J. Patera, *Verma bases for representations of classical simple Lie algebras*, J. Math. Phys. **27** (1986), no. 3, 668–677.
14. O. Mathieu, *Good bases for G-modules*, Geom. Dedicata **36** (1990), 51–66.
15. V. S. Retakh and A. V. Zelevinsky, *Base affine space and canonical basis in irreducible representations of Sp_4*, Dokl. Akad. Nauk USSR **300** (1988), no. 1, 31–35; English transl., Soviet Math. Dokl. **37** (1988), no. 3, 618–622.

DEPARTMENT OF MATHEMATICS, NORTHEASTERN UNIVERSITY, BOSTON, MASSACHUSETTS 02115

ADVANCES IN SOVIET MATHEMATICS
Volume 16, Part 1, 1993

Limit Distributions of Orbits of Unipotent Flows and Values of Quadratic Forms

S. G. DANI AND G. A. MARGULIS

§1. Introduction

Let G be a Lie group and Γ be a discrete subgroup of G. Consider the homogeneous space G/Γ. On such a space we have a natural class of dynamical systems arising from actions of one-parameter subgroups of G, acting on the left. Study of dynamics of these flows, and more generally of actions of subgroups of G on G/Γ, has attracted considerable attention for over half a century now, on account of their role as a natural class of dynamical systems. Some of the remarkable dynamical features discovered in the early years in special cases are connections with geometry, the structure of discrete subgroups, and, more pertinently from the current standpoint, applications to Diophantine approximation. In recent years a long-standing conjecture due to A. Oppenheim on values of quadratic forms at integral points was proved (cf. [24, 25, 26]) and also various strengthenings were obtained (cf. [15, 16, 17]) via such a study. The reader is referred to [27] for a general perspective of the area (see also [6, 13, 23] and references therein).

Specifically, the proofs of the Oppenheim conjecture and related problems involve flows which are Ad-unipotent (an element g of G is said to be Ad-unipotent if $\operatorname{Ad} g$ is a unipotent linear transformation of the Lie algebra of G, and a flow is said to be Ad-unipotent if the corresponding one-parameter subgroup consists of Ad-unipotent elements) and, more generally, actions of subgroups of G which are generated by Ad-unipotent elements. While the above mentioned applications depend on certain particular Ad-unipotent flows (and actions as above) having the property that the closure of any orbit is a homogeneous subset (namely an orbit of a possibly larger subgroup) it was expected that a similar result holds for all Ad-unipotent flows and actions as above, whenever Γ is a lattice in G (namely G/Γ carries a G-invariant probability measure); this was conjectured by M. S. Raghunathan (cf. [11]) for the case of flows and by G. A. Margulis (cf. [24]) in the general case.

1991 *Mathematics Subject Classification*. Primary 11H55, 58E11; Secondary 22E40, 58E17.

Major progress has now been made in the area by Marina Ratner who, through a series of papers (cf. [30, 31, 32]), obtained a classification of all finite invariant measures of W-actions on G/Γ, where G and Γ are as set up at the outset and W is a closed connected subgroup of G generated by Ad-unipotent elements, and later, in [33], deduced the above mentioned conjecture on orbit closures of such a W, for the case of lattices; the actual results are somewhat more general (cf. [32, Theorem 3]), but we shall not go into it as it is not relevant to the present context; on the question of classification of invariant measures the reader may also see [28] for some more recent results, where p-adic groups are also considered. Ratner also proved that if Γ is a lattice in a Lie group G and $\{u_t\}$ is an Ad-unipotent one-parameter subgroup of G then for any $x \in G/\Gamma$ the $\{u_t\}$-orbit of x is uniformly distributed in its closure in the sense that for any bounded continuous function φ on G/Γ,

$$\frac{1}{T}\int_0^T \varphi(u_t x)\,dt \to \int_{G/\Gamma} \varphi\,d\pi_x, \tag{i}$$

where π_x is a (uniquely defined) probability measure which is invariant under the action of a closed subgroup H of G which acts transitively on the closure of the orbit, namely such that $\overline{\{u_t x \mid t \in \mathbb{R}\}} = Hx$. In particular, for any x in G/Γ which is not contained in any subset of the form Hx with a finite H-invariant measure, for any proper closed subgroup H of G, we have

$$\frac{1}{T}\int_0^T \varphi(u_t x)\,dt \to \int_{G/\Gamma} \varphi\,d\mu, \tag{ii}$$

for all bounded continuous functions φ on G/Γ, where μ is the G-invariant probability measure on G/Γ; thus all such orbits are uniformly distributed with respect to the canonical probability measure on G/Γ.

The main question which we now want to address is how the convergences in (i) and (ii) depend on the initial point x and the one-parameter subgroup $\{u_t\}$; for instance, in (ii) is it uniform when we vary x and $\{u_t\}$ over compact subsets on which the convergence holds pointwise? In (i) does π_x come close to μ as x approaches a point where the corresponding measure is μ? etc. These questions, apart from their intrinsic interest, are of relevence in studying certain Diophantine problems relating to values of quadratic forms.

It turns out that certain ideas involved in [16], and developed further in [36], in both of which partial results were obtained towards Raghunathan's conjecture, can be applied to this problem. The ideas pertain to the behaviour of orbits of Ad-unipotent flows near "singular sets", considered simultaneously for different one-parameter subgroups and initial points. In this paper we prove a general result in this regard (see Theorem 1) which together with Ratner's classification of invariant measures enables us to provide certain answers to the questions posed above (see Theorems 2 and 3) and, in turn, to prove some new results pertaining to lower estimates for the number of integral solutions for certain quadratic inequalities, in large enough euclidean balls (see Corollary 5). Most of the results in this paper were first announced in [19] together with a sketch of the ideas involved.

We now introduce some notation and give the formal statements of the results. Let G be a connected Lie group and Γ be a discrete subgroup of G. For any closed subgroup W of G we denote by $\mathscr{S}(W)$ the set of all x in G/Γ for which there exists a proper closed subgroup H of G containing W such that Hx admits a finite H-invariant measure; under this condition Hx is automatically a proper closed subset of G/Γ (cf. [29, Theorem 1.13]); we also put $\mathscr{G}(W) = G/\Gamma - \mathscr{S}(W)$. Observe that if $\{u_t\}$ is an Ad-unipotent one-parameter subgroup of G then (ii) holds for any $x \in \mathscr{G}(\{u_t\})$ and thus any $x \in \mathscr{G}(\{u_t\})$ is a generic point for the flow, in the usual sense of ergodic theory. Conversely, it is true that any generic point belongs to $\mathscr{G}(\{u_t\})$, but it should be borne in mind that this is only an a posteriori fact, following from Ratner's results in [33] or the results of the present paper.

We now state the main general result about the behaviour of Ad-unipotent flows near "singular sets" $\mathscr{S}(W)$ for certain subgroups W. As usual l will denote the Lebesgue measure on $\mathbb{R}$.

1. THEOREM. *Let G be a connected Lie group and let Γ be a discrete subgroup of G. Let W be any closed connected subgroup of G which is generated by the* Ad-*unipotent elements contained in it. Let F be a compact subset of $\mathscr{G}(W)$. Then for any $\varepsilon > 0$ there exists a neighbourhood Ω of $\mathscr{S}(W)$ such that for any* Ad-*unipotent one-parameter subgroup $\{u_t\}$ of G, any $x \in F$, and any $T \geqslant 0$,*

$$l(\{t \in [0, T] \mid u_t x \in \Omega\}) \leqslant \varepsilon T.$$

Our proof of this result is independent of Ratner's work. We next use Ratner's classification of invariant measures of Ad-unipotent flows and deduce from Theorem 1 the following result which may be seen to provide an answer to the general question posed above, concerning (ii). For this, W as in Theorem 1 is chosen to be the limit one-parameter subgroup $\{u_t\}$ as in the statement below.

2. THEOREM. *Let G be a connected Lie group and let Γ be a lattice in G. Let μ be the G-invariant probability measure on G/Γ. Let $\{u_t^{(i)}\}$ be a sequence of* Ad-*unipotent one-parameter subgroups converging to an* Ad-*unipotent one-parameter subgroup $\{u_t\}$; that is, $u_t^{(i)} \to u_t$ for all t. Let $\{x_i\}$ be a sequence in G/Γ converging to a point in $\mathscr{G}(\{u_t\})$ and let $\{T_i\}$ be a sequence in $\mathbb{R}^+$, tending to infinity. Then for any bounded continuous function φ on G/Γ,*

$$\frac{1}{T_i}\int_0^{T_i} \varphi(u_t^{(i)} x_i)\, dt \to \int_{G/\Gamma} \varphi\, d\mu.$$

Specialising to $\{u_t^{(i)}\}$ and $\{x_i\}$ being constant sequences in Theorem 2 yields Ratner's distribution rigidity theorem for Ad-unipotent flows (cf. [32, Theorem B]) and hence Raghunathan's conjecture on closures of orbits. Since we use only her classification of invariant measures as in [32] and not her deduction in [33] of the distribution rigidity theorem and Raghunathan's conjecture, our method provides another way (simpler in our view) of deducing the above mentioned results from the classification of invariant measures. We also deduce from Theorem 2 a result (cf. Corollary 6.2) on the distribution of closed orbits under certain closed subgroups, which implies in particular an assertion made by M. Burger and P. Sarnak, related to the distribution of Hecke points.

We next study the variation in the averages without limiting ourselves to generic points. In this we fix an Ad-unipotent one-parameter subgroup; for the phenomenon being observed this seems to be a natural restriction.

For any closed subgroups H and W of G let $X(H, W) = \{g \in G \mid Wg \subseteq gH\}$. Clearly for any proper closed subgroup H such that $H \cap \Gamma$ is a lattice in H, $X(H, W)\Gamma/\Gamma$ is contained in $\mathscr{S}(W)$. The following result says that the ergodic averages of a given bounded continuous function from two initial points are comparable if the initial points are outside certain *compact* subsets of $\mathscr{S}(W)$ contained in a union of *finitely many* subsets as above. It also means that for x outside such a compact subset of $\mathscr{S}(W)$, π_x as in (i) is close to the G-invariant probability measure μ.

3. THEOREM. *Let G be a connected Lie group and Γ be a lattice in G. Let μ be the G-invariant probability measure on G/Γ. Let $U = \{u_t\}$ be an* Ad-*unipotent one-parameter subgroup of G and let φ be a bounded continuous function on G/Γ. Let K be a compact subset of G/Γ and let $\varepsilon > 0$ be given. Then there exist finitely many proper closed subgroups $H_1, \dots, H_k$ such that $H_i \cap \Gamma$ is a lattice in H_i for all i, and compact subsets $C_1, \dots, C_k$ of $X(H_1, U), \dots, X(H_k, U)$ respectively, for which the following holds: For any compact subset F of $K - \bigcup C_i\Gamma/\Gamma$ there exists a $T_0 \geqslant 0$ such that for all $x \in F$ and $T > T_0$,*

$$\left|\frac{1}{T}\int_0^T \varphi(u_t x)\, dt - \int_{G/\Gamma} \varphi\, d\mu\right| < \varepsilon.$$

We also prove the following.

4. THEOREM. *Let G be a connected Lie group and Γ be a lattice in G. Let $U = \{u_t\}$ be an* Ad-*unipotent one-parameter subgroup of G. Let $\{x_i\}$ be a convergent sequence in G/Γ. Suppose that for any proper closed connected subgroup H of G such that $H \cap \Gamma$ is a lattice in H and any compact subset C of $X(H, U)$ the set of all i such that $x_i \in C\Gamma/\Gamma$ is finite. Let R be a subset of $\mathbb{R}^+$ for which there exists an $\alpha > 0$ such that $l(R \cap [0, T]) \geqslant \alpha T$ for all $T \geqslant 0$. Then there exists a sequence $\{t_i\}$ in R such that $\{u_{t_i} x_i\}$ has a subsequence converging to a point in $\mathscr{G}(U)$.*

We now describe an application of the results to studying the set of values of quadratic forms at integral points. Let $n \geqslant 3$ and $1 \leqslant p < n$ be fixed and let $\mathscr{Q}(p, n)$ denote the space of quadratic forms on $\mathbb{R}^n$ with discriminant ± 1 and signature $(p, n-p)$; the signature determines the sign of the discriminant. The space $\mathscr{Q}(p, n)$ has a natural locally compact topology given by pointwise convergence as functions on $\mathbb{R}^n$. We denote by $\|.\|$ and λ respectively the usual Hilbert norm and the Lebesgue measure on $\mathbb{R}^n$. For any subset E of $\mathbb{R}^n$ we denote by $|E|$ the cardinality of E. Recall that by the Oppenheim conjecture proved by Margulis, for any $Q \in \mathscr{Q}(p, n)$ which is not a scalar multiple of a rational form and any interval I of $\mathbb{R}$ of positive length there exist integral n-tuples x such that $Q(x) \in I$. The first assertion below gives, in particular, a lower bound for the number of such solutions in euclidean balls of radius r centered at 0, for all large r; also, we consider a compact set of forms simultaneously and the result requires exclusion of only finitely many forms which are multiples of rational forms, rather than all such forms. By Meyer's theorem one knows that if $n \geqslant 5$ any rational form in $\mathscr{Q}(p, n)$ has a (nontrivial) rational zero. This suggests that for $n \geqslant 5$ no exclusion of rational forms may be necessary for an interval of the form $(-\varepsilon, \varepsilon)$. The second assertion in the corollary upholds such an expectation though, naturally, the constant involved is smaller. It can be verified that the volumes appearing on the right-hand side in the expressions below are proportional to r^{n-2}. We mention here that for a single Q a similar estimate, with the same exponent of r but only a positive constant rather than one arbitrarily close to 1, was obtained earlier by the first named author jointly with Shahar Mozes, and also independently by M. Ratner (both unpublished) using only Ratner's theorem on uniform distribution.

5. Corollary. *Let $\mathscr{K}$ be a compact subset of $\mathscr{Q}(p, n)$. Let ν be a continuous positive function on the sphere $\{v \in \mathbb{R}^n \mid \|v\| = 1\}$ and let $\Omega = \{v \in \mathbb{R}^n \mid \|v\| < \nu(v/\|v\|)\}$. Then we have the following:*

i) *for any interval I in $\mathbb{R}$ and $\theta > 0$ there exists a finite subset $\mathscr{S}$ of $\mathscr{K}$ such that each $Q \in \mathscr{S}$ is a scalar multiple of a rational quadratic form and for any compact subset $\mathscr{C}$ of $\mathscr{K} - \mathscr{S}$ there exists $r_0 \geqslant 0$ such that for all Q in $\mathscr{C}$ and $r \geqslant r_0$,*

$$|\{z \in r\Omega \cap \mathbb{Z}^n \mid Q(z) \in I\}| \geqslant (1-\theta)\lambda(\{v \in r\Omega \mid Q(v) \in I\});$$

ii) *if $n \geqslant 5$, for any $\varepsilon > 0$ there exist $c > 0$ and $r_0 \geqslant 0$ such that for all $Q \in \mathscr{K}$ and $r \geqslant r_0$,*

$$|\{z \in r\Omega \cap \mathbb{Z}^n \mid |Q(z)| < \varepsilon\}| \geqslant c\lambda(\{v \in r\Omega \mid |Q(v)| < \varepsilon\}).$$

The paper is organised as follows. Section 2 is devoted to preliminaries on the structure of $\mathscr{S}(W)$. In Section 3 we relate the sets $X(H, W)$ to algebraic subvarieties in certain (finite-dimensional) representation spaces. The proof of Theorem 1 is contained in Section 4; while the basic ideas involved in the proof are developed in that section itself, we use the preliminaries from earlier sections and also a result, Theorem 3.4, stated in Section 3 together with a proof in the particular case of arithmetic subgroups. Section 5 contains a "finiteness theorem" for homogeneous subsets of G/Γ of volume bounded by a given constant, which would be of independent interest; the theorem is then used to deduce the general case of Theorem 3.4. In Section 6 we prove Theorem 2, using the results of the earlier sections and Ratner's classification of invariant measures; the section also contains a proof of an assertion of Burger and Sarnak, alluded to above, related to the distribution of Hecke points. In Section 7 we refine the ideas involved in the proof of Theorem 1 and prove some technical results which are then used in Section 8 to complete the proofs of Theorems 3 and 4. The proof of Corollary 5 is the subject of the last two sections.

ACKNOWLEDGMENTS. The authors would like to thank Nimish Shah for numerous comments on a preliminary version of the paper, contributing substantially in improving the text. Thanks are also due to Shahar Mozes and Marina Ratner for useful suggestions. The authors would like to gratefully acknowledge the hospitality of the Sonderforschungsbereich-170, Göttingen, and the Mathematical Sciences Research Institute, Berkeley, during the course of the work.

Also, S. G. Dani would like to thank the University of Chicago and Yale University, New Haven, and G. A. Margulis would like to thank the Max-Planck-Institut, Bonn, for hospitality while the work was going on.

G. A. Margulis was partially supported by NSF (grant DMS 9204270).

§2. Structure of singular sets

Recall that for a connected Lie group G, a discrete subgroup Γ, and a closed connected subgroup W of G the "singular" set $\mathscr{S}(W)$ is defined to be the set of all $x \in G/\Gamma$ such that there exists a proper closed subgroup H containing W such that Hx (is closed and) has finite H-invariant measure. In this section we discuss the structure of this set.

A subgroup of $GL(V)$, where V is a finite-dimensional real vector space, is said to be almost algebraic if it is an open subgroup of an algebraic subgroup of $GL(V)$; since any algebraic subgroup of $GL(V)$ has only finitely many connected components (cf. [5, Section 14]), any almost algebraic subgroup is of finite index in an algebraic subgroup; for any subgroup S of $GL(V)$ we denote by $\tilde{S}$ the smallest almost algebraic subgroup of $GL(V)$ containing S. We note that if S is connected so is $\tilde{S}$.

To begin with we note below a general fact concerning subgroups intersecting a given discrete subgroup in a lattice. The reader may like to compare the result with Theorem 1.1 of [32]; it may be pointed out that the Zariski-density condition here is related to the condition of existence of an Ad-unipotent element acting ergodically on $H/H \cap \Gamma$ involved there (cf. [36, Section 2]). We mention also that later we will actually be proving a stronger result (see Theorem 5.1 and especially step iv) there); however it seems convenient to content ourselves with the following for the present.

2.1. PROPOSITION. *Let G be a connected Lie group and $\rho: G \to GL(V)$ be a representation of G over a finite-dimensional real vector space V, such that* $\ker \rho$ *is contained in the center of G. Let Γ be a discrete subgroup of G and let $\mathscr{C}$ be the class of all closed connected subgroups H of G such that $H \cap \Gamma$ is a lattice in H and $\rho(H \cap \Gamma)$ is Zariski-dense in $\rho(H)$. Then $\mathscr{C}$ is countable.*

PROOF. We first consider the case when $\rho(G)$ is closed. For any $H \in \mathscr{C}$ clearly $\widetilde{\rho(H)} = \widetilde{\rho(H \cap \Gamma)}$ and by dimension considerations there exists a finitely generated subgroup Δ of Γ such that $\widetilde{\rho(H \cap \Gamma)} = \widetilde{\rho(\Delta)}$. Since Γ is countable, it has only countably many finitely generated subgroups. Hence to prove the proposition it is enough to prove that given any finitely generated subgroup Δ of Γ such that $\widetilde{\rho(\Delta)}$ is connected, there are only countably many $H \in \mathscr{C}$ such that $\widetilde{\rho(H)} = \widetilde{\rho(\Delta)}$. Let such a Δ be given and let $S = \widetilde{\rho(\Delta)}$. Then $S/[S, S]$ is an abelian connected Lie group and hence has the form $\mathbb{R}^a \times \mathbb{T}^b$ for some $a, b \geqslant 0$. Let $\eta: S \to \mathbb{R}^a \times \mathbb{T}^b$ be the canonical quotient homomorphism under the identification. In $\mathbb{R}^a \times \mathbb{T}^b$ there are only countably many closed connected subgroups L such that $L/\overline{\eta(\rho(H \cap \Gamma))}$ is compact; let $\mathscr{K}$ be the class of all closed subgroups of S of the form $\eta^{-1}(L)$ where L is such a subgroup. Now let $H \in \mathscr{C}$ be such that $\widetilde{\rho(H)} = \widetilde{\rho(\Delta)} = S$. For any connected Lie subgroup C of $GL(V)$ one knows that $[C, C] = [\tilde{C}, \tilde{C}]$ (cf. [8, Chapter II, Theorem 13]). Hence $\rho(H)$ as above contains $[S, S]$. Then $\eta(\overline{\rho(H)})$ is a closed connected subgroup of $\mathbb{R}^a \times \mathbb{T}^b$. Further, since $H \cap \Gamma$ is a lattice in H, $\eta(\overline{\rho(H)})/\overline{\eta(\rho(H \cap \Gamma))}$ is compact. This implies that $\overline{\rho(H)} \in \mathscr{K}$. (It may be observed that if ρ is a faithful representation and also $\rho(G)$ is closed as assumed above, then the preceding conclusion already implies that $\mathscr{C}$ is countable, thus completing the proof in that case.) Since $\mathscr{K}$ is countable, to prove the proposition it is now enough to prove that for any $K \in \mathscr{K}$ there are only countably many H in $\mathscr{C}$ such that $\overline{\rho(H)} = K$. Let $K \in \mathscr{K}$ be given and let H be any such subgroup. Since $\rho(G)$ is closed, K is contained in $\rho(G)$ and hence $\rho^{-1}(K) = \overline{H(\ker \rho)}$. Let A be the connected component of the identity in $\rho^{-1}(K)$. Then H is contained in A. Since $\ker \rho$ is contained in the center of G, it follows that H contains $\overline{[A, A]}$. Now, $H/\overline{[A, A]}$ is a closed subgroup of the abelian Lie group $A/\overline{[A, A]}$, containing $\overline{(H \cap \Gamma)[A, A]}/\overline{[A, A]}$ as a cocompact subgroup. As noted earlier this means that it has only countably many possibilities and hence so does H. This proves the proposition in the case at hand, namely when $\rho(G)$ is closed.

Now suppose that $\rho(G)$ is not closed. We note that $[\rho(G), \rho(G)]$ is an almost algebraic subgroup (cf. [8, Chapter II, Theorem 13]) and hence in particular it is closed. Then $\rho(G)/[\rho(G), \rho(G)]$ is a connected abelian Lie group (where $\rho(G)$ is considered with its Lie group topology as the quotient of G by $\ker \rho$ and not the induced topology from $GL(V)$). Hence it is Lie isomorphic to $\mathbb{R}^a \times \mathbb{T}^b$ for some $a, b \geqslant 0$. Therefore there exists a representation $\sigma : \rho(G)/[\rho(G), \rho(G)] \to GL(V')$, for a suitable finite-dimensional vector space V', such that the image of σ is closed. Let $\tau : G \to GL(V) \times GL(V')$ (the latter may be viewed canonically as a subgroup of $GL(V \oplus V')$) be the representation defined by $\tau(g) = (\rho(g), \sigma(\rho(g)[\rho(G), \rho(G)]))$ for all $g \in G$. It is then easy to verify that $\tau(G)$ is closed (cf. [20, Proposition 2.5], for details, if necessary). Also if H is a closed subgroup of G such that $\rho(H \cap \Gamma)$ is Zariski-dense in $\rho(H)$ then $\tau(H \cap \Gamma)$ is Zariski-dense in $\tau(H)$; this follows from the fact that if S is a Zariski-closed subgroup of $GL(V) \times GL(V')$ containing $\tau(H \cap \Gamma)$ then its projection on $GL(V)$ is a Zariski-closed subgroup (cf. [2, Corollary 1.4]) containing $\rho(H \cap \Gamma)$ and hence $\rho(H)$. The general case of the proposition now follows from application of the above special case to the representation τ.

2.2. LEMMA. *Let G be a connected Lie group and Γ be a lattice in G. Let $\rho : G \to GL(V)$ be a representation of G over a finite-dimensional real vector space V. Let W be any connected Lie subgroup of G such that $\rho(W)$ is generated by the unipotent elements contained in it. Then there exists a closed connected subgroup H of G such that $W \subseteq H$, $H \cap \Gamma$ is a lattice in H, and $\rho(H \cap \Gamma)$ is Zariski-dense in $\rho(H)$.*

PROOF. Let H_1 be the smallest closed subgroup containing all one-parameter subgroups $\{g_t\}$ of G such that $\{\rho(g_t)\}$ is unipotent. Clearly, H_1 is a connected normal subgroup of G. Let $H = (\overline{H_1\Gamma})^0$, the connected component of the identity in $\overline{H_1\Gamma}$; then $H\Gamma = \overline{H_1\Gamma}$ and it is a closed subgroup of G. Since Γ is a lattice in G it follows that Γ is a lattice in $H\Gamma$. Hence $H \cap \Gamma$ is a lattice H. Now let L be the Zariski closure of $\rho(H \cap \Gamma)$. Let $\{g_t\}$ be any one-parameter subgoup of G such that $\{\rho(g_t)\}$ is unipotent. Then $\{g_t\}$ is contained in H_1 and hence also in H. Since $H \cap \Gamma$ is a lattice in H, by Borel's density theorem (cf. [10]), we get that $\{\rho(g_t)\}$ is contained in L. Since H_1 is the smallest closed subgroup containing all $\{g_t\}$ as above, it follows that $\rho(H_1)$ is contained in L. It is easy to see that $H = \overline{H_1(H \cap \Gamma)}$ and hence the preceding conclusion implies that $\rho(H)$ is contained in L. Hence $\rho(H \cap \Gamma)$ is Zariski-dense in $\rho(H)$. Any connected Lie subgroup of $GL(V)$ which is generated by the unipotent elements contained in it is also generated by the unipotent one-parameter subgroups contained in it; this can be deduced using Theorem 13 of Chapter II in [8]. Hence any subgroup W as in the hypothesis is contained in the subgroup H_1 defined above, and hence also in H. This proves the lemma.

2.3. PROPOSITION. *Let* G, Γ, *and* W *be as in Theorem* 1. *Let* $\mathscr{H}$ *be the class of all proper closed connected subgroups* H *of* G *such that* $H \cap \Gamma$ *is a lattice in* H *and* $\operatorname{Ad}(H \cap \Gamma)$ *is Zariski-dense in* $\operatorname{Ad} H$, *where* Ad *denotes the adjoint representation of* G. *Then* $\mathscr{S}(W) = \bigcup_{H \in \mathscr{H}} X(H, W)\Gamma/\Gamma$.

PROOF. Let $x \in \mathscr{S}(W)$ and let $g \in G$ be such that $x = g\Gamma$. Then there exists a proper closed connected subgroup F containing W such that Fx has a finite F-invariant measure; hence $g^{-1}Fg \cap \Gamma$ is a lattice in $g^{-1}Fg$. Then by Lemma 2.2 (applied to $g^{-1}Fg$ in the place of G and the restriction of Ad in the place of ρ there) there exists a closed connected subgroup H of $g^{-1}Fg$ containing $g^{-1}Wg$ and such that $H \cap \Gamma$ is a lattice in H and $\operatorname{Ad}(H \cap \Gamma)$ is Zariski-dense in $\operatorname{Ad} H$. Then $H \in \mathscr{H}$ and since $g^{-1}Wg \subseteq H$ we have $g \in X(H, W)$ and hence $x = g\Gamma \in X(H, W)\Gamma/\Gamma$. Conversely it is clear that $X(H, W)\Gamma/\Gamma \subseteq \mathscr{S}(W)$, for any $H \in \mathscr{H}$. This proves the proposition.

§3. Linearisation of singular submanifolds

In this section we relate subsets of the form $X(H, W)$ to affine subvarieties in certain vector spaces.

Let G be a connected Lie group and let Γ be a discrete subgroup of G. Let W be a closed connected subgroup of G which is generated by the Ad-unipotent elements contained in it. As before we denote by $\mathscr{H}$ the class of all proper closed connected subgroups H of G such that $H \cap \Gamma$ is a lattice in H and $\operatorname{Ad}(H \cap \Gamma)$ is Zariski-dense in $\operatorname{Ad} H$; here and in the sequel Ad denotes the adjoint representation of G. We denote by $\mathfrak{G}$ the Lie algebra of G. Also, for any subgroup S of G, S^0 will denote the connected component of the identity in S.

To each $H \in \mathscr{H}$ we associate a representation ρ_H of G as follows. Let $H \in \mathscr{H}$ and h be the dimension of H. Let $V_H = \bigwedge^h \mathfrak{G}$, the hth exterior power of $\mathfrak{G}$, as a vector space, and let $\rho_H : G \to GL(V_H)$ be the hth exterior power of the adjoint representation of G. Let $p_H \in V_H$ be a nonzero element of the form $\xi_1 \wedge \xi_2 \wedge \cdots \wedge \xi_h$, where $\xi_1, \xi_2, \ldots, \xi_h$ belong to the Lie subalgebra of $\mathfrak{G}$ corresponding to H. Also let $\eta_H : G \to V_H$ be the orbit map defined by $\eta_H(g) = \rho_H(g)p_H$ for all $g \in G$.

For any $H \in \mathscr{H}$ let $\Gamma_H = N(H) \cap \Gamma$, where $N(H)$ denotes the normaliser of H in G. We note the following:

3.1. LEMMA. *Let* $H \in \mathscr{H}$. *Then* $\Gamma_H = \{\gamma \in \Gamma \mid \rho_H(\gamma)p_H = \pm p_H\}$.

PROOF. If $\gamma \in \Gamma$ and $\rho_H(\gamma)p_H = \pm p_H$ then it is clear from the definition of ρ_H and p_H that the Lie subalgebra of $\mathfrak{G}$ corresponding to H is invariant under the adjoint action of γ; since H is connected, this implies that γ normalises H and hence $\gamma \in \Gamma_H$. Conversely, let $\gamma \in \Gamma_H$. Then γ normalises H and hence also $H \cap \Gamma$. Since $H \cap \Gamma$ is a lattice in H, this implies that the conjugation automorphism $h \mapsto \gamma h \gamma^{-1}$ preserves the Haar measure on H;

note that H is unimodular, as it admits a lattice. The assertion now follows from the well-known relation between the Haar measures and exterior forms.

3.2. **Proposition.** *Let $H \in \mathscr{H}$ and let A_H be the Zariski closure of $\eta_H(X(H, W))$ in V_H. Then $X(H, W) = \eta_H^{-1}(A_H) = \{g \in G \mid \eta_H(g) \in A_H\}$.*

Proof. Let $\mathfrak{H}$ and $\mathfrak{W}$ be the Lie subalgebras corresponding to H and W. Since W is a connected Lie subgroup we see that for $g \in G$, $g \in X(H, W)$ if and only if $\mathfrak{W} \subseteq (\operatorname{Ad} g)(\mathfrak{H})$ or equivalently if and only if $\rho_H(g)p_H \wedge w = 0$ for all $w \in \mathfrak{W}$. Thus $\rho_H(X(H, W))p_H$ is contained in the set $A_0 = \{v \in V_H \mid v \wedge w = 0 \ \forall w \in \mathfrak{W}\}$. Since A_0 is clearly a Zariski-closed subset of V_H the preceding observation implies that A_H is contained in A_0. Now let $x \in \eta_H^{-1}(A_H)$ be arbitrary. Then we have $\eta_H(x) = \rho_H(x)p_H \in A_0$ and hence $\rho_H(x)p_H \wedge w = 0$ for all $w \in \mathfrak{W}$. But this implies that $\mathfrak{W} \subseteq (\operatorname{Ad} x)(\mathfrak{H})$ and hence by the above observation $x \in X(H, W)$. This shows that $X(H, W) = \eta_H^{-1}(A_H)$, thus proving the proposition.

(Note: The above simple proof was pointed out by Shahar Mozes. The authors had earlier deduced the result using Proposition 6.7 of [2].)

Let $H \in \mathscr{H}$ and X be a subset of G. Consider the quotient map π_X : $X\Gamma_H/\Gamma_H \to X\Gamma/\Gamma$ defined by $g\Gamma_H \mapsto g\Gamma$ for all $g \in X$; $x \in X$ is said to be a point of (H, Γ)-*self-intersection* of X if $\pi_X^{-1}(\{x\Gamma\})$ has more than one point, namely if there exists a $\gamma \in \Gamma - \Gamma_H$ such that $x\gamma \in X$.

3.3. **Proposition.** *Let $H \in \mathscr{H}$ and let h be the dimension of H. Then the set of points of (H, Γ)-self-intersection of $X(H, W)$ is contained in the union of $X(H', W)$ over $H' \in \mathscr{H}$ of dimension at most $h - 1$.*

Proof. Let $H \in \mathscr{H}$ and $\gamma \in \Gamma$ be arbitrary. Since $H\Gamma$ and $\gamma H\Gamma$ are closed it follows that so is $(H \cap \gamma H \gamma^{-1})\Gamma$. Now let H_γ be the smallest closed (connected) subgroup of H containing all Ad-unipotent elements in $(H \cap \gamma H\gamma^{-1})^0$ and such that $H_\gamma\Gamma$ is closed; clearly H_γ is contained in $(H \cap \gamma H \gamma^{-1})^0$. The results in Section 2 of [36] then show that $H_\gamma \cap \Gamma$ is a lattice in H_γ and $\operatorname{Ad}(H_\gamma \cap \Gamma)$ is Zariski-dense in $\operatorname{Ad} H_\gamma$. Thus $H_\gamma \in \mathscr{H}$.

Now let g be any point of (H, Γ)-self-intersection of $X(H, W)$. Then there exists a $\gamma \in \Gamma - \Gamma_H$ such that $g^{-1}Wg \subseteq H \cap \gamma H \gamma^{-1}$. Since γ does not normalise H, $H \cap \gamma H \gamma^{-1}$ is a proper subgroup of H and hence so is the subgroup H_γ as defined above; since H is connected, this shows that H_γ is of dimension at most $h - 1$. Since $g^{-1}Wg \subseteq H \cap \gamma H\gamma^{-1}$, the definition of H_γ shows that $g^{-1}Wg \subseteq H_\gamma$ and hence $g \in X(H_\gamma, W)$. This proves the proposition.

3.4. **Theorem.** *Let the notation be as above. Then for any $H \in \mathscr{H}$, $\eta_H(\Gamma) = \rho_H(\Gamma)p_H$ is a closed (and hence discrete) subset of V_H.*

We defer the proof of this until Section 5. We note however that if G is a closed subgroup of $GL(n, \mathbb{R})$, for some $n \geqslant 2$, and Γ is contained in

$GL(n, \mathbb{Z})$, which is the case in our application to Diophantine approximation, the assertion follows readily from well-known results: In this case, for any $H \in \mathscr{H}$, $\tilde{H}$ is the connected component of the identity in an algebraic subgroup of $GL(n, \mathbb{R})$ defined over $\mathbb{Q}$ and hence so is $[\tilde{H}, \tilde{H}]$; since commutator subgroups of algebraic groups do not admit nontrivial characters it follows from a theorem of Borel and Harish-Chandra (cf. [4] and also [23, Theorem 3.11 and Remark 3.12 (II)]) that $[\tilde{H}, \tilde{H}] \cap \Gamma$ is a lattice in $[\tilde{H}, \tilde{H}]$. Using this together with the fact that $[\tilde{H}, \tilde{H}] = [H, H]$ (cf. [8, Chapter II, Theorem 13]) and considering the images of H and $H \cap \Gamma$ modulo that subgroup we see that the Lie subalgebra corresponding to any H in $\mathscr{H}$ is a rational subspace of $\mathscr{M}(n, \mathbb{R})$. It then follows that p_H is a multiple of a rational vector while $\rho_H(\Gamma)$ is given by integral matrices and hence the orbit $\rho_H(\Gamma)p_H$ is discrete.

3.5. Corollary. *Let $H \in \mathscr{H}$ and let A_H be the closure of $\eta_H(X(H, W))$ in V_H (in Zariski topology). Let D be a compact subset of A_H. Let Y_H be the set of points of (H, Γ)-self-intersection of $\eta_H^{-1}(D)$ and let K be a compact subset of $G - Y_H\Gamma$. Then there exists a neighbourhood Ω of D in V_H such that the quotient map of $(\eta_H^{-1}(\Omega)\Gamma_H \cap K\Gamma)/\Gamma_H$ onto $(\eta_H^{-1}(\Omega)\Gamma \cap K\Gamma)/\Gamma$ is injective; that is, $\eta_H^{-1}(\Omega) \cap K\Gamma$ has no points of (H, Γ)-self-intersection.*

Proof. We have to show that there exists a neighbourhood Ω of D in V such that for any $g, g' \in G$ such that $\eta_H(g), \eta_H(g') \in \Omega$ and $g\Gamma = g'\Gamma \in K\Gamma/\Gamma$ we have $g\Gamma_H = g'\Gamma_H$. Suppose this is not true. Then there exist sequences $\{g_i\}$ and $\{g_i'\}$ in G such that $g_i\Gamma = g_i'\Gamma \in K\Gamma/\Gamma$ and $g_i\Gamma_H \neq g_i'\Gamma_H$ for all i and $\{\eta_H(g_i)\}$ and $\{\eta_H(g_i')\}$ converge to some elements, say d and d' respectively, of D. Since K is compact, without loss of generality we may assume $\{g_i\Gamma\}$ to be convergent. Hence there exist a convergent sequence $\{h_i\}$ in G, say $h_i \to h \in G$, and two sequences $\{\gamma_i\}$ and $\{\gamma_i'\}$ in Γ such that $g_i = h_i\gamma_i$ and $g_i' = h_i\gamma_i'$ for all i. Now, we have $\rho_H(\gamma_i)p_H = \rho_H(h_i^{-1})\eta_H(g_i) \to \rho_H(h^{-1})d$ and similarly $\rho_H(\gamma_i')p_H \to \rho_H(h^{-1})d'$. Since by Theorem 3.4 $\rho_H(\Gamma)p_H$ is discrete, this implies that for all large i, $\rho_H(\gamma_i)p_H = \rho_H(h^{-1})d$ and $\rho_H(\gamma_i')p_H = \rho_H(h^{-1})d'$. Since $d, d' \in D$, it follows that $h\gamma_i, h\gamma_i' \in \eta_H^{-1}(D)$ for all large i. Since $h\Gamma \in K\Gamma/\Gamma \subseteq (G - Y_H\Gamma)/\Gamma$, the last condition implies that for all large i, $\gamma_i^{-1}\gamma_i' \in \Gamma_H$. But then we get that $g_i\Gamma_H = h_i\gamma_i\Gamma_H = h_i\gamma_i'\Gamma_H = g_i'\Gamma_H$ for all large i, which contradicts the choice of the sequences. This shows that there must exist a neighbourhood Ω as above.

§4. Proof of Theorem 1

In this section we give a proof of Theorem 1. It may be worth noting that the arguments in this section itself are of elementary nature; hence, in cases where the results referred to in earlier sections, in the course of the proof, are either known or can be obtained by elementary means, one gets an elementary proof of the theorem. We now begin the proof of Theorem 1 with a simple observation about values of polynomials, on which it is based.

4.1. LEMMA. *Let $n \in \mathbb{N}$ and $\varepsilon > 0$ be given. Let $\delta = \varepsilon^n/(n+1)n^n$. Let $k \geqslant 1$ and $\varphi_1, \dots, \varphi_k$ be polynomials of degree at most n such that $\sup|\varphi_i(0)| \geqslant 1$. Then for any $T > 0$,*

$$l(\{t \in [0, T] \mid |\varphi_i(t)| < \delta \ \forall i\}) < \varepsilon l(\{t \in [0, T] \mid |\varphi_i(t)| < 1 \ \forall i\}).$$

PROOF. We first note that for any polynomial φ of degree at most n and any $n+1$ distinct real numbers $\alpha_0, \alpha_1, \dots, \alpha_n$, by the Lagrange interpolation formula, we have

$$\varphi(x) = \sum_{m=0}^{n} \varphi(\alpha_m) \frac{(x-\alpha_0)\cdots(x-\alpha_{m-1})(x-\alpha_{m+1})\cdots(x-\alpha_n)}{(\alpha_m-\alpha_0)\cdots(\alpha_m-\alpha_{m-1})(\alpha_m-\alpha_{m+1})\cdots(\alpha_m-\alpha_n)}$$

and hence if $\alpha_m \in [0, 1]$ and $|\varphi(\alpha_m)| \leqslant \delta$ for all m and $|\alpha_p - \alpha_q| \geqslant \varepsilon/n$ for all distinct p and q then for all $x \in [0, 1]$ we have

$$|\varphi(x)| < (n+1)\delta(\varepsilon/n)^{-n} = 1.$$

Now let $\varphi_1, \dots, \varphi_k$ be k polynomials, where $k \geqslant 1$, as in the hypothesis. Let $T > 0$ be given. Put

$$E = \{t \in (0, T) \mid |\varphi_i(t)| < \delta \ \forall i\} \quad \text{and} \quad F = \{t \in (0, T) \mid |\varphi_i(t)| < 1 \ \forall i\}.$$

Let (a, b) be any connected component of F. Then there exists an i_0 such that $|\varphi_{i_0}(a)| = 1$; this follows from continuity and the fact that if $a = 0$ then by hypothesis $|\varphi_i(a)| \geqslant 1$ for some i. Let φ be the polynomial defined by $\varphi(t) = \varphi_{i_0}(a + (b-a)t)$ for all t. Then $|\varphi(0)| = 1$ and hence the above observation implies that there do not exist $(n+1)$ points $\alpha_0, \alpha_1, \dots, \alpha_n$ in $[0, 1]$ such that $|\varphi(\alpha_m)| \leqslant \delta$ for all m and $|\alpha_p - \alpha_q| \geqslant \varepsilon/n$ for all distinct p, q. Then the set $\{t \in [0, 1] \mid |\varphi(t)| < \delta\}$ must have Lebesgue measure less than ε. This yields that

$$l(\{t \in (a, b) \mid |\varphi_{i_0}(t)| < \delta\}) < \varepsilon(b-a).$$

Thus we get that $l(E \cap (a, b)) < \varepsilon(b-a)$ for any connected component (a, b) of F and hence $l(E) < \varepsilon l(F)$, which proves the lemma.

4.2. PROPOSITION. *Let V be a finite-dimensional real vector space and let A be an algebraic subvariety of V. Then for any compact subset C of A and any $\varepsilon > 0$ there exists a compact subset D of A such that the following holds: for any neighbourhood Φ of D in V there exists a neighbourhood Ψ of C in V such that for any unipotent one-parameter subgroup $\{u_t\}$ of $GL(V)$, any $v \in V - \Phi$, and any $T > 0$, we have*

$$l(\{t \in [0, T] \mid u_t v \in \Psi\}) < \varepsilon l(\{t \in [0, T] \mid u_t v \in \Phi\}).$$

PROOF. Since A is an algebraic subvariety of V there exists a polynomial function θ on V (namely a function given by a polynomial in the coordinate variables with respect to any fixed basis) such that $A = \{v \in V \mid \theta(v) = 0\}$; we note that though a priori a subvariety is in general defined as the set zeros of a finite set of polynomials, in the case of real vector spaces the sum of squares of the polynomials is a single polynomial which has the same set of zeros. Let d be the dimension of V, m be the degree of θ, and $n = 2dm$. Let ε be as in the hypothesis and let $\delta = \varepsilon^n/(n+1)n^n$.

We now equip V with a Hilbert norm $\|\cdot\|$. Let $r > 0$ be such that the ball of radius r with center at 0 contains C. We choose $D = \{a \in A \mid \|a\| \leqslant r/\sqrt{\delta}\}$. Now let Φ be any neighbourhood of D in V. Then there exists $\alpha > 0$ such that

$$\{v \in V \mid \|v\| < (r+\alpha)/\sqrt{\delta} \text{ and } \theta(v)| < \alpha\} \subseteq \Phi.$$

Choose

$$\Psi = \{v \in V \mid \|v\| < (r+\alpha) \text{ and } |\theta(v)| < \alpha\delta\}.$$

We show that the contention of the proposition holds for this choice. Let a unipotent one-parameter subgroup $\{u_t\}$ of $GL(V)$, a $v \in V - \Phi$, and a $T > 0$ be given. Let $\varphi_1(t) = \delta\|u_t v\|^2/(r+\alpha)^2$ and $\varphi_2(t) = \theta(u_t v)/\alpha$ for all $t \in \mathbb{R}$; since $\{u_t\}$ is a unipotent one-parameter subgroup it follows that φ_1 and φ_2 are polynomials in t. Both of them are of degree at most n; specifically, they are of degree at most $2d$ and dm respectively. Since $v \in V - \Phi$ we see that either $\varphi_1(0) = \delta\|v\|^2/(r+\alpha)^2 \geqslant 1$ or $|\varphi_2(0)| = |\theta(v)/\alpha| \geqslant 1$. We can therefore apply Lemma 4.1 to φ_1, φ_2 and the conclusion from the lemma then evidently yields the assertion as in the proposition.

Given a real vector space V and compact subsets C and D of V, with $C \subseteq D$, C is said to be of *relative size less than* ε *in* D if for any neighbourhood Φ of D in V there exists a neighbourhood Ψ of C in V such that for any unipotent one-parameter subgroup $\{u_t\}$ of $GL(V)$, any $v \in V - \Phi$, and any $T > 0$, we have

$$l(\{t \in [0, T] \mid u_t v \in \Psi\}) < \varepsilon l(\{t \in [0, T] \mid u_t v \in \Phi\}).$$

Proposition 4.2 asserts that given an $\varepsilon > 0$ and a compact subset C of an algebraic subvariety A there exists a compact subset D of A such that C is of relative size less than ε in D.

PROOF OF THEOREM 1. Let the notation G, Γ, etc. be as in the statement of the theorem. As before let $\mathscr{H}$ be the class of all proper closed connected subgroups of G such that $H \cap \Gamma$ is a lattice in H and $\mathrm{Ad}(H \cap \Gamma)$ is Zariski-dense in $\mathrm{Ad}\, H$. Also let $\mathscr{U}$ denote the class of all Ad-unipotent one-parameter subgroups of G. We will now carry out the proof in three steps.

i) *Some reductions*

Recall that by Propositions 2.1 and 2.3 $\mathscr{S}(W)$ is the union of the countably many subsets $X(H, W)\Gamma/\Gamma$, $H \in \mathscr{H}$. Further, each $X(H, W)$ is σ-compact and therefore $\mathscr{S}(W)$ can be expressed as $\bigcup C_i\Gamma/\Gamma$, where $\{C_i\}$

is a sequence of compact subsets such that each C_i is contained in $X(H, W)$ for some $H \in \mathscr{H}$. Let $\{X_j\}$ be a sequence of compact subsets whose interiors cover G/Γ. If we show that for each i and j there exists a neighbourhood Ω_{ij} of C_i such that $l(\{t \in [0, T] \mid u_t x \in \Omega_{ij} \cap X_j\}) \leqslant \varepsilon T/2^{i+j}$ for all $\{u_t\} \in \mathscr{U}$, $x \in F$, and $T \geqslant 0$, then the assertion in the theorem would follow for $\Omega = \bigcup(\Omega_{ij} \cap X_j)$, the union being taken over both i and j. Therefore to prove Theorem 1 it is enough to show that for any compact subsets K of G/Γ and C of $X(H, W)$, where $H \in \mathscr{H}$, and any $\varepsilon > 0$ there exists a neighbourhood Ω of $C\Gamma/\Gamma$ in G/Γ such that

$$l(\{t \in [0, T] | u_t x \in \Omega \cap K\}) \leqslant \varepsilon T \quad \forall \{u_t\} \in \mathscr{U}, \ x \in F, \ \text{and } T \geqslant 0. \quad (4.3)$$

We shall do this by induction on the dimension of H. For low dimensions (less than that of W) the assertion follows trivially. Now assume it to be true when the dimension is at most $n - 1$ and let C be a compact subset of $X(H, W)$, where $H \in \mathscr{H}$ is of dimension n. Let K be any compact subset of G/Γ. Let Y be the set of points of (H, Γ)-self-intersection of $X(H, W)$. Then by the induction hypothesis and Proposition 3.3 (and a countable union argument) there exists an open neighbourhood Ω' of $Y\Gamma/\Gamma$ such that

$$l(\{t \in [0, T] | u_t x \in \Omega' \cap K\}) \leqslant \varepsilon T/2 \quad \forall \{u_t\} \in \mathscr{U}, \ x \in F, \ \text{and } T \geqslant 0. \quad (4.4)$$

ii) *Linearisation*

Now let $V = V_H$, $\rho = \rho_H$, $p = p_H$, $\eta = \eta_H$, and $A = A_H$, where the right-hand side terms are as defined in Section 3; we recall in particular that A is the Zariski-closure of $\eta(X(H, W))$ in V and by Proposition 3.2 $X(H, W) = \eta^{-1}(A)$. By Proposition 4.2 there exists a compact subset D of A such that $\eta(C)$ is of relative size less than $\varepsilon/4$ in D. We now choose a neighbourhood Φ of D as follows.

Let F' be a compact subset of G such that $F'\Gamma/\Gamma = F$. By Theorem 3.4 $\eta(\Gamma) = \rho(\Gamma)p$ is closed and hence so is $\eta(F'\Gamma) = \rho(F'\Gamma)p = \rho(F')\rho(\Gamma)p$. Since $F \subseteq \mathscr{G}(W)$, $F'\Gamma$ is disjoint from $X(H, W)$. Since $X(H, W) = \eta^{-1}(A)$ this implies that $\eta(F'\Gamma)$ is disjoint from A and hence from D. Since $\eta(F'\Gamma)$ is closed and D is compact, there exists a neighbourhood Φ' of D in V, which is disjoint from $\eta(F'\Gamma)$.

Next let $K_1 = K - \Omega'$. Then K_1 is a compact subset of $G/\Gamma - Y\Gamma/\Gamma$. Let K' be a compact subset of G such that $K'\Gamma/\Gamma = K_1$. Then by Corollary 3.5 there exists a neighbourhood Φ of D in V such that the quotient map of $(\eta^{-1}(\Phi)\Gamma_H \cap K'\Gamma)/\Gamma_H$ onto $(\eta^{-1}(\Phi)\Gamma \cap K'\Gamma)/\Gamma$ is injective; we choose an open neighbourhood Φ of D contained in Φ', for which this condition holds.

We note that since $\rho = \rho_H$ is an exterior power of the adjoint representation of G, for any $\{u_t\} \in \mathscr{U}$, $\{\rho(u_t)\}$ is a unipotent one-parameter subgroup of $GL(V)$. Since $\eta(C)$ is of relative size less than $\varepsilon/4$ in D there exists a neighbourhood Ψ of $\eta(C)$ in V such that

$$l(\{t \in [0, T] \mid \rho(u_t)v \in \Psi\}) < \varepsilon\, l(\{t \in [0, T] \mid \rho(u_t)v \in \Phi\})/4 \tag{4.5}$$

for all $\{u_t\} \in \mathscr{U}$, $v \in V - \Phi$, and $T > 0$. Now let $\Omega = \eta^{-1}(\Psi)\Gamma/\Gamma$. We shall show that (4.3) holds for this Ω.

iii) *Completion of the proof*

Let $\{u_t\} \in \mathscr{U}$, $x \in F$, and $T \geqslant 0$ be given. Let $g \in F'$ be such that $x = g\Gamma$. For any $q \in \eta(\Gamma) = \rho(\Gamma)p$ let

$$I(q) = \{t \in (0, T) | \rho(u_t g)q \in \Phi\} \text{ and}$$
$$J(q) = \{t \in (0, T) | \rho(u_t g)q \in \Psi \text{ and } u_t g\Gamma \in K_1\}.$$

Also for each $q \in \eta(\Gamma)$ let $I'(q)$ be the subset of $I(q)$ consisting of all t for which there exists $a \geqslant 0$ such that $[t, t+a] \subseteq I(q)$ and $u_{t+a}g\Gamma \in K_1$. Clearly, $J(q) \subseteq I'(q)$ for all $q \in \eta(\Gamma)$.

We shall show that for $q_1, q_2 \in \eta(\Gamma)$ the sets $I'(q_1), I'(q_2)$ are disjoint unless $q_1 = \pm q_2$. Let $q_1, q_2 \in \eta(\Gamma)$ and suppose that there exists a $t \in I'(q_1) \cap I'(q_2)$. Then there exists $\alpha \geqslant 0$ such that $[t, t+\alpha] \subseteq I(q_1) \cap I(q_2)$ and $u_{t+\alpha}g\Gamma \in K_1$. Let $\gamma_1, \gamma_2 \in \Gamma$ be such that $q_1 = \eta(\gamma_1)$ and $q_2 = \eta(\gamma_2)$. Then from the definition of $I(q)$'s we get that $\eta(u_{t+\alpha}g\gamma_1), \eta(u_{t+\alpha}g\gamma_2) \in \Phi$. Since $u_{t+\alpha}g\Gamma \in K_1 = K'\Gamma/\Gamma$ and the quotient map of $(\eta^{-1}(\Phi)\Gamma_H \cap K'\Gamma)/\Gamma_H$ onto $(\eta^{-1}(\Phi)\Gamma \cap K'\Gamma)/\Gamma$ is injective, we get that $\gamma_1\Gamma_H = \gamma_2\Gamma_H$. Hence by Lemma 3.1 $q_1 = \rho(\gamma_1)p = \pm\rho(\gamma_2)p = \pm q_2$.

Now let $q \in \eta(\Gamma)$ and (a, b) be any connected component of $I(q)$. We note that $\rho(u_a g)q \notin \Phi$; this is obvious if $a > 0$ and for $a = 0$ it follows from the fact that Φ is disjoint from $\eta(F'\Gamma)$. If $J(q) \cap (a, b)$ is nonempty then $u_t g\Gamma \in K_1$ for some $t \in (a, b)$ and hence $I'(q)$ contains an interval of the form (a, c), where $a < c \leqslant b$, such that $J(q) \cap (a, b)$ is contained in $(a, c]$. Hence by (4.5), applied to $\rho(u_a g)q$ and $c - a$ in the place of v and T there, respectively, we have

$$l(J(q) \cap (a, b)) \leqslant \varepsilon l(I'(q) \cap (a, b))/4.$$

Since this holds for all connected components of $I(q)$ it follows that $l(J(q)) \leqslant \varepsilon l(I'(q))/4$. Then we have

$$\begin{aligned} &l(\{t \in [0, T] \mid u_t g\Gamma \in \Omega \cap K_1\}) \\ &\quad = l\Big(\bigcup_{q \in \eta(\Gamma)} J(q)\Big) \leqslant \varepsilon l\Big(\bigcup_{q \in \eta(\Gamma)} I'(q)\Big)/4 \leqslant \varepsilon T/2, \end{aligned}$$

where the last step follows from the fact that for $q_1, q_2 \in \eta(\Gamma)$, $I'(q_1)$, $I'(q_2)$ are disjoint unless $q_1 = \pm q_2$. Since $K_1 = K - \Omega'$, (4.4) and the last conclusion together imply (4.3), for the given $\{u_t\} \in \mathscr{U}$, $x = g\Gamma \in F$, and $T \geqslant 0$. As noted earlier, this proves the theorem.

§5. Volumes of homogeneous subsets

In this section we prove a result on the volumes of homogeneous subsets and deduce Theorem 3.4.

Let G be a connected Lie group and Γ be a discrete subgroup of G. We equip G with a Riemannian metric invariant under the G-action on the right. This yields a Riemannian metric on G/Γ. In the following discussion all distances, volumes, etc. on submanifolds of G or G/Γ are meant to be the ones arising from the Riemannian metrics as above. We note that if H is a closed subgroup of G such that $H \cap \Gamma$ is a lattice in H then $H\Gamma/\Gamma$ is a Riemannian submanifold of finite Riemannian volume and the Riemannian measure on $H\Gamma/\Gamma$ is H-invariant for the H-action on the left; this follows from the unimodularity of H.

5.1. THEOREM. *Let G and Γ be as above. Let $c > 0$ and let $\mathscr{V}_c$ be the class of all closed connected subgroups H of G such that $H\Gamma/\Gamma$ is closed and has volume at most c. Then there are only finitely many subgroups of the form $H \cap \Gamma$ with $H \in \mathscr{V}_c$ (that is, $\{H \cap \Gamma | H \in \mathscr{V}_c\}$ is finite). Further, given a representation $\rho : G \to GL(V)$ (over a finite-dimensional real vector space) such that $\ker \rho$ is contained in the center of G, there are only finitely many closed subgroups H in $\mathscr{V}_c$ for which $\rho(H \cap \Gamma)$ is Zariski-dense in $\rho(H)$.*

PROOF. We present the proof in four steps.

i) *A limit subgroup*

Let, if possible, $\{H_i\}$ be a sequence in $\mathscr{V}_c$ such that $H_i \cap \Gamma$, $i = 1, 2, \dots$, are distinct subgroups. We shall show that this leads to a contradiction. Let $\mathfrak{G}$ be the Lie algebra of G and for any $i \geqslant 1$ let $\mathfrak{H}_i$ be the Lie subalgebra of $\mathfrak{G}$ corresponding to H_i. Passing to a subsequence if necessary, we may assume that $\mathfrak{H}_i$ are all of the same dimension, say h, and that, as $i \to \infty$, they converge in the Grassmannian manifold of h-dimensional subspaces of $\mathfrak{G}$, say $\mathfrak{H}_i \to \mathfrak{L}$. It is easy to see that $\mathfrak{L}$ is a Lie subalgebra of $\mathfrak{G}$. Let L be the connected Lie subgroup of G with Lie subalgebra $\mathfrak{L}$.

Passing to a subsequence again, we may also assume that the sequence $\{H_i\}$ converges in the Hausdorff topology on the space of closed subsets of G, namely, that there exists a closed subset H_∞ of G such that for any compact subset K of G, the Hausdorff distance between $H_i \cap K$ and $H_\infty \cap K$ tends to 0. It is clear that H_∞ is a subgroup of G. Since the exponential map $\exp\colon \mathfrak{G} \to G$ restricts to a diffeomorphism in a neighbourhood of 0 in $\mathfrak{G}$, it follows that L is contained in H_∞. We claim that L is the connected component of the identity in H_∞.

Let e be the identity of G. Since Γ is a discrete subgroup of G there exists a neighbourhood Ω of e such that $\Omega^{-1}\Omega \cap \Gamma = \{e\}$. We choose an open relatively compact exponential neighbourhood Ω for which this holds; by an exponential neighbourhood we mean one of the form $\exp \Omega_0$, where

Ω_0 is a neighbourhood of 0 in $\mathfrak{G}$ on which the exponential map is a diffeomorphism. Let Ψ be a compact neighbourhood of e contained in Ω and let $\delta = d(\Psi, G - \Omega)/2 > 0$, where d is the Riemannian distance function on G. There exists an $\alpha > 0$ such that for any complete Riemannian submanifold M of G and any $x \in \Psi \cap M$ the volume of the δ-neighbourhood of x in M (with respect to the induced metric) is at least α. Now let $i \geqslant 1$ and let E be any set of points in $\Psi \cap H_i$ such that for any two distinct points of E the δ-neighbourhoods in H_i are disjoint. If Δ is the union of the δ-neighbourhoods of points in E then $\Delta \subseteq \Omega$ and, since $\Omega^{-1}\Omega \cap \Gamma = \{e\}$, the (total) volume of Δ is at most $\operatorname{vol} H_i\Gamma/\Gamma \leqslant c$. It follows that the cardinality of any such set E is at most c/α. For each i let E_i be a maximal subset of $\Psi \cap H_i$ with the above property. Then the 2δ-neighbourhood of E_i contains $\Psi \cap H_i$. Passing to a subsequence, without loss of generality we may assume that $\{E_i\}$ converges to a set, say E. Clearly E is a finite set with cardinality at most c/α. Since Ω is contained in the exponential neighbourhood and L is the Grassmannian limit of $\{H_i\}$ it follows that the Hausdorff limit of the sequence of 2δ-neighbourhoods of e in H_i is contained in L. Since the 2δ-neighbourhoods of E_i's contain $\Psi \cap H_i$ the preceding two observations imply that $\Psi \cap H_\infty$ is contained in LE. Since E is finite it follows that L is the connected component of the identity in H_∞.

Passing to a subsequence once more, we may also assume that for any compact subset K of G, $K \cap H_i \cap \Gamma$ is the same subset for all large i; namely that there exists an i_0 depending on K such that $K \cap H_i \cap \Gamma = K \cap H_{i_0} \cap \Gamma$ for all $i \geqslant i_0$. Let $\Gamma_1 = \{\gamma \in L \mid \gamma \in H_i \cap \Gamma \text{ for all large } i\}$. We now prove the following:

ii) Γ_1 *is a lattice in* L

Since H_i, $i = 1, 2, \ldots$, admit lattices, they are unimodular subgroups of G. Let λ_i, $i = 1, 2, \ldots$, and λ be the Haar measures on H_i and H_∞ respectively, where the appropriate choice up to scalars is made so that they correspond to the volume elements on the subgroups, induced by the metric on G. Then $\lambda_i \to \lambda$, in the weak* topology on the space of measures on G.

By a standard argument (see the proof of Chabouty's theorem [29, Theorem 1.20]) one can see that there exists an open subset Φ of L such that $\Phi\gamma$, $\gamma \in \Gamma_1$, are mutually disjoint and $\lambda(L - \Phi\Gamma_1) = 0$. Now let K be any compact subset of Φ. Since L is open in H_∞ there exists a compact neighbourhood Θ of K in G such that $\Theta^{-1}\Theta \cap H_\infty \subseteq \Phi^{-1}\Phi$. Since $\Phi\gamma$, $\gamma \in \Gamma_1$, are mutually disjoint, it now follows that for all large i, $\Theta\gamma$, $\gamma \in H_i \cap \Gamma$, are mutually disjoint; otherwise there would exist a sequence $\{\gamma_i\}$ consisting of nontrivial elements such that $\gamma_i \in (H_i \cap \Gamma) \cap \Theta^{-1}\Theta$ for all i; by the compactness of Θ it has only finitely many distinct elements and the assumption made at the end of step i) shows that the element occurring infinitely many times must belong to H_∞; the condition in the choice of Θ

then implies that it belongs to $\Phi^{-1}\Phi$, contradicting the disjointness of $\Phi\gamma$, $\gamma \in \Gamma_1$. Since $\lambda_i \to \lambda$ and Θ contains a neighbourhood of K, it follows that $\lambda(K) \leqslant \liminf \lambda_i(\Theta)$. Since $\mathrm{vol}(H_i\Gamma/\Gamma) \leqslant c$ and $\Theta\gamma$, $\gamma \in H_i \cap \Gamma$, are mutually disjoint, it follows that $\lambda_i(\Theta) \leqslant c$, for all i. Thus, $\lambda(K) \leqslant c$. Since this holds for any compact subset K of Φ, we get that $\lambda(\Phi) \leqslant c$. Since $\lambda(L - \Phi\Gamma_1) = 0$, it now follows that the L-invariant measure on L/Γ_1 is finite. Thus Γ_1 is a lattice in L.

iii) *Proof of the first assertion*

Since Γ_1 is a lattice in L, in particular, it is a finitely generated group (cf. [29, Remark 13.21] and [9, Lemma 9.1]). Since any $\gamma \in \Gamma_1$ is contained in H_i for all large i, this implies that Γ_1 is contained in H_i for all large i, say for all $i \geqslant i_0$. Let S be a connected component of the identity in the Zariski-closure of $\mathrm{Ad}(\Gamma_1)$ in $GL(\mathfrak{G})$, where as before Ad denotes the adjoint representation of G. Let $i \geqslant i_0$ be arbitrary. We then have $[S, S] \subseteq [\mathrm{Ad}(\tilde{H_i}), \mathrm{Ad}(\tilde{H_i})] = [\mathrm{Ad}(H_i), \mathrm{Ad}(H_i)]$, where the latter relation follows from Theorem 13 of Chapter II in [8]. In particular, $[S, S] \subseteq \mathrm{Ad}(H_i) \subseteq \mathrm{Ad}(G)$ (though, a priori, S may not be contained in $\mathrm{Ad}(G)$). Let D be the smallest closed subgroup of G such that $\Gamma_1 \subseteq D$ and $[S, S] \subseteq \mathrm{Ad}(D)$. Clearly D is contained in H_i for all $i \geqslant i_0$. By a similar argument we also see that it is contained in L. Since Γ_1 is a lattice in L, by Borel's density theorem (cf. [10]) every connected noncompact simple Lie subgroup of $\mathrm{Ad}(L)$ is contained in S and hence also in $[S, S]$. Since $[S, S]$ is contained in $\mathrm{Ad}(D)$ this implies that D contains every connected noncompact simple Lie subgroup of L. Let M be the smallest closed subgroup containing all connected noncompact simple Lie subgroups of L. Then M is a normal subgroup of L contained in D. Since Γ_1 is a lattice in L, L/D admits a finite L-invariant measure. Now, L/M is a Lie group with no noncompact simple Lie subgroups, acting transitively on L/D and therefore the last observation implies that L/D is compact (cf. [29, Corollary 8.25]). Since L is the connected component of the identity in the Hausdorff limit of $\{H_i\}$, this implies that there exists a compact subset of G/D containing H_i/D for all large i. Since $\Gamma_1 \subseteq D \subseteq L$ and Γ_1 is a lattice in L it is also a lattice in D. Since $D \subseteq H_i$ for all $i \geqslant i_0$ we see that $D \cap \Gamma = \Gamma_1$ and since it is a lattice in D, in particular ΓD is closed. Now let $\{\gamma_i\}$ be any sequence such that $\gamma_i \in H_i \cap \Gamma$ for all i. Then $\{\gamma_i D\}$ is contained in a compact subset of G/D and therefore has a subsequence, say $\{\gamma_{i_j} D\}$, converging to an element in L/D. Since ΓD is closed, and hence a discrete union of cosets of D, there must then exist a $\gamma \in L \cap \Gamma$ such that for each j, $\gamma_{i_j} = \gamma d_j$ for some $d_j \in D$. Then $\gamma = \gamma_{i_j} d_j^{-1} \in H_{i_j}$ for all large j. Recall that any element of Γ which belongs to infinitely many H_i's belongs to all of them after some stage (see step i)). Hence $\gamma \in \Gamma_1$. Since $d_j \in D \cap \Gamma = \Gamma_1$ this implies that the subsequence $\{\gamma_{i_j}\}$ is contained in Γ_1. We have thus shown that any sequence $\{\gamma_i\}$ such that $\gamma_i \in H_i \cap \Gamma$ for all i has a subsequence contained in Γ_1. This is impossible unless $H_i \cap \Gamma = \Gamma_1$ for infinitely many i. But this contradicts the assumption that $H_i \cap \Gamma$, $i \geqslant 1$, are all distinct. This proves the first part of the theorem.

iv) *Completion of the proof*

To prove the remaining part it is now enough to show that for any subgroup Δ of Γ the class of all closed subgroups H in $\mathscr{V}_c$ such that $\rho(\Delta)$ is Zariski-dense in $\rho(H)$ and $H \cap \Gamma = \Delta$ is finite; fix such a subgroup Δ and let $\mathscr{F}$ be the corresponding class subgroups H. Let $T = \widetilde{\rho(\Delta)}$. Let $H \in \mathscr{F}$ be arbitrary. Then we have $\widetilde{\rho(H)} = T$. Let C be the connected component of the identity in $\rho^{-1}(T)$. Then H is contained in C. Since H is connected, $[\rho(H), \rho(H)] = [T, T]$ (cf. [8, Chapter II, Theorem 13]) and hence we also have $[\rho(H), \rho(H)] = [\rho(C), \rho(C)]$. Let $C_1 = \overline{[C, C]}$ and $C_2 = \overline{[C, C_1]}$. Then by the preceding conclusion C_1 is contained in $\overline{[H, H](\ker \rho)}$ and hence in $\overline{HZ}$, where Z is the center of G, and consequently C_2 is contained in H. Since C/C_2 is a nilpotent Lie group, there exists a unique closed connected subgroup B of C such that $\Delta C_2 \subset B$, $B/\overline{\Delta C_2}$ is compact, and B/C_2 contains the (unique) maximal compact (central) subgroup of C/C_2; this can be deduced by applying Theorem 2.1 of [29] to the quotient of C/C_2 by the maximal compact subgroup; the uniqueness implies in particular that the image of H in the same quotient must coincide with that of B and hence we get that H is contained in B. Now let $N = \overline{\Delta C_2}^0$. Then, again by Theorem 2.1 of [29], N is a normal subgroup of B. Now B/N is a nilpotent Lie group, $\Delta' := \Delta N/N$ is a lattice in B/N, and H/N is a closed subgroup of B/N containing Δ'. This implies that the image of H/N in the simply connected quotient group of B/N by its maximal compact central subgroup must be the entire quotient group. Therefore H/N contains $[B/N, B/N]$ and hence H contains $\Delta[B, B]$. One can see that the latter is a closed subgroup (alternatively, one can argue using its closure). It is easy to see, using the Fubini-Weil formula (cf. [38, Chapter II, Section 9]), that if we fix a Riemannian metric on $B/\Delta[B, B]$ appropriately then for any $H \in \mathscr{F}$ the volume of $H/\Delta[B, B]$ is bounded by a fixed constant depending on c. The finiteness of $\mathscr{F}$ therefore follows from the fact that in the torus $B/\Delta[B, B]$ there are only finitely many closed subgroups with volumes bounded by a given constant.

(Note: The authors originally proved the first assertion in Theorem 5.1 and deduced Theorem 3.4 for a class of subgroups H, including all almost algebraic subgroups, which was large enough to be adequate in proving the other theorems where it is applied. In this context it was observed by Nimish Shah that the first part implied the second, for the adjoint representations. The proof of the second assertion in step iv) is a variation of his argument.)

PROOF OF THEOREM 3.4. Let $H \in \mathscr{H}$ and $\gamma \in \Gamma$ be arbitrary. Let $\rho_H : G \to GL(V_H)$ be the representation associated to H. Let $x_1, x_2, \ldots, x_n$

be an orthonormal basis of $\mathfrak{G}$, with respect to the inner product corresponding to the Riemannian metric on G. We equip V_H with the norm for which $\{x_{i_1} \wedge \cdots \wedge x_{i_h} \mid 1 \leqslant i_1 < i_2 < \cdots < i_h \leqslant n\}$ is an orthonormal basis; the norm on V_H is independent of the choice of the orthonormal basis $x_1, x_2, \ldots, x_n$. It can be verified that for any $g \in G$, $\|\rho_H(g)p_H\| = \delta_g \|p_H\|$, where δ_g is the factor by which volumes of subsets get multiplied under the transformation $h \mapsto ghg^{-1}$ for all $h \in H$. If D is a fundamental domain for $H \cap \Gamma$ in H then $\gamma D \gamma^{-1}$ is a fundamental domain for $\gamma H \gamma^{-1} \cap \Gamma$ in $\gamma H \gamma^{-1}$. It follows therefore that for any $\gamma \in \Gamma$ as above, $\operatorname{vol} \gamma H\Gamma/\Gamma = \delta_\gamma \operatorname{vol} H\Gamma/\Gamma = \|\rho_H(\gamma)p_H\| \|p_H\|^{-1} \operatorname{vol} H\Gamma/\Gamma$. Now let B be any ball in V_H. Then there exists a $c > 0$ such that if $\gamma \in \Gamma$ and $\rho_H(\gamma)p_H \in B$ then $\operatorname{vol} \gamma H\Gamma/\Gamma \leqslant c$. Since $\operatorname{Ad}(H \cap \Gamma)$ is Zariski-dense in $\operatorname{Ad} H$ by Theorem 5.1 we get that $\{\gamma H \gamma^{-1} \mid \rho_H(\gamma)p_H \in B\}$ is finite. If γ_1 and γ_2 are such that $\gamma_1 H \gamma_1^{-1} = \gamma_2 H \gamma_2^{-1}$ and $\gamma = \gamma_1^{-1}\gamma_2$, then γ normalises H and further, since $H \cap \Gamma$ is a lattice in H, the automorphism $h \mapsto \gamma h \gamma^{-1}$ is volume-preserving. These conditions imply that $\rho_H(\gamma)p_H = \pm p_H$ and in turn that $\rho_H(\gamma_1)p_H = \pm \rho_H(\gamma_2)p_H$. In view of the finiteness assertion above this implies that $B \cap \rho_H(\Gamma)p_H$ is finite. This implies the contention in the theorem.

§6. Proof of Theorem 2

In this section we deduce Theorem 2 from Theorem 1, using Ratner's classification of invariant measures of unipotent flows (cf. [32, Theorem 1]) and the following variation of the results in [12] and [14].

6.1. Theorem. *Let G be a connected Lie group and Γ be a lattice in G. Let F be a compact subset of G/Γ and let $\varepsilon > 0$ be given. Then there exists a compact subset K of G/Γ such that for any* Ad*-unipotent one-parameter subgroup $\{u_t\}$ of G, any $x \in F$, and $T \geqslant 0$,*

$$l(\{t \in [0, T] \mid u_t x \in K\}) \geqslant (1 - \varepsilon)T.$$

Proof. In view of Lemma 9.1 of [9] and theorem 5.22 of [29], in proving the theorem we may without loss of generality assume G to be a semisimple Lie group with trivial center and no (nontrivial) compact factors and Γ to be an irreducible lattice in G. Since the assertion of the theorem is trivial if G/Γ is compact, using the arithmeticity theorem in the form as in [22] and arguing as in [16, Appendix], we may further assume that either G is of $\mathbb{R}$-rank 1 or Γ is arithmetic with respect to a $\mathbb{Q}$-structure on G. In the latter case G can be realised as a closed subgroup of $SL(n, \mathbb{R})$, for some n, in such a way that a subgroup Γ' of finite index in Γ is contained in $SL(n, \mathbb{Z})$ and the induced map of G/Γ' into $SL(n, \mathbb{R})/SL(n, \mathbb{Z})$ is a proper map (cf. [29, Proposition 10.15]). Hence it is enough to prove the theorem when either $G = SL(n, \mathbb{R})$ and $\Gamma = SL(n, \mathbb{Z})$, for some $n \geqslant 2$, or G is of $\mathbb{R}$-rank 1. In the former case the theorem then follows from Proposition 2.7

of [14] together with the Mahler criterion (cf. [29, Corollary 10.9]), if the choices involved in the proposition are made as follows: $h > 1$ and $k > 1$ satisfying condition (2.3) of [14] for the given ε; $S = \varnothing$ (the empty set); $\theta > 0$ such that (in the notation of [14]) $d^2(\Delta) \geqslant \theta$ for any subgroup Δ of $g\mathbb{Z}^n$ for any $g \in G = SL(n, \mathbb{R})$ such that $g\Gamma \in F$ (where $\Gamma = SL(n, \mathbb{Z})$ and F is the given compact subset of G/Γ); $\alpha > 0$ and $\beta > 0$ arbitrary; $a = 0$ and $b = k^{-1}T$. With these choices the proposition shows that the contention as in the statement of the theorem holds for the set K consisting of all $g\Gamma$ where $g \in G$ is such that $g\mathbb{Z}^n$ has no nonzero point in the δ_0-neighbourhood of 0, for the $\delta_0 > 0$ as in the proposition; by the Mahler criterion the set is compact. This proves the theorem for the case at hand. A study analogous to that in [14] for lattices in $\mathbb{R}$-rank 1 groups was carried out in [12] and though there is no explicit statement there to which we could refer to for our present purpose, an examination of the proof of Proposition 1.2 there readily shows that the contention of the theorem holds in this case as well; we omit the details.

PROOF OF THEOREM 2. Let the notation be as in the hypothesis. We shall first show that without loss of generality we may assume that $x_i \in \mathscr{G}(\{u_t\})$ for all i. Let φ be any bounded continuous function on G/Γ. Let $x \in \mathscr{G}(\{u_t\})$ be the limit point of $\{x_i\}$ and let $\{\Omega_i\}$ be a fundamental system of open neighbourhoods of x such that $x_i \in \Omega_i$ for all i. For each i let $\Theta_i = \{y \mid |\varphi(u_t^{(i)}y) - \varphi(u_t^{(i)}x_i)| < 2^{-i}\ \forall t \in [0, T_i]\}$. Then for each i, $\Theta_i \cap \Omega_i$ is a neighbourhood of x_i. Since $x \in \mathscr{G}(\{u_t\})$, in particular the latter is nonempty and hence by Propositions 2.1 and 2.3 it is dense in G/Γ. Therefore for any i we can find a $y_i \in \mathscr{G}(\{u_t\})$ such that $y_i \in \Theta_i \cap \Omega_i$. Since $\{\Omega_i\}$ is a fundamental system of neighbourhoods of x it follows that $y_i \to x$. Also since $y_i \in \Theta_i$ for all i, we have

$$\left|\frac{1}{T_i}\int_0^{T_i} \varphi(u_t^{(i)}x_i)\,dt - \frac{1}{T_i}\int_0^{T_i} \varphi(u_t^{(i)}y_i)\,dt\right| < 2^{-i}.$$

Therefore the assertion as in the theorem holds for $\{x_i\}$ if and only if it holds for $\{y_i\}$ and hence in proving the theorem we may without loss of generality assume that $x_i \in \mathscr{G}(\{u_t\})$ for all i.

Now let $C_c(G/\Gamma)$ be the space of continuous functions with compact support on G/Γ. We define a sequence $\{\lambda_i\}$ of probability measures on G/Γ by setting

$$\int \varphi\, d\lambda_i = \frac{1}{T_i}\int_0^{T_i} \varphi(u_t^{(i)}x_i)\,dt$$

for all $i = 1, 2, \ldots$ and $\varphi \in C_c(G/\Gamma)$; in other words, λ_i is the normalised Lebesgue measure along the orbit segment $\{u_t^{(i)}x_i \mid t \in [0, T_i]\}$. We shall show that $\{\lambda_i\}$ converges to μ, in the usual weak* topology on the space of bounded measures on G/Γ. This implies the assertion as in the theorem for continuous functions with compact support and a standard measure-theoretic argument then shows that it must hold for all bounded continuous functions.

Let Z denote the space G/Γ if the latter is compact and the one-point compactification of G/Γ if it is not compact; the point at infinity, if any, will be denoted by ∞. Any finite measure on G/Γ can be viewed canonically as a measure on Z by assigning 0 mass to the point at infinity, if any. Since λ_i, $i = 1, 2, \ldots$, are probability measures and the space of probability measures on a compact space is compact with respect to the weak* topology, to prove that $\lambda_i \to \mu$ it is enough to show that any measure λ on Z which is a limit point of the sequence $\{\lambda_i\}$ is none other than μ. Let λ be such a limit point. Then there exists a subsequence of $\{\lambda_i\}$ converging to λ. We may now assume that $\lambda_i \to \lambda$; there is no loss of generality in this, since in proving that λ as above coincides with μ, we may ignore the elements of the sequence outside the convergent subsequence as above.

Since $\{x_i\}$ is a convergent sequence, by Theorem 6.1, for any $j = 1, 2, \ldots$ there exists a compact subset K_j of G/Γ such that for all $i = 1, 2, \ldots$ we have

$$l(\{t \in [0, T_i] \mid u_t^{(i)} x_i \in K_j\}) \geqslant (1 - 2^{-j})T_i.$$

This implies that $\lambda_i(K_j) \geqslant (1 - 2^{-j})$ for all i and j and in turn $\lambda(K_j) \geqslant (1 - 2^{-j})$ for all j. Therefore $\lambda(G/\Gamma) = 1$; that is, the point at infinity, if any, has zero mass under λ and the latter may be viewed as a probability measure on G/Γ.

We see also that λ is invariant under the action of $\{u_t\}$. Let $s > 0$ and $\varphi \in C_c(G/\Gamma)$ be arbitrary. Then for any $i = 1, 2, \ldots$ we have

$$\begin{aligned} \left|\int \varphi(u_s y)\, d\lambda(y) - \int \varphi\, d\lambda\right| \leqslant & \left|\int \varphi(u_s y)\, d\lambda(y) - \int \varphi(u_s y)\, d\lambda_i(y)\right| \\ & + \left|\int \varphi(u_s y)\, d\lambda_i(y) - \int \varphi(u_s^{(i)} y)\, d\lambda_i(y)\right| \\ & + \left|\int \varphi(u_s^{(i)} y)\, d\lambda_i(y) - \int \varphi\, d\lambda_i\right| \\ & + \left|\int \varphi\, d\lambda_i - \int \varphi\, d\lambda\right|. \end{aligned}$$

Let $\varepsilon > 0$ be arbitrary. Since $\lambda_i \to \lambda$, for all large i the first and last terms on the right-hand side are at most $\varepsilon/4$. Since λ_i are all probability measures, $u_s^{(i)} \to u_s$, and φ is uniformly continuous, it follows that the second term is less than $\varepsilon/4$ for all large i. The third term is clearly the same, by the definition of λ_i's, as $T_i^{-1}\left|\int_{T_i}^{T_i+s} \varphi(u_t^{(i)} y)\, dt - \int_0^s \varphi(u_t^{(i)} y)\, dt\right|$, which is majorised by $2sT_i^{-1}M$, where M is a bound for $|\varphi|$ on G/Γ. Since $T_i \to \infty$ it follows that the third term is also less than $\varepsilon/4$ for all large i. Thus we get that for any $\varepsilon > 0$ the right-hand side of the above inequality is less than ε for all large i and hence it follows that the left-hand side, which is independent of i, must be 0. Since this holds for any $s > 0$ and $\varphi \in C_c(G/\Gamma)$ it follows that λ is $\{u_t\}$-invariant.

We next show, using Theorem 2, that $\lambda(\mathscr{S}(\{u_t\})) = 0$. In Theorem 1 choose W to be $\{u_t\}$ and F to be the set consisting of the sequence $\{x_i\}$ together with the limit point x; recall that by a choice made in the beginning of the proof, $x_i \in \mathscr{G}(\{u_t\})$ and since by hypothesis $x \in \mathscr{G}(\{u_t\})$ it follows that F is contained in $\mathscr{G}(\{u_t\})$ as required in the hypothesis of the theorem. The theorem then yields that for any $j = 1, 2, \ldots$ there exists a neighbourhood Ω_j of $\mathscr{S}(\{u_t\})$ such that for any $i = 1, 2, \ldots$ and $T > 0$, $l(\{t \in [0, T] \mid u_t^{(i)} x_i \in \Omega_j\}) < T/2^j$. In particular this implies that $\lambda_i(\Omega_j) < 2^{-j}$ for all $i, j = 1, 2, \ldots$. Hence $\lambda(\Omega_j) \leqslant 2^{-j}$ for all j and in turn $\lambda(\mathscr{S}(\{u_t\})) = 0$.

Since λ is a $\{u_t\}$-invariant measure, by Ratner's classification of invariant measures of unipotent flows (cf. [32, Theorem 1]) and the ergodic decomposition theorem for finite invariant measures of flows (cf. [34]) the preceding observation implies that $\lambda = \mu$ and, as noted earlier, this completes the proof.

We now deduce from Theorem 2 a result on the asymptotics of closed orbits of certain subgroups, which includes in particular an assertion of M. Burger and P. Sarnak, which they apply to uniform distribution of Hecke points (cf. [7, "Theorem 5.2"]); their assertion follows as a particular case of the following corollary if, given a noncompact simple Lie group S and a lattice Λ in S, we choose $G = S \times S$, $\Gamma = \Lambda \times \Lambda$, and $H = \Delta(S)$, the diagonal subgroup in $S \times S$. For the corollary we have chosen a set-up which is general enough to put the above special case in perspective, but have not striven for complete attainable generality, since it is intended only as an illustration.

6.2. COROLLARY. *Let G be a connected semisimple Lie group with no (nontrivial) compact factors and let Γ be a lattice in G. Let H be a maximal connected semisimple subgroup of G, having no compact factors. Let $\{x_i\}$ be a sequence in G/Γ such that for each i, Hx_i (is closed and) admits a (unique) H-invariant probability measure, say μ_i, and $x_i \to x$, where $x \in G/\Gamma$ is such that Hx is not closed. Then $\{\mu_i\}$ converges to the G-invariant probability measure on G/Γ (in the weak * topology on the space of probability measures on G/Γ).*

PROOF. Let $U = \{u_t\}$ be an Ad-unipotent one-parameter subgroup contained in H but not contained in any proper closed normal subgroup of H (such a subgroup exists since H is a semisimple subgroup with no compact factors). Then, by Moore's ergodicity theorem (see [39, Chapter II]; see also [6], for more general results), for each i, U acts ergodically on Hx_i with respect to μ_i and hence for almost all $h \in H$ (with respect to the Haar measure) the U-orbit of hx_i is uniformly distributed in Hx_i. We also claim that $hx \in \mathscr{G}(U)$ for almost all $h \in H$. If the claim does not hold then by Propositions 2.1 and 2.3 there exists a proper closed connected subgroup L of G such that $L \cap \Gamma$ is a lattice in L and $\{h \in H \mid hx \in X(L, U)\Gamma/\Gamma\}$ has positive Haar measure; hence so does the set $\{h \in H \mid hg \in X(L, U)\}$ for some (fixed) g such that $x = g\Gamma$. Since $X(L, U)$ is a real analytic submanifold, such a condition implies that $hg \in X(L, U)$, and hence $h^{-1}Uh \subseteq gLg^{-1}$, for all $h \in H$. Since U is not contained in any proper normal subgroup of H and since H is a maximal connected subgroup of G, that in turn implies that $H = gLg^{-1}$ and hence $Hx = Hg\Gamma/\Gamma = gL\Gamma/\Gamma$ is closed, contradicting the hypothesis. Therefore the claim must hold.

Putting together the observations above we get that there exists an $h \in H$ such that for each i the $h^{-1}Uh$-orbit of x_i is uniformly distributed in Hx_i (with respect to μ_i) and $x \in \mathscr{G}(h^{-1}Uh)$. It is now easy to deduce the contention in the corollary by applying Theorem 2 with $u_t^{(i)} = h^{-1}u_t h$, for all i, $\{x_i\}$ as above, and $\{T_i\}$ chosen suitably (depending on the test function involved in the verification of the contention). We omit the details.

§7. Trajectories from singular points

In this section we prove a stronger, but somewhat technical version of Theorem 1 (see Theorem 7.3 below) which will be used in the next section to deduce Theorems 3 and 4.

Let G be a connected Lie group and Γ be a discrete subgroup of G. Let W be a closed connected subgroup which is generated by the Ad-unipotent elements contained in it. We shall follow the notation introduced in Section 3. Also for $H \in \mathscr{H}$ let $I_H = \{g \in G \mid \rho_H(g)p_H = p_H\}$.

7.1. Proposition. *Let $H \in \mathscr{H}$ and C be a compact subset of V_H. Let K be a compact subset of G/Γ. Then there exists a compact subset C' of $\eta_H^{-1}(C)$ such that $K \cap \eta_H^{-1}(C)\Gamma/\Gamma = C'\Gamma/\Gamma$.*

Proof. Let K_0 be a compact subset of G such that $K_0\Gamma/\Gamma = K$. For any $\gamma \in \Gamma$ let $K_\gamma = \{g \in K_0 \mid g\gamma \in \eta_H^{-1}(C)\}$. Then clearly $K \cap \eta_H^{-1}(C)\Gamma/\Gamma = \bigcup_{\gamma\in\Gamma} K_\gamma\Gamma/\Gamma$. Let $\gamma \in \Gamma$ be such that K_γ is nonempty; say $g \in K_\gamma$. Then $\eta_H(g\gamma) = \rho_H(g)\eta_H(\gamma) \in C$ and hence $\eta_H(\gamma) \in \rho_H(K_\gamma^{-1})C \subseteq \rho_H(K_0^{-1})C$. Since the latter is a compact subset of V_H and $\eta_H(\Gamma) = \rho_H(\Gamma)p_H$ is discrete (cf. Theorem 3.4) the preceding observation implies that there exists a finite subset Δ of Γ such that K_γ is nonempty if and only if $\gamma \in \Delta(I_H \cap \Gamma)$. Let $C' = \bigcup_{\gamma\in\Delta} K_\gamma\gamma$. Then C' is a compact subset of $\eta_H^{-1}(C)$ and

$$C'\Gamma/\Gamma = \bigcup_{\gamma\in\Delta} K_\gamma\Gamma/\Gamma = K \cap \eta_H^{-1}(C)\Gamma/\Gamma;$$

we use here the fact that $K_\gamma = K_{\gamma\theta}$ if $\gamma \in \Gamma$ and $\theta \in I_H \cap \Gamma$ and that K_γ is empty if $\gamma \notin \Delta(I_H \cap \Gamma)$.

7.2. PROPOSITION. *Let* $H \in \mathscr{H}$ *and* D *be a compact subset of* A_H. *Let* K *be a compact subset of* G/Γ. *Then* $\{K \cap (\eta_H^{-1}(D) \cap \eta_H^{-1}(D)\gamma)\Gamma/\Gamma\}_{\gamma \in \Gamma}$ *is a family of sets with only finitely many distinct elements. Further, for each* $\gamma \in \Gamma$ *there exists a compact subset* C_γ *of* $\eta_H^{-1}(D) \cap \eta_H^{-1}(D)\gamma$ *such that*

$$K \cap (\eta_H^{-1}(D) \cap \eta_H^{-1}(D)\gamma)\Gamma/\Gamma = C_\gamma\Gamma/\Gamma.$$

PROOF. We note that if $\gamma, \gamma' \in \Gamma$ are such that $\gamma' \in (I_H \cap \Gamma)\gamma(I_H \cap \Gamma)$ then the elements of the family of sets as above corresponding to γ and γ' are the same. Therefore it is enough to show that there exists a finite subset Δ of Γ such that the set corresponding to $\gamma \in \Gamma$ is empty unless $\gamma \in (I_H \cap \Gamma)\Delta(I_H \cap \Gamma)$. Let $\gamma \in \Gamma$ be such that $K \cap (\eta_H^{-1}(D) \cap \eta_H^{-1}(D)\gamma)\Gamma/\Gamma$ is nonempty. Let K_0 be a compact subset of G such that $K_0\Gamma/\Gamma = K$. Then there exist $g \in K_0$ and $\gamma' \in \Gamma$ such that $\eta_H(g\gamma'), \eta_H(g\gamma'\gamma^{-1}) \in D$. Since $\rho_H(K_0^{-1})D$ is compact and $\eta_H(\Gamma) = \rho_H(\Gamma)p_H$ is discrete, it follows that there exists a finite subset Δ' of Γ such that $\eta_H^{-1}(\rho_H(K_0^{-1})D) \cap \Gamma$ is contained in $\Delta'(I_H \cap \Gamma)$. Again since $\rho_H((K_0\Delta')^{-1})D$ is compact, there exists a finite subset Δ of Γ such that $\eta_H^{-1}(\rho_H((K_0\Delta')^{-1})D) \cap \Gamma$ is contained in $\Delta^{-1}(I_H \cap \Gamma)$. Now, for γ' as above $\eta_H(\gamma') \in \rho_H((K_0)^{-1})D$ and hence $\gamma' \in \Delta'(I_H \cap \Gamma)$. Let $\delta \in \Delta'$ and $\gamma_0 \in I_H \cap \Gamma$ be such that $\gamma' = \delta\gamma_0$. Since $\eta_H(g\gamma'\gamma^{-1}) \in D$, we now have $\eta_H(\gamma_0\gamma^{-1}) \in \rho_H((K_0\Delta')^{-1})D$ and therefore $\gamma_0\gamma^{-1} \in \Delta^{-1}(I_H \cap \Gamma)$ and in turn, $\gamma \in (I_H \cap \Gamma)\Delta(I_H \cap \Gamma)$. This shows that there exists a finite subset of Γ with the desired property.

Now consider the Cartesian square action of Γ on $V_H \times V_H$, the action on each component being given by ρ_H. The Γ-orbit of $(p_H, \rho_H(\gamma)p_H)$ is a discrete subset of $V_H \times V_H$ and since $\rho_H(K_0^{-1})D \times \rho_H(K_0^{-1})D$ is compact, it follows that the latter contains only finitely many elements of the orbit. Observe that the isotropy subgroup of $(p_H, \rho_H(\gamma)p_H)$ for the Γ-action is $I_H \cap (\gamma I_H \gamma^{-1}) \cap \Gamma$. Hence the preceding observation implies that there exists a finite subset Φ of Γ such that $\{\theta \in \Gamma \mid \rho_H(\theta)p_H, \rho_H(\theta\gamma)p_H \in \rho_H(K_0^{-1})D\} = \Phi(I_H \cap (\gamma I_H \gamma^{-1}) \cap \Gamma)$. Applying this to the element γ' as above and arguing as in the proof of Proposition 7.1 we see that there exists a compact subset C_γ of $\eta_H^{-1}(D) \cap \eta_H^{-1}(D)\gamma$ such that $K \cap (\eta_H^{-1}(D) \cap \eta_H^{-1}(D)\gamma)\Gamma/\Gamma = C_\gamma\Gamma/\Gamma$. This proves the proposition.

We denote by $\mathscr{E}$ the class of subsets of G for which there exist $H_1, \dots, H_r \in \mathscr{H}$ and compact subsets $D_1, \dots, D_r$ of $A_{H_1}, \dots, A_{H_r}$ respectively such that $E = \eta_{H_1}^{-1}(D_1) \cup \cdots \cup \eta_{H_r}^{-1}(D_r)$; for such a subset E, considered together with the decomposition, we denote by $\mathscr{N}(E)$ the class of all neighbourhoods of E of the form $\Phi = \eta_{H_1}^{-1}(\Theta_1) \cup \cdots \cup \eta_{H_r}^{-1}(\Theta_r)$, where $\Theta_1, \dots, \Theta_r$ are neighbourhoods of $D_1, \dots, D_r$ in $V_{H_1}, \dots, V_{H_r}$ respectively; the subsets $\eta_{H_i}^{-1}(\Theta_i)$, $i = 1, \dots, r$, will be called the components of Φ.

7.3. THEOREM. *Let the notation be as above. Let K be a compact subset of G/Γ and let $\varepsilon > 0$ be given. Then for any $E \in \mathscr{E}$ there exists $E' \in \mathscr{E}$ such that the following holds: given any $\Phi \in \mathscr{N}(E')$ there exists a neighbourhood Ω of $E\Gamma/\Gamma$ such that for any* Ad*-unipotent one-parameter subgroup $\{u_t\}$ of G, any $g \in G$, and $T_0 \geqslant 0$ either*

i) *there exist a $\gamma \in \Gamma$ and a component of Φ containing $u_t g\gamma$ for all $t \in [0, T_0]$, or*
ii) *$l(\{t \in [T_0, T] | u_t g\Gamma \in \Omega \cap K\}) \leqslant \varepsilon T$ for all $T \geqslant T_0$.*

PROOF. The proof is similar to that of Theorem 1. However some of the arguments need to be modified; this applies especially to the induction argument there, as it involves all $H' \in \mathscr{H}$ of dimension at most $n-1$ and the whole of $X(H', W)$ in each case, which is not allowable in the present context. We shall use notation compatible with that in the proof of Theorem 1 and draw from the arguments there wherever convenient.

Clearly, it is enough to prove the assertion when $E = \eta_H^{-1}(C)$, where $H \in \mathscr{H}$ and C is a compact subset of A_H. We proceed by induction on the dimension of H. For low dimensions A_H is empty and hence the assertion is obvious. Now assume the result when the dimension is at most $n-1$ and let C be a compact subset of A_H for some $H \in \mathscr{H}$ of dimension n. Let $V = V_H$, $\rho = \rho_H$, $p = p_H$, $\eta = \eta_H$, and $A = A_H$. By Proposition 4.2 there exists a compact subset D of A such that C is of relative size less than $\varepsilon/4$ in D. If there exists a $\gamma \in \Gamma$ such that $\rho(\gamma)p = -p$ then $A = -A$ and in that case, by enlarging D if necessary, we may assume that $-D = D$. Let $B = \eta^{-1}(D)$. By Proposition 7.2 the family of sets $\{K \cap (B \cap B\gamma^{-1})\Gamma/\Gamma\}_{\gamma\in\Gamma}$ has only finitely many elements, say $K\cap(B\cap B\gamma_1^{-1})\Gamma/\Gamma, \dots, K\cap(B\cap B\gamma_k^{-1})\Gamma/\Gamma$, where $\gamma_1, \gamma_2, \dots, \gamma_k \in \Gamma$ and $\gamma_1 = e$, the identity element of G. Further, each $K\cap(B\cap B\gamma_j^{-1})\Gamma/\Gamma$ is of the form $C_j\Gamma/\Gamma$, where C_j is a compact subset of $B\cap B\gamma_j^{-1} \subseteq X(H, W)\cap X(\gamma_j H\gamma_j^{-1}, W) \subseteq X(H\cap\gamma_j H\gamma_j^{-1}, W)$. Let $j \geqslant 2$ be arbitrary. Then $K\cap(B\cap B\gamma_j^{-1})\Gamma/\Gamma \neq K\cap B\Gamma/\Gamma$ and hence, using Lemma 3.1 and the assumption on D as above we get that $\gamma_j \notin \Gamma_H$. Therefore γ_j does not normalise H and hence $H \cap \gamma_j H\gamma_j^{-1}$ is a proper subgroup of H. Arguing as in the proof of Proposition 3.3 we see that there exists a subgroup $H_j \in \mathscr{H}$ contained in $(H \cap \gamma_j H\gamma_j^{-1})^0$ and containing all Ad-unipotent elements of the latter subgroup. Then H_j is of dimension at most $n-1$ and C_j is a compact subset of $X(H_j, W)$. Let $E_j = \eta_{H_j}^{-1}(\eta_{H_j}(C_j))$. Then by the induction hypothesis there exists $E_j' \in \mathscr{E}$ such that given $\Phi_j \in \mathscr{N}(E_j')$ there exists a neighbourhood Ω_j of E_j such that for any Ad-unipotent one-parameter subgroup $\{u_t\}$ of G, any $g \in G$, and $T_0 \geqslant 0$ either there exists a $\gamma \in \Gamma$ such that $\{u_t g\gamma \mid t \in [0, T_0]\}$ is contained in a component of Φ_j, or $l(\{t \in [T_0, T] \mid u_t g\Gamma \in \Omega_j \cap K\}) \leqslant \varepsilon T/2k$, for all $T \geqslant T_0$.

Let $E' = B \cup E_2' \cup \cdots \cup E_k'$. Then $E' \in \mathscr{E}$. We shall verify the assertion in the theorem for this E'. Let $\Phi \in \mathscr{N}(E')$ be given. Then by the above observation there exists a neighbourhood Ω' of $(E_2 \cup \cdots \cup E_k)\Gamma/\Gamma$ such that for any Ad-unipotent one-parameter subgroup $\{u_t\}$ of G, any $g \in G$, and $T_0 \geqslant 0$ either there exists a $\gamma \in \Gamma$ such that $\{u_t g\gamma \mid t \in [0, T_0]\}$ is contained in a component of Φ, or

$$l(\{t \in [T_0, T] \mid u_t g\Gamma \in \Omega' \cap K\}) \leqslant \varepsilon T/2 \quad \forall T \geqslant T_0. \tag{7.4}$$

This corresponds to (4.4) in the proof of Theorem 1. Now let $K_1 = K - \Omega'$ and let K' be a compact subset of G such that $K_1 = K'\Gamma/\Gamma$. As in the proof of Theorem 1, applying Corollary 3.5 we find a neighbourhood Φ_1 of D in V such that $\eta^{-1}(\Phi_1) \subseteq \Phi$ and the quotient map of $(\eta^{-1}(\Phi_1)\Gamma_H \cap K'\Gamma)/\Gamma_H$ onto $(\eta^{-1}(\Phi_1)\Gamma \cap K'\Gamma)/\Gamma$ is injective. Then, as in the proof of Theorem 1, there exists a neighbourhood Ψ of C in V such that

$$l(\{t \in [0, T] \mid \rho(u_t)v \in \Psi\}) < \varepsilon l(\{t \in [0, T] \mid \rho(u_t)v \in \Phi_1\})/4 \tag{7.5}$$

for any Ad-unipotent one-parameter subgroup $\{u_t\}$ in G, $v \notin \Phi_1$, and $T > 0$. Let $\Omega = \eta^{-1}(\Psi)\Gamma/\Gamma$. Now let a Ad-unipotent one-parameter subgroup $\{u_t\}$ of G, a $g \in G$, and a $T_0 \geqslant 0$ be given. Suppose assertion i) as in the statement of the theorem does not hold for this data; that is, for any $\gamma \in \Gamma$ there exists a $t \in [0, T_0]$ such that $u_t g\gamma \notin \Phi$. For any $q \in \eta(\Gamma) = \rho(\Gamma)p$ let

$$I(q) = \{t \in (T_0, T) \mid \rho(u_t g)q \in \Phi_1\},$$
$$J(q) = \{t \in (T_0, T) \mid \rho(u_t g)q \in \Psi \text{ and } u_t g\Gamma \in K_1\},$$

and $I'(q)$ be the subset of $I(q)$ consisting of all t for which there exists $a \geqslant 0$ such that $[t, t+a] \subseteq I(q)$ and $u_{t+a}g\Gamma \in K_1$. An argument as in the proof of Theorem 1 shows that, for $q_1, q_2 \in \eta(\Gamma)$, $I'(q_1)$, $I'(q_2)$ are disjoint unless $q_1 = \pm q_2$. Also for any connected component (a, b) of any $I(q)$ for which $a > T_0$, the argument there shows that $l(J(q) \cap (a, b)) \leqslant \varepsilon l(I'(q) \cap (a, b))/4$. Now suppose that there exists a $q_0 \in \eta(\Gamma)$ such that $I(q_0)$ has a connected component of the form (T_0, b), for some $b > T_0$, intersecting $J(q_0)$ nontrivially; then $-q_0$ can be the only other element of $\eta(\Gamma)$ for which this can hold. Let $\gamma \in \Gamma$ be such that $q_0 = \eta(\gamma)$. By our assumption, there exists $a \in [0, T_0]$ such that $u_a g\gamma \notin \eta^{-1}(\Phi_1)$; in particular $\eta(u_a g\gamma) = \rho(u_a g)q_0 \notin \Phi_1$. Applying (7.5) with $v = \rho(u_a g)q_0$ and $T = c - a$, where c is the right-hand endpoint of the interval in $I'(q_0)$ starting from T_0, we get that $l(J(q) \cap (T_0, b)) \leqslant \varepsilon l((a, T_0) \cup I'(q) \cap (T_0, b))/4$. Now summing the terms as in the proof of Theorem 1 we get that

$$l(\{t \in [T_0, T] \mid u_t g\Gamma \in \Omega \cap K_1\}) \leqslant \varepsilon T/2.$$

Together with (7.4) this shows that condition ii) as in the theorem holds. This completes the proof.

§8. Proofs of Theorems 3 and 4

We now apply the results of the last section to lattices. We shall first make some simple observations following the previous notation, and then deduce Theorems 3 and 4. Henceforth W will be a (Ad-unipotent) one-parameter subgroup $U = \{u_t\}$. We shall continue to use the notations depending on W, such as A_H, meaning U in the place of W.

8.1. PROPOSITION. *Let G be a connected Lie group, Γ be a lattice in G, and $U = \{u_t\}$ be an* Ad*-unipotent one-parameter subgroup of G. Let $H_1, \dots, H_k \in \mathscr{H}$ and for each $i = 1, \dots, k$ let D_i be a compact subset of A_{H_i} and Θ_i be a compact neighbourhood of D_i in V_{H_i}. Let K be a compact subset of G/Γ. Then there exist $H_1', \dots, H_k' \in \mathscr{H}$ and compact subsets $D_1', \dots, D_k'$ of $A_{H_1'}, \dots, A_{H_k'}$ respectively such that given any compact subset F of $K - \bigcup(\eta_H^{-1}(D_i) \cup \eta_{H'}^{-1}(D_i'))\Gamma/\Gamma$ there exists an $T_0 \geqslant 0$ such that for any $g \in G$ with $g\Gamma \in F$ and $i = 1, \dots, k$ there exists a $t \in [0, T_0]$ such that $u_t g \notin \eta_{H_i}^{-1}(\Theta_i)$.*

PROOF. We first observe that for any $H \in \mathscr{H}$ there exists a $H' \in \mathscr{H}$ such that $X(I_H, U)$ is contained in $X(H', U)$; here, as before, I_H denotes the isotropy subgroup of p_H: Since $\rho_H(\Gamma)p_H$ is closed in V_H, ΓI_H is a closed subset of G and hence so is $I_H\Gamma$. Let H' be the smallest closed (connected) subgroup of G containing all Ad-unipotent elements in I_H and such that $H'\Gamma$ is closed. Then by the results of Section 2 of [36] $H' \cap \Gamma$ is a lattice in H' and $\operatorname{Ad}(H' \cap \Gamma)$ is Zariski-dense in $\operatorname{Ad} H'$; thus $H' \in \mathscr{H}$. Observe that if $g \in X(I_H, U)$ then $g^{-1}Ug \subseteq I_H$ and hence, by our choice of H', $g^{-1}Ug \subseteq H'$. Hence $X(I_H, U)$ is contained in $X(H', U)$.

Now let $H_1', \dots, H_k' \in \mathscr{H}$ be such that, for each $i = 1, \dots, k$, $X(I_{H_i}, U)$ is contained in $X(H_i', U)$. Also for each i let $Q_i = \{v \in \Theta_i \mid \rho_{H_i}(u_t)v = v \text{ for all } t \in \mathbb{R}\}$. By Proposition 7.1 there exist compact subsets C_i, $i = 1, \dots, k$, of $\eta_{H_i}^{-1}(Q_i)$ such that $K \cap \eta_{H_i}^{-1}(Q_i)\Gamma/\Gamma = C_i\Gamma/\Gamma$. Then clearly $\eta_{H_i}^{-1}(Q_i) \subseteq X(I_{H_i}, U) \subseteq X(H_i', U)$, for all i. Thus each C_i is a compact subset of $X(H_i', U)$. For each i let $D_i' = \eta_{H_i'}(C_i)$; it is a compact subset of $A_{H_i'}$.

Now let a compact subset F of $K - \bigcup(\eta_{H_i}^{-1}(D_i) \cup \eta_{H_i'}^{-1}(D_i'))\Gamma/\Gamma$ be given. Let F' be a compact subset of G such that $F'\Gamma/\Gamma = F$. Since $\rho_{H_i}(\Gamma)p_{H_i} = \eta_{H_i}(\Gamma)$ is closed (and hence discrete) in V_{H_i}, for each i, there exist only finitely many $\gamma \in \Gamma$ such that $\eta_{H_i}(\gamma) \in \rho_{H_i}(F')^{-1}\Theta_i$ for some $i = 1, \dots, k$. Therefore to prove the proposition it is enough to show that for any $i =$

$1, \ldots, k$ and $\gamma \in \Gamma$ there exists a $T \geqslant 0$ such that for all $t \geqslant T$ the set $\rho_{H_i}(u_t)(\rho_{H_i}(F'\gamma)p_{H_i} \cap \Theta_i)$ is disjoint from Θ_i. Let $1 \leqslant i \leqslant k$ and $\gamma \in \Gamma$ be given. We note that $\rho_{H_i}(F'\gamma)p_{H_i} \cap \Theta_i$ is a compact subset of V_{H_i} not containing any fixed point of $\{\rho_{H_i}(u_t)\}$; if $g \in F'$ is such that $\rho_{H_i}(g\gamma)p_{H_i}$ is contained in Θ_i and fixed by $\rho_{H_i}(\{u_t\})$ then $g\gamma \in \eta_{H_i}^{-1}(Q_i)$ and hence $g\Gamma \in K \cap \eta_{H_i}^{-1}(Q_i)\Gamma/\Gamma = C_i\Gamma/\Gamma \subseteq \eta_{H_i}^{-1}(D_i')\Gamma/\Gamma$, which contradicts F being disjoint from $\eta_{H_i}^{-1}(D_i')\Gamma/\Gamma$. Since $\{\rho_{H_i}(u_t)\}$ is a unipotent one-parameter subgroup of $GL(V_{H_i})$ it now follows that there exists a $T \geqslant 0$ with the desired property. This proves the proposition.

8.2. Corollary. *Let G, Γ, and $\{u_t\}$ be as in Proposition* 8.1. *Let a compact subset K of G/Γ and $\varepsilon > 0$ be given. Then for any $E \in \mathscr{E}$ there exist a neighbourhood Ω of $E\Gamma/\Gamma$ in G/Γ and a $E^* \in \mathscr{E}$ such that the following holds: for any compact subset F of $K - E^*\Gamma/\Gamma$ there exists a $T_0 \geqslant 0$ such that for all $x \in F$ and $T \geqslant T_0$,*

$$l(\{t \in [0, T] \mid u_t x \in \Omega \cap K\}) \leqslant \varepsilon T.$$

Proof. Let $E \in \mathscr{E}$ be given and let $E' \in \mathscr{E}$ be such that the conclusion of Theorem 7.3 holds for this E and $\varepsilon/2$ in the place of ε there. Let $H_1, \ldots, H_k \in \mathscr{H}$ and $D_1, \ldots, D_k$ be compact subsets of $A_{H_1}, \ldots, A_{H_k}$ respectively such that $E' = \eta_{H_1}^{-1}(D_1) \cup \cdots \cup \eta_{H_k}^{-1}(D_k)$. Let Φ be a neighbourhood of E' of the form $\eta_{H_1}^{-1}(\Phi_1) \cup \cdots \cup \eta_{H_k}^{-1}(\Phi_k)$, where $\Phi_1, \ldots, \Phi_k$ are compact neighbourhoods of $D_1, \ldots, D_k$ respectively. Let Ω be a neighbourhood of $E\Gamma/\Gamma$ for which the conclusion of Theorem 7.3 holds, for this Φ. By Proposition 8.1 there exists $E'' \in \mathscr{E}$ such that for any compact subset F of $K - (E' \cup E'')\Gamma/\Gamma$ there exists a $T_0' \geqslant 0$ such that for any $g \in G$ with $g\Gamma \in F$ and any i there exists a $t \in [0, T_0']$ such that $u_t g \notin \Phi_i$. The assertion in the corollary is then immediate from Theorem 7.3 if we choose $E^* = E' \cup E''$ and $T_0 = 2\varepsilon^{-1}T_0'$.

8.3. Remark. For any closed subgroup L of G let $\check{L}$ be the smallest closed subgroup containing all the Ad-unipotent elements in L and such that $\check{L}\Gamma$ is closed. A scrutiny of the proofs of Proposition 7.3 and Corollary 8.2 readily shows the following: If $\mathscr{H}'$ is a subclass of $\mathscr{H}$ such that i) for any $H \in \mathscr{H}'$ and $\gamma \in \Gamma$, $(H \cap \gamma H \gamma^{-1})\check{} \in \mathscr{H}'$ and ii) for any $H \in \mathscr{H}'$, $\check{I}_H \in \mathscr{H}'$ and if $\mathscr{E}'$ is the class of subsets of the form $\eta_{H_1}^{-1}(D_1) \cup \cdots \cup \eta_{H_k}^{-1}(D_k)$, where $H_1, \ldots, H_k \in \mathscr{H}'$ and each D_i is a compact subset of A_{H_i}, then for any $E \in \mathscr{E}'$ the set E^* as in the conclusion of Corollary 8.2 can also be chosen to be in $\mathscr{E}'$; this will be used in the sequel, in the proof of the second part of Corollary 5; we have avoided incorporating it in the original statements, in the interest of keeping them relatively simple.

PROOF OF THEOREM 3. Suppose that the conclusion of the theorem is not true. Then there exist a bounded continuous function φ on G/Γ, a compact subset K_1 of G/Γ, and $\varepsilon > 0$ such that for any proper closed subgroups $H_1, \ldots, H_k$ of G such that $H_j \cap \Gamma$ is a lattice in H_j for each j, and any compact subsets $C_1, \ldots, C_k$ of $X(H_1, U), \ldots, X(H_k, U)$ respectively, there exists a compact subset F of $K_1 - \bigcup C_j\Gamma/\Gamma$ such that for any $T_0 \geqslant 0$,

$$\left\{x \in F \mid \left|\frac{1}{T}\int_0^T \varphi(u_t x)\,dt - \int_{G/\Gamma} \varphi\,d\mu\right| > \varepsilon \text{ for some } T > T_0\right\} \neq \varnothing.$$

It is easy to see that in view of Theorem 6.1, by altering the funtion φ outside a suitable compact subset of G/Γ and modyfying ε suitably, we may assume that φ has compact support.

By Theorem 6.1 there exists a compact subset K of G/Γ such that for all $x \in K_1$ and $T \geqslant 0$,

$$l(\{t \in [0, T] \mid u_t x \notin K\}) < T/3. \tag{8.4}$$

Applying Corollary 8.2 we can construct a sequence $\{E_i\}$ in $\mathscr{E}$ such that i) $E_1 \subseteq E_2 \subseteq \cdots$ and $\bigcup E_i = \mathscr{S}(U)$ and ii) for each i there exists an open neighbourhood Ω_i of $E_i\Gamma/\Gamma$ such that for any compact subset F of $K - E_{i+1}\Gamma/\Gamma$ there exists a $T_i' \geqslant 0$ such that for all $x \in F$ and $T \geqslant T_i'$,

$$l(\{t \in [0, T] \mid u_t x \in \Omega_i \cap K\}) \leqslant T/4^i.$$

Recall that by Proposition 7.1 each set $K \cap E_i\Gamma/\Gamma$ is of the form $\bigcup C_j\Gamma/\Gamma$, where $C_1, \ldots, C_k$ are compact subsets of $X(H_1, U), \ldots, X(H_k, U)$, for some $H_1, \ldots, H_k \in \mathscr{H}$. Hence by the above condition for each i there exists a compact subset F_i of $K_1 - E_{i+1}\Gamma/\Gamma$ such that for any $T_0 \geqslant 0$,

$$\left\{x \in F_i \mid \left|\frac{1}{T}\int_0^T \varphi(u_t x)\,dt - \int_{G/\Gamma} \varphi\,d\mu\right| > \varepsilon \text{ for some } T > T_0\right\} \neq \varnothing.$$

Then by the defining condition of the sequence $\{E_i\}$ there exist, for all i, $T_i \geqslant 0$ such that

$$l(\{t \in [0, T] \mid u_t x \in \Omega_i \cap K\}) \leqslant T/4^i \quad \forall x \in F_i \text{ and } T \geqslant T_i. \tag{8.5}$$

Without loss of generality we may assume $\{T_i\}$ to be an increasing sequence. By the preceding condition then there exist $x_i \in F_i$ and $s_i \geqslant 3(1+\varepsilon^{-1})T_i\|\varphi\|$ (where $\|\varphi\|$ is the supremum norm of φ) such that

$$\left|\frac{1}{s_i}\int_0^{s_i} \varphi(u_t x_i)\,dt - \int_{G/\Gamma} \varphi\,d\mu\right| > \varepsilon.$$

In view of (8.4) and (8.5) there exist, for each $j \geqslant 1$, $t_j \in [0, T_j]$ such that $u_{t_j} x_j \in K - \bigcup_{i=1}^j \Omega_i$. For each j let $y_j = u_{t_j} x_j$. Then clearly, we have

$$\left|\frac{1}{s_j}\int_0^{s_j} \varphi(u_t y_j)\,dt - \int_{G/\Gamma} \varphi\,d\mu\right| \geqslant \varepsilon - 2t_j s_j^{-1}\|\varphi\| \geqslant \frac{\varepsilon}{3},$$

for all $j \geqslant 1$. Since $\{y_j\}$ is contained in the compact set K it has a limit point, say y. Clearly $y \notin \Omega_j$ for any $j \geqslant 1$. Since $\mathscr{S}(U) = \bigcup E_i\Gamma/\Gamma \subseteq \bigcup \Omega_i$ this implies that y is not contained in $\mathscr{S}(U)$. In other words, $y \in \mathscr{G}(\{u_t\})$. But then, applying Theorem 2 to a subsequence of $\{y_j\}$ converging to y and the corresponding subsequence of $\{s_j\}$, in the place of $\{x_i\}$ and $\{T_i\}$ respectively, we see that an estimate as above cannot hold for all j. The contradiction shows that the theorem is true.

PROOF OF THEOREM 4. By Theorem 6.1 there exists a compact subset K of G/Γ such that for all $i \geqslant 1$, $l(\{t \in [0, T] \mid u_t x_i \notin K\}) \leqslant \alpha/3$. Using Corollary 8.2 we can construct a sequence $\{E_j\}$ in $\mathscr{E}$ such that the following conditions are satisfied: a) $E_1 \subseteq E_2 \subseteq \cdots$ and $\bigcup E_j\Gamma/\Gamma = \mathscr{S}(U)$ and b) for each j there exists an open neighbourhood Ω_j of $E_j\Gamma/\Gamma$ such that for any $x \notin E_{j+1}\Gamma/\Gamma$ there exists a $T_0 \geqslant 0$ such that

$$l(\{t \in [0, T] \mid u_t x \in \Omega_j \cap K\}) \leqslant \alpha T/4^j \quad \forall T \geqslant T_0.$$

In view of Proposition 7.1 and the condition in the hypothesis any $E \in \mathscr{E}$ contains only finitely many terms of the sequence $\{x_i\}$. We can therefore find a sequence $\{i_k\}$ of natural numbers such that $i_1 < i_2 < \cdots$ and $x_{i_k} \notin E_{j+1}\Gamma/\Gamma$ whenever $j \leqslant k$. Hence for each $k \geqslant 1$ there exists an $s_k \geqslant 0$ such that

$$l(\{t \in [0, T] \mid u_t x_{i_k} \in \Omega_j \cap K\}) \leqslant \alpha T/4^j \quad \forall k \geqslant 1,\ 1 \leqslant j \leqslant k, \text{ and } T \geqslant s_k.$$

Since $l(R \cap [0, T]) \geqslant \alpha T$ and $l(\{t \in [0, T] \mid u_t x_{i_k} \notin K\}) \leqslant \alpha T/3$ for all $T > 0$ and $k \geqslant 1$, the above condition implies that for each k there exists $t_k \in R \cap [0, s_k]$ such that $u_{t_k} x_{i_k} \in K - \bigcup_{j=1}^{k} \Omega_j$. Clearly, the sequence $\{u_{t_k} x_{i_k}\}$ has convergent subsequences and any of its limit points is contained in the complement of $\bigcup_{j=1}^{\infty} \Omega_j$. Since $\mathscr{S}(U) \subseteq \bigcup E_j\Gamma/\Gamma \subseteq \bigcup_{j=1}^{\infty} \Omega_j$, this proves the theorem.

§9. Values of quadratic forms at integral points. I

In this and the following section we relate the results on the dynamics of unipotent flows to studying values of quadratic forms and prove Corollary 5.

Let $n \geqslant 3$ and $1 \leqslant p < n$. We denote by $\mathscr{Q}(p, n)$ the space of all quadratic forms on $\mathbb{R}^n$ with signature $(p, n-p)$ and discriminant ± 1; the signature determines the sign of the discriminant. We equip $\mathscr{Q}(p, n)$ with its natural locally compact topology given by pointwise convergence as functions on $\mathbb{R}^n$.

Let $\{e_1, e_2, \ldots, e_n\}$ be the standard basis of $\mathbb{R}^n$. Let Q_0 be the quadratic form defined by

$$Q_0\Big(\sum_{i=1}^{n} v_i e_i\Big) = 2v_1 v_n + \sum_{i=2}^{p} v_i^2 - \sum_{i=p+1}^{n-1} v_i^2 \quad \forall v_1, \ldots, v_n \in \mathbb{R}.$$

It is straightforward to verify that $Q_0 \in \mathscr{Q}(p, n)$. Let $G = SL(n, \mathbb{R})$, the group of $n \times n$ matrices with determinant 1, with its usual Lie group structure. For each quadratic form Q on $\mathbb{R}^n$ and $g \in G$ let Q^g denote the quadratic form defined by $Q^g(v) = Q(gv)$ for all $v \in \mathbb{R}^n$. By the well-known classification of quadratic forms (cf. [1], for instance) it follows that for any $Q \in \mathscr{Q}(p, n)$ there exists a $g \in G$ such that $Q = Q_0^g$. For any quadratic form Q, $SO(Q)$ denotes the special orthogonal group corresponding to Q; namely $\{g \in G \mid Q^g = Q\}$. Let $H = SO(Q_0)$. Then we have a one-one correspondence of $H\backslash G$ and $\mathscr{Q}(p, n)$ associating to each coset Hg, $g \in G$, the quadratic form Q_0^g. Since $\mathscr{Q}(p, n)$ is locally compact it follows that the correspondence is a homeomorphism, when $H\backslash G$ is equipped with the quotient topology.

Let U be the subgroup of H consisting of all $h \in H$ such that $he_i = e_i$ for all $i \leqslant n-2$ and he_{n-1} and he_n are contained in the subspaces spanned by $\{e_1, e_{n-1}\}$ and $\{e_1, e_{n-1}, e_n\}$ respectively. It is easy to verify that U is a unipotent one-parameter subgroup in G; it corresponds to $\{v_1(t)\}$ as in [16] if $n = 3$ and $p = 1$.

Let M be the subgroup of H consisting of all h leaving invariant the subspace spanned by $\{e_1 + e_n, e_2, \dots, e_p\}$. We note that the latter is a maximal subspace on which Q_0 is positive. Therefore M consists of orthogonal matrices in the usual sense, of transpose being equal to the inverse, and hence it is compact. We denote by σ the normalised Haar measure on M.

Let $\Gamma = SL(n, \mathbb{Z})$, the subgroup of G consisting of integral matrices. It is well known that Γ is a lattice in G (cf. [29, Corollary 10.5]); let μ denote the G-invariant probability measure on G/Γ. There is a well-defined one-one correspondence of G/Γ with the space of lattices in $\mathbb{R}^n$ with discriminant 1 associating to each $g\Gamma \in G/\Gamma$ the lattice $g\mathbb{Z}^n$. Using the correspondence one associates to each (not necessarily continuous) function ψ on $\mathbb{R}^n$ vanishing outside a compact subset, a function $\tilde{\psi}$ on G/Γ by setting

$$\tilde{\psi}(g\Gamma) = \sum_{v \in g\mathbb{Z}^n} \psi(v); \tag{9.1}$$

note that since ψ vanishes outside a compact subset, the expression on the right-hand side is actually a finite sum. It is clear that if ψ is measurable on $\mathbb{R}^n$, then $\tilde{\psi}$ is measurable on G/Γ. Further it is well known that if ψ is integrable on $\mathbb{R}^n$ then $\tilde{\psi}$ is integrable on G/Γ and $\int_{\mathbb{R}^n} \psi \, d\lambda = \int_{G/\Gamma} \tilde{\psi} \, d\mu$; cf. [37]. The proof of Corollary 5 is based on the following identity which is immediate from the above definition:

$$\int_T^{\kappa T} \int_B \sum_{v \in g\mathbb{Z}^n} \psi(u_t m v) \, d\sigma(m) \, dt = \int_T^{\kappa T} \int_B \tilde{\psi}(u_t m g\Gamma) \, d\sigma(m) \, dt \tag{9.2}$$

for any bounded measurable function ψ on $\mathbb{R}^n$ vanishing outside a compact subset, any Borel subset B of M, $\kappa > 1$, and $T > 0$. Applying Theorem 3 to suitable functions we get lower estimates for the right-hand side which involve volumes and compare them with certain upper estimates for the left-hand side, involving the number of integral points in the regions of interest in the corollary. We shall go over to the details after proving some preliminary results.

9.3. PROPOSITION. *Let* $\mathscr{K}$ *be a compact subset of* $\mathscr{Q}(p, n)$. *Let* K *be a compact subset of* G *such that for any* $Q \in \mathscr{K}$ *there exists a* $g \in \operatorname{Int} K$ *(the interior of* K*) such that* $Q = Q_0^g$. *Let* $H_1, \ldots, H_k$ *be proper closed connected subgroups of* G *such that* $H_i \cap \Gamma$ *is a lattice in* H_i, *for all* i, *and let* $C_1, \ldots, C_k$ *be compact subsets of* $X(H_1, U), \ldots, X(H_k, U)$ *respectively. Let* $\mathscr{S} = \{Q \in \mathscr{K} \mid \exists \gamma \in \Gamma$ *and* $1 \leqslant i \leqslant k$ *such that* $SO(Q)^0 = \gamma H_i \gamma^{-1}\}$. *Then we have the following:*

i) $\mathscr{S}$ *is finite and any* $Q \in \mathscr{S}$ *is a scalar multiple of a rational form and*

ii) *for any compact subset* $\mathscr{C}$ *of* $\mathscr{K} - \mathscr{S}$ *there exists a compact subset* C *of* $K - \bigcup C_i \Gamma$ *such that any* $Q \in \mathscr{C}$ *is of the form* Q_0^g *for some* $g \in C$.

PROOF. i) It is well known that any $Q \in \mathscr{Q}(p, n)$ such that $SO(Q) \cap \Gamma$ is a lattice in $SO(Q)$ is a multiple of a rational form (cf. [17, Proposition 9], for instance). In particular this holds for any $Q \in \mathscr{S}$. For any rational form $Q \in \mathscr{Q}(p, n)$ (and hence also for any scalar multiple of such a form) $\{Q^\gamma \mid \gamma \in \Gamma\}$ is a discrete subset of $\mathscr{Q}(p, n)$; for rational forms this follows from the fact that the denominators of the coefficients of the forms from the family are bounded. If $Q \in \mathscr{S}$ and $SO(Q)^0 = \gamma^{-1} H_i \gamma$ for some $\gamma \in \Gamma$ then $Q = Q_i^\gamma$, where $Q_i \in \mathscr{Q}(p, n)$ is such that $SO(Q_i)^0 = H_i$. Further, for $Q, Q' \in \mathscr{Q}(p, n)$, $SO(Q)^0 = SO(Q')^0$ only if $Q' = \pm Q$. Thus if we choose a form $Q_i \in \mathscr{Q}(p, n)$ such that $SO(Q)^0 = H_i$, whenever such a form exists, then any form in $\mathscr{S}$ can be expressed as $\pm Q_i^\gamma$ for some $\gamma \in \Gamma$ and $i = 1, \ldots, k$. Since $\mathscr{K}$ is compact and $\{Q_i^\gamma \mid \gamma \in \Gamma\}$ is discrete for all Q_i it now follows that $\mathscr{S}$ is finite.

ii) Let $\mathscr{C}$ be a compact subset of $\mathscr{K} - \mathscr{S}$. We first note that it is enough to show that each $Q \in \mathscr{C}$ is of the form Q_0^g for some $g \in \operatorname{Int} K - \bigcup C_i \Gamma$; if $g \in \operatorname{Int} K - \bigcup C_i \Gamma$ is such that $Q_0^g \in \mathscr{C}$ and Φ is a compact neighbourhood of g in $\operatorname{Int} K - \bigcup C_i \Gamma$ then $\{Q_0^x \mid x \in \Phi\}$ is a neighbourhood of Q_0^g in $\mathscr{Q}(p, n)$ and since, by the compactness of $\mathscr{C}$, finitely many such neighbourhoods cover $\mathscr{C}$ we would get a compact subset C of $\operatorname{Int} K - \bigcup C_i \Gamma$ as desired. Now let, if possible, $Q \in \mathscr{K} - \mathscr{S}$ be such that the above assertion is not true. Let $g \in \operatorname{Int} K$ be such that $Q = Q_0^g$. Clearly $Q = Q_0^{hg}$ for all $h \in H$ and hence the condition on Q implies that $Hg \cap \operatorname{Int} K \subseteq \bigcup C_i \Gamma \subseteq \bigcup X(H_i, U)\Gamma$. Since H and each $X(H_i, U)$ are real analytic submanifolds of G the condition implies that there exist $1 \leqslant i \leqslant k$ and $\gamma \in \Gamma$ such that $H^0 g \subseteq X(H_i, U)\gamma$. Hence $(hg\gamma^{-1})^{-1} U (hg\gamma^{-1}) \subseteq H_i$ for all $h \in H^0$;

in other words, $h^{-1}Uh \subseteq g\gamma^{-1}H_i\gamma g^{-1}$ for all $h \in H^0$. But the subgroup generated by $\{h^{-1}Uh \mid h \in H^0\}$ is clearly a normal subgroup of H^0 of positive dimension and since H^0 is a simple Lie group it must be the whole. Thus $H^0 \subseteq g\gamma^{-1}H_i\gamma g^{-1}$. But H^0 is a maximal connected Lie subgroup of G; this may be seen by considering the complexification and using Cartan decomposition (cf. [21, Chapter III]). Hence the preceding conclusion implies that $g\gamma^{-1}H_i\gamma g^{-1} = H^0$. But $g^{-1}H^0 g = SO(Q)^0$ and since $Q \notin \mathcal{S}$ it is not of the form $\gamma^{-1}H_i\gamma$ for any $\gamma \in \Gamma$. The contradiction shows that there must exist a g as desired.

9.4. PROPOSITION. *Let K be a compact subset of G such that $MK = K$. Let φ be a bounded continuous nonnegative function on G/Γ and B be a Borel subset of M. Let $\delta > 0$ and $\kappa > 1$ be given. Then there exist proper closed subgroups $H_1, \dots, H_k$ of G such that $H_i \cap \Gamma$ is a lattice in H_i, for all i, and compact subsets $D_1, \dots, D_k$ of $X(H_1, U), \dots, X(H_k, U)$ respectively such that for any compact subset C of $K - \bigcup_i D_i\Gamma$ there exists a $T' \geqslant 0$ such that for all $g \in C$ and $T > T'$ we have*

$$\frac{1}{(\kappa-1)T}\int_T^{\kappa T}\int_B \varphi(u_t mg\Gamma)\,d\sigma(m)\,dt \geqslant (1-\delta)\sigma(B)\int_{G/\Gamma}\varphi\,d\mu.$$

PROOF. We may assume without loss of generality that B is contained in M^0; if M has a nontrivial connected component, it contains an element m_0 normalising U and to get the result for a subset B of m_0M^0 one only needs to apply the proposed special case to a translate of the function φ as above.

Put $\alpha = \int_{G/\Gamma}\varphi\,d\mu$; we may assume $\alpha > 0$. By Theorem 3 there exist proper closed subgroups $H_1, \dots, H_k$ of G such that $H_i \cap \Gamma$ is a lattice in H_i for all i and compact subsets $C_1, \dots, C_k$ of $X(H_1, U), \dots, X(H_k, U)$ respectively such that for any compact subset F of $K - \bigcup_i C_i\Gamma$ there exists a T_0 such that for all $x \in F$ and $T \geqslant T_0$ we have

$$\left|\frac{1}{T}\int_0^T \varphi(u_t x\Gamma)\,dt - \alpha\right| < \frac{\delta\alpha(\kappa-1)}{2(\kappa+1)}. \tag{9.5}$$

For each $i = 1, \dots, k$ let $D_i = M^0C_i \cap X(H_i, U)$. Now let C be any compact subset of $K - \bigcup D_i\Gamma$. For each $g \in C$ let $M_i(g) = \{m \in M^0 \mid mg \in C_i\Gamma\}$. Suppose $\sigma(M_i(g)) > 0$ for some $g \in C$ and $1 \leqslant i \leqslant k$. Then there exists a $\gamma \in \Gamma$ such that $\sigma(\{m \in M^0 \mid mg \in C_i\gamma\}) > 0$. Since $C_i \subseteq X(H_i, U)$ and the latter is a real analytic submanifold, the condition implies that $M^0 g \subseteq X(H_i, U)\gamma$. There exists an $m \in M^0$ such that $mg \in C_i\gamma$ and hence $g\gamma^{-1} \in m^{-1}C_i \cap X(H_i, U) \subseteq D_i$. But then $g \in D_i\Gamma$, which is a contradiction since C is disjoint from $D_i\Gamma$. Therefore $\sigma(M_i(g)) = 0$ for all $g \in C$ and $i = 1, \dots, k$. Hence there exists a compact subset $B(g)$ of $B - \bigcup M_i(g)$ such that $\sigma(B(g)) \geqslant (1-\delta/2)\sigma(B)$. Then for each

$g \in C$, $B(g)g \subseteq K - \bigcup C_i\Gamma$ and since $B(g)$ is compact and $\bigcup C_i\Gamma$ is closed there exists a compact neighbourhood $\Psi(g)$ of g in K such that $B(g)\Psi(g) \subseteq K - \bigcup C_i\Gamma$. Since C is compact we can now extract finitely many compact subsets $B_1, \ldots, B_l$ of B and $\Psi_1, \ldots, \Psi_l$ of C respectively such that $C = \bigcup \Psi_j$, $\sigma(B_j) > (1-\delta/2)\sigma(B)$, and $B_j\Psi_j \subseteq K - \bigcup C_i\Gamma$ for all $j = 1, \ldots, l$. Then for each j there exists a $T_j > 0$ such that (9.5) holds for all $x \in B_j\Psi_j$ and $T \geqslant T_j$. Put $T' = \max\{T_1, \ldots, T_l\}$. Now let $g \in C$ be given. Then there exists a j, $1 \leqslant j \leqslant l$, such that $g \in \Psi_j$ and we have

$$\frac{1}{(\kappa-1)T}\int_T^{\kappa T}\int_B \varphi(u_t m g\Gamma)\,d\sigma(m)\,dt \geqslant \frac{1}{(\kappa-1)T}\int_T^{\kappa T}\int_{B_j} \varphi(u_t m g\Gamma)\,d\sigma(m)\,dt$$

$$\geqslant \int_{B_j}\left(\alpha - \left|\frac{1}{(\kappa-1)T}\int_T^{\kappa T}\varphi(u_t x\Gamma)\,dt - \alpha\right|\right)d\sigma(m)$$

$$\geqslant \int_{B_j}\left(\alpha - \frac{\alpha\delta}{2}\right)d\sigma(m) = \alpha\left(1-\frac{\delta}{2}\right)\sigma(B_j) \geqslant \alpha(1-\delta)\sigma(B).$$

This proves the proposition.

For any subset I of $\mathbb{R}$ and $Q \in \mathcal{Q}(p, n)$ let $V_Q(I) = \{v \in \mathbb{R}^n \mid Q(v) \in I\}$; for convenience $V_{Q_0}(I)$ will also be denoted by $V_0(I)$. Clearly, if $Q = Q_0^g$ for some $g \in G$ then $V_Q(I) = g^{-1}V_0(I)$ for any subset I of $\mathbb{R}$.

Let $S^{n-1} = \{v \in \mathbb{R}^n \mid \|v\| = 1\}$. For any $\alpha, \beta \geqslant 0$ we denote by $A(\alpha, \beta)$ the 'annular region' of $\mathbb{R}^n$ given by $\{sv \mid \alpha \leqslant s < \beta, v \in S^{n-1}\}$. For each subset B of S^{n-1} let $c(B)$ denote the cone over B with vertex at 0; namely, $c(B) = \{sb \mid s \geqslant 0, b \in B\}$. It can be verified that for any closed interval $I_0 = [a, b]$, where $0 < a < b$, and $0 < \varrho < 1 < \kappa$ there exist $0 < \varsigma < 1$ such that for any interval I contained in I_0 and any open subset Φ of S^{n-1} there exists an $r_0 \geqslant 0$ such that for all $r \geqslant r_0$,

$$\begin{aligned}\varrho^{-1}\lambda(V_0(I) \cap A(\varsigma^{-1}r, \varsigma\kappa r) \cap c(\Phi)) &\geqslant \lambda(V_0(I) \cap A(r, \kappa r) \cap c(\Phi)) \\ &\geqslant \varrho\lambda(V_0(I) \cap A(\varsigma r, \varsigma^{-1}\kappa r) \cap c(\Phi)).\end{aligned} \tag{9.6}$$

Now let $v_0 = (e_1 + e_n)/\sqrt{2}$. Then $\|v_0\| = Q_0(v_0) = 1$. Also for any $v \in S^{n-1}$ and $\delta > 0$ let $B(v, \delta)$ denote the open δ-neighbourhood of v in S^{n-1} (with respect to the induced metric).

9.7. PROPOSITION. *Let $0 < a < \alpha < \beta < b$ and $0 < \varepsilon < 1$ be given. Let $I = [\alpha, \beta)$. Let Δ be a compact neighbourhood of Iv_0 in $V_0(\bar{I}) \cap A(a, b)$ not containing any fixed point of U. Let Φ be an open subset S^{n-1}. Let $\Psi = \{w \in \Phi \mid B(w, 2\varepsilon) \subseteq \Phi\}$ and $B = \{m \in M \mid B(e_1, \varepsilon) \subseteq m\Phi\}$. Then there exists a $T_0 \geqslant 0$ such that for all $\kappa > 1$ and $T \geqslant T_0$ the set $\{v \in \mathbb{R}^n \mid u_t m v \in \Delta$ for some $T \leqslant t \leqslant \kappa T$ and $m \in B\}$ contains $V_0(I) \cap$*

$A((1+\varepsilon)\beta T^2/2,\ \kappa^2(1-\varepsilon)\alpha T^2/2)\cap c(\Psi)$ and is contained in

$$A((1-\varepsilon)aT^2/2,\ \kappa^2(1+\varepsilon)bT^2/2)\cap c(\Phi).$$

PROOF. It is straightforward to verify by writing the matrix form of $\{u_t\}$ and using the fact that Δ is a compact subset not containing any fixed point of U that there exists a $T_0 \geqslant 0$ such that for all $w \in \Delta$ and $t \in \mathbb{R}$ such that $|t| \geqslant T_0$, $u_tw/\|u_tw\| \in B(e_1, \varepsilon)$ and $(1-\varepsilon)t^2\|w\|/2 \leqslant \|u_tw\| \leqslant (1+\varepsilon)t^2\|w\|/2$. Now suppose that $u_tmv = w \in \Delta$ for some $T \leqslant t \leqslant \kappa T$ and $m \in B$, where $T \geqslant T_0$. Then $\|v\| = \|mv\| = \|u_{-t}w\|$ and the latter is between $(1-\varepsilon)aT^2/2$ and $\kappa^2(1+\varepsilon)bT^2/2$. Also $mv/\|v\| = mv/\|mv\| = u_{-t}w/\|u_{-t}w\| \in B(e_1, \varepsilon)$. Hence $v/\|v\| \in m^{-1}B(e_1, \varepsilon) \subseteq \Phi$. This proves the second part of the assertion.

Now let T_0 be also chosen to be greater than $2/\sqrt{\varepsilon}$. Let $v \in V_0(I) \cap A((1+\varepsilon)\beta T^2/2, \kappa^2(1-\varepsilon)\alpha T^2/2) \cap c(\Psi)$, for some $T \geqslant T_0$. Let $w \in Iv_0$ be such that $Q_0(v) = Q_0(w)$. Then $\|v\| \geqslant (1+\varepsilon)\beta T_0^2/2 \geqslant \beta \geqslant \|w\|$ and since $\|u_{-t}w\|$ increases to infinity it follows that there exists a $t \geqslant 0$ such that $\|u_{-t}w\| = \|v\|$. Hence there exists $m \in M$ such that $u_{-t}w = mv$. Since $\|v\| \leqslant \kappa^2(1-\varepsilon)\alpha T^2/2$ and since for $t \geqslant T_0$, $\|u_{-t}w\| \geqslant (1-\varepsilon)\alpha t^2/2$ we get that $t \leqslant \kappa T$. Similarly since $\|v\| \geqslant (1+\varepsilon)\beta T^2/2$ and $\|u_{-t}w\|^2 \leqslant \beta((1+t^2/2)^2+1)$ a simple calculation using the assumption that $T \geqslant T_0 \geqslant 2/\sqrt{\varepsilon}$ shows that $t \geqslant T$. Since $mv = u_{-t}w$ and $t \geqslant T$, $mv/\|v\| \in B(e_1, \varepsilon)$ and hence $m^{-1}e_1 \in B(v/\|v\|, \varepsilon)$. Since $v \in c(\Psi)$ this implies that $m^{-1}B(e_1, \varepsilon) = B(m^{-1}e_1, \varepsilon) \subseteq \Phi$ and hence $m \in B$. This proves the proposition.

Let W_0 and W_1 be the subspaces of $\mathbb{R}^n$ defined by $W_q = \{\sum v_ie_i \in \mathbb{R}^n \mid v_{n-q} = 0\}$, for $q = 0$ or 1. The one-parameter subgroup $U = \{u_t\}$ yields a map $f: \mathbb{R} \times (W_1 - W_0) \to \mathbb{R}^n - W_0$, defined by $f(t, v) = u_tv$, for all $t \in \mathbb{R}$ and $v \in (W_1 - W_0)$. It is straightforward to verify that f is a diffeomorphism. In particular it follows that if Δ_0 is a subset of $W_1 - W_0$ and $\Delta = f([-c, c] \times \Delta_0)$ then the intersection of any $\{u_t\}$-orbit in $\mathbb{R}^n$ with Δ is either empty or a segment of length $2c$, along the flow parameter.

For any $I = [\alpha, \beta)$, where $0 < \alpha < \beta$, and $0 < \tau < \alpha/2$ let $I(\tau)$ be the τ-neighbourhood of Iv_0 in $(W_1 - W_0) \cap V_0(\bar{I})$ and let $\Delta(I, \tau) = f([-\tau, \tau] \times I(\tau))$.

9.8. LEMMA. *Let $\kappa > 1$, $0 < \delta < \kappa - 1$, and $0 < \alpha < \beta$ such that $\beta/\alpha < (1+\delta)$ be given. Then there exist $0 < \tau < \delta$ and $T_0 \geqslant 0$ such that the following holds: if, for $T > 0$, $E(T) = \{v \in \mathbb{R}^n \mid u_tv \in \Delta(I, \tau)$ for some $t \in [T, \kappa T]\}$ then for all $v \in \mathbb{R}^n$ and $T \geqslant T_0$ we have*

i) $\sigma(\{m \in M \mid mv \in E(T)\}) \leqslant \kappa^2(1+2\delta)\lambda(E(T))/\lambda(ME(T))$.

ii) *If* Θ *is a subset of* S^{n-1} *with piecewise smooth boundary and* $B = \{m \in M \mid e_1 \in m\Theta\}$ *then*

$$\lambda(\{v \in \mathbb{R}^n \mid mv \in E(T) \text{ for some } m \in B\}) \leqslant (1+\delta)\sigma(B)\lambda(ME(T)).$$

PROOF. It can be verified that if $\tau > 0$ is small then for $T \geqslant T_0$, where T_0 is as in Proposition 9.7, and $v_1, v_2 \in A((1+\tau)\beta T^2/2, \kappa^2(1-\tau)\alpha T^2/2)$, the measures $\sigma(\{m \in M \mid mv_1 \in E(T)\})$ and $\sigma(\{m \in M \mid mv_2 \in E(T)\})$ are comparable in the sense that their ratios are at most $\kappa^2(1+\delta)$; for v outside the annular region the corresponding measures can be larger than these only by a factor of at most $1+\tau$. Since $\sigma(\{m \in M \mid mv \in E(T)\})$ are the conditional measures of $E(T)$ for the restriction of the measure λ to $ME(T)$, with respect to the partition into the M-orbits, and by Proposition 9.7 and (9.6) most of $ME(T)$ is contained in the annular region as above, this implies that condition i) holds for all $T \geqslant T_0$. Condition ii) follows from the fact that the angle subtended at 0 by the set of points of $E(T)$ with norm r, tends to 0 as r tends to infinity. We omit the details.

PROOF OF COROLLARY 5, (i). Since we can simultaneously consider, along with $\mathscr{K} \subseteq \mathscr{Q}(p, n)$, the subset $\{-Q \mid Q \in \mathscr{K}\}$ of $\mathscr{Q}(n-p, n)$, there is no loss of generality in assuming that the interval I is contained in $[0, \infty)$. Further, evidently it may be assumed to be of the form $[a, b)$, for some $0 \leqslant a < b < \infty$. Also, if $I = [0, b)$ for some $b > 0$, it can be verified that there exist $a > 0$ and $r_0 \geqslant 0$ such that for all $Q \in \mathscr{K}$ and $r \geqslant r_0$, $\lambda(V_Q([a, b)) \cap r\Omega) \geqslant (1-\theta/2)\lambda(V_Q(0, b) \cap r\Omega)$. In view of this, modifying the given θ if necessary, we may assume I to be of the form $[a, b)$ where $0 < a < b < \infty$. Now let θ be as in the hypothesis and let $\kappa > 1$ be such that $\kappa^{-4} > (1-\theta/4)$. Let $0 < \varsigma < 1$ be such that (9.6) holds for $I_0 = [a, b]$, with κ^2 and κ^{-1} in the place of κ and ϱ there, respectively. Let $0 < \omega < 1$ be such that $\omega^6 > \max\{\varsigma, (1-\theta/4)\}$. Let $\delta > 0$ be such that $2\delta < 1-\omega$ and $\delta < \kappa - 1$. It is enough to prove the contention in the corollary for $[\alpha, \beta)$, where α and β are such that $a \leqslant \alpha < \beta \leqslant b$ and $\beta/\alpha < 1+\delta$, since summing over finitely many such intervals yields the result for the interval $[a, b)$. Now fix $\alpha, \beta \in [a, b]$ such that $\alpha < \beta$ and $\beta/\alpha < 1+\delta$. Let $J = [\alpha, \beta)$. Let $\tau > 0$ be such that the conclusion of Lemma 9.8 holds for the κ, δ, α, and β as above, for some $T_0 \geqslant 0$, and let $\Delta = \Delta(J, \tau)$, in the notation as before. Let χ be the characteristic function of Δ, $\tilde{\chi}$ be the function on G/Γ associated to χ by (9.1), and let φ be a bounded continuous function on G/Γ such that $0 \leqslant \varphi(x) \leqslant \tilde{\chi}(x)$ for all $x \in G/\Gamma$ and $\int \varphi d\mu \geqslant \omega \int \tilde{\chi} d\mu$. Recall that $\int \tilde{\chi} d\mu = \int \chi d\lambda = \lambda(\Delta)$; hence $\int \varphi \geqslant \omega\lambda(\Delta)$. Now let K be a compact subset of G such that $MK = K$ and each $Q \in \mathscr{K}$ is of the form Q_0^g for some $g \in \operatorname{Int} K$. Let $H_1, \dots, H_k$ be proper closed subgroups of G and $D_1, \dots, D_k$ be compact subsets of $X(H_1, U), \dots, X(H_k, U)$ respectively for which the conclusion of Proposition 9.4 holds for this function φ (for K and δ as above). Let

$\mathcal{S} = \{Q \in \mathcal{K} \mid SO(Q)^0 = \gamma H_i \gamma^{-1}$ for some $i = 1, \ldots, k$ and $\gamma \in \Gamma\}$. By Proposition 9.3 $\mathcal{S}$ is finite and consists of scalar multiples of rational forms.

Now let $\mathcal{C}$ be any compact subset of $\mathcal{K} - \mathcal{S}$. By Proposition 9.3 there exists a compact subset C of $K - \bigcup D_i \Gamma$ such that any $Q \in \mathcal{C}$ is of the form Q_0^g for some $g \in C$. We next choose (finitely many) subsets $B_1, \ldots, B_l$ of M as follows. Let ν be the function as in the hypothesis and for each $g \in C$ let $\nu_g : S^{n-1} \to \mathbb{R}$ be the function defined by $\nu_g(v) = \|g^{-1}v\|^{-1}\nu(g^{-1}v/\|g^{-1}v\|)$; namely ν_g is the function such that $g\Omega = \{v \in \mathbb{R}^n \mid \|v\| < \nu_g(v/\|v\|)\}$. Since ν is a continuous positive function and C is compact it follows that the family of functions $\{\log \nu_g\}_{g\in C}$ is equicontinuous. For any $d \in \mathbb{R}$ let $E_d = \{v \in S^{n-1} \mid Q_0(v) = d\}$. Clearly each E_d is M-invariant. Further it is straightforward to verify that M acts transitively on E_d for all $d \in \mathbb{R}$. Let σ_d be the M-invariant probability measure on E_d. We now choose open subsets $\Phi_0, \ldots, \Phi_l$ of S^{n-1} satisfying the following conditions: i) $\Phi_1, \ldots, \Phi_l$ are mutually disjoint, $\bigcup_0^l \overline{\Phi}_j = S^{n-1}$ and $\bigcup_1^l \overline{\Phi}_j$ contains a neighbourhood of E_0, ii) each Φ_j, $j = 1, \ldots, l$, is bounded by a piecewise smooth hypersurface which is transverse to E_0 at all points of intersection, and iii) for any $j = 1, \ldots, l$ and any $v, v' \in \Phi_j$, $\nu_g(v')/\nu_g(v) < \omega^{-1}$ for all $g \in C$. In view of condition ii) there exist $d_0 > 0$ and compact subsets $\Psi_1, \ldots, \Psi_l$ of $\Phi_1, \ldots, \Phi_l$ respectively such that for all $j = 1, \ldots, l$ and $d \in [0, d_0]$, $\sigma_d(\Psi_j) > \omega\sigma_d(\Phi_j)$. By replacing the original choice suitably, if necessary, we assume that Φ_0 contains E_d for all $d \geqslant d_0$ and that its closure is disjoint from E_0. Let $\varepsilon > 0$ be such that $B(w, 2\varepsilon) \subseteq \Phi_j$ for all $w \in \Psi_j$, $j = 1, \ldots, l$. Finally, let $B_1, \ldots, B_l$ be the subsets of M defined by $B_j = \{m \in M \mid B(e_1, \varepsilon) \subseteq m\Phi_j\}$, $j = 1, \ldots, l$.

Applying Proposition 9.4 with the choices as above we get that there exists a $T_1 \geqslant T_0$ such that for all $j = 1, \ldots, l$, $g \in C$, and $T > T_1$,

$$\frac{1}{(\kappa - 1)T}\int_T^{\kappa T}\int_{B_j} \varphi(u_t mg\Gamma)\, d\sigma(m)\, dt \geqslant (1-\delta)\sigma(B_j)\int_{G/\Gamma}\varphi\, d\mu \geqslant \omega^2\sigma(B_j)\lambda(\Delta).$$

Recalling that $\varphi \leqslant \tilde{\chi}$ and using the identity (9.2) with $B = B_j$ we deduce from this that for all $j = 1, \ldots, l$, $g \in C$, and $T \geqslant T_1$,

$$\frac{1}{(\kappa - 1)T}\int_T^{\kappa T}\int_{B_j}\sum_{v\in g\mathbb{Z}^n}\chi(u_t mv)\, d\sigma(m)\, dt \geqslant \omega^2\sigma(B_j)\lambda(\Delta). \tag{9.9}$$

Let $g \in C$ be arbitrary (but considered fixed). For any $T \geqslant 0$ let $E(T) = \{v \in \mathbb{R}^n \mid u_t v \in \Delta$ for some $T \leqslant t \leqslant \kappa T\}$, namely the set as in Lemma 9.8

for Δ and κ as above, and for $1 \leqslant j \leqslant l$ and $T \geqslant T_1$ and put $S_j(T) = \{v \in \mathbb{R}^n \mid mv \in E(T) \text{ for some } m \in B_j\}$. Since χ is the characteristic function of Δ it follows that the left-hand side of (9.9) is bounded by

$$|S_j(T) \cap g\mathbb{Z}^n|(2\tau/(\kappa - 1)T) \sup_{v \in S_j(T)} \sigma(\{m \in M \mid mv \in E(T)\}).$$

The latter is at most $\kappa^2\omega^{-1}|S_j(T) \cap g\mathbb{Z}^n|(2\tau/(\kappa-1)T)\lambda(E(T))/\lambda(ME(T))$, by Lemma 9.8. Clearly $\lambda(E(T)) = ((\kappa - 1)T + 2\tau)\lambda(\Delta)/2\tau$. Hence for $T > (\kappa - 1)^{-1}$ the left-hand side of (9.9) is bounded by

$$\kappa^2\omega^{-1}(1 + 2\tau)|S_j(T) \cap g\mathbb{Z}^n|\lambda(\Delta)/\lambda(ME(T)).$$

Comparing with the right-hand side and recalling that $\tau < \delta$ we now get that

$$|S_j(T) \cap g\mathbb{Z}^n| \geqslant \kappa^{-2}\omega^4\sigma(B_j)\lambda(ME(T)) \geqslant \kappa^{-2}\omega^5\lambda(S_j(T)).$$

By Proposition 9.7 there exists a $T_2 \geqslant T_1 + (\kappa - 1)^{-1}$ such that for all $T \geqslant T_2$, $V_0(J) \cap A((1+\delta)\beta T^2/2,\ \kappa^2(1-\delta)\alpha T^2/2) \cap c(\Psi_j) \subseteq S_j(T) \subseteq A((1-\delta)\omega\alpha T^2/2,\ \kappa^2(1+\delta)\omega^{-1}\beta T^2/2) \cap c(\Phi_j)$. Hence for all $j = 1, \dots, l$ and $T \geqslant T_2$ we have

$$\begin{aligned}
|V_0(J) \cap A((1-\delta)\omega\alpha T^2/2,\ &\kappa^2(1+\delta)\omega^{-1}\beta T^2/2) \cap c(\Phi_j) \cap g\mathbb{Z}^n| \\
&\geqslant |S_j(T) \cap g\mathbb{Z}^n| \geqslant \kappa^{-2}\omega^5\lambda(S_j(T)) \\
&\geqslant \kappa^{-2}\omega^5\lambda(V_0(J) \cap A((1+\delta)\beta T^2/2,\ \kappa^2(1-\delta)\alpha T^2/2) \cap c(\Psi_j)).
\end{aligned}$$

There exists $T_3 \geqslant T_2$ such that $V_0(J) \cap A(\beta T_3^2/2, \infty)$ is contained in the cone $c(\bigcup_{0<d\leqslant d_0} E_d)$. Then the choice of Ψ_j's shows that for $T \geqslant T_3$ the last term above is at least

$$\kappa^{-2}\omega^6\lambda(V_0(J) \cap A((1+\delta)\beta T^2/2,\ \kappa^2(1-\delta)\alpha T^2/2) \cap c(\Phi_j)).$$

Then (9.6) and the choice of δ imply that there exists $T_4 \geqslant T_3$ such that for all $T \geqslant T_4$ the last term is at least

$$\kappa^{-4}\omega^6\lambda(V_0(J) \cap A((1-\delta)\omega\alpha T^2/2,\ \kappa^2(1+\delta)\omega^{-2}\beta T^2/2) \cap c(\Phi_j)).$$

Now let $r' = \kappa^2(1+\delta)\omega^{-1}\beta T_4^2/2$ and let $r \geqslant r'$ be arbitrary. We can then construct a finite sequence $r_1, \dots, r_q$ such that $r = r_1 > r_2 > \cdots > r_q$, $r_q \leqslant r'$, and each interval (r_{i+1}, r_i) is of the form $((1-\delta)\omega\alpha T^2/2,\ \kappa^2(1+\delta)\omega^{-1}\beta T^2/2)$ for some $T \geqslant T_4$. Then by the above argument and the fact that κ^{-4} and ω^6 are at least $(1 - \theta/4)$ we get that

$$|V_0(J) \cap A(r_{i+1}, r_i) \cap c(\Phi_j) \cap g\mathbb{Z}^n| \geqslant (1-\theta/2)\lambda(V_0(J) \cap A(r_{i+1}, \omega^{-1}r_i) \cap c(\Phi_j))$$

for all $i = 1, \dots, q-1$ and hence

$$\begin{aligned}
|V_0(J) \cap A(0, r) \cap c(\Phi_j) \cap g\mathbb{Z}^n| &\geqslant (1-\theta/2)\lambda(V_0(J) \cap A(r_q, \omega^{-1}r) \cap c(\Phi_j)) \\
&\geqslant (1-\theta/2)\lambda(V_0(J) \cap A(r', \omega^{-1}r) \cap c(\Phi_j)).
\end{aligned}$$

Since $\nu_g(v')/\nu_g(v) < \omega^{-1}$ for all $g \in C$ and $v, v' \in \Phi_j$, $j = 1, \dots, l$, and $g\Omega = \{v \in \mathbb{R}^n \mid \|v\| < \nu_g(v/\|v\|)\}$ there exist $\rho_0, \dots, \rho_l > 0$ such that

$$\bigcup_{j=1}^{l} A(0, \rho_j) \cap c(\Phi_j) \subseteq g\Omega \subseteq \bigcup_{j=0}^{l} A(0, \omega^{-1}\rho_j) \cap c(\Phi_j)$$

for all $g \in C$. Hence for any $r \geqslant \max\{r'\rho_j^{-1} | 0 \leqslant j \leqslant l\}$, we have

$$\begin{aligned}|V_0(J) \cap r(g\Omega) \cap g\mathbb{Z}^n| &\geqslant \sum_{j=1}^{l} |V_0(J) \cap A(0, r\rho_j) \cap c(\Phi_j) \cap g\mathbb{Z}^n| \\ &\geqslant (1-\theta/2) \sum_{j=1}^{l} \lambda(V_0(J) \cap A(r', \omega^{-1} r\rho_j) \cap c(\Phi_j)).\end{aligned}$$

Since the closure of Φ_0 is disjoint from E_0, there exists a $\rho' \geqslant r'$ such that $V_0(J) \cap c(\Phi_0)$ is contained in $A(0, \rho')$. Therefore together with $A(0, \rho')$ the sets in the right-hand expression above cover $V_0(J) \cap r(g\Omega)$ and hence the term is at least $(1-\theta/2)\{\lambda(V_0(J) \cap r(g\Omega)) - \lambda(V_0(J) \cap A(0, \rho'))\}$. There exists an $r_0 \geqslant \max\{r'\rho_j^{-1} | 0 \leqslant j \leqslant l\}$ such that for all $g \in C$, $\lambda(V_0(J) \cap A(0, \rho')) \leqslant \theta\lambda(V_0(J) \cap r_0(g\Omega))/2$. Then for any $g \in C$ and $r \geqslant r_0$ we have

$$|V_0(J) \cap r(g\Omega) \cap g\mathbb{Z}^n| \geqslant (1-\theta)\lambda(V_0(J) \cap r(g\Omega)).$$

Now if $Q \in \mathscr{C}$ and if $g \in C$ is such that $Q = Q_0^g$ then $V_0(J) = gV_Q(J)$ and hence the above relation implies that for all $r \geqslant r_0$,

$$\begin{aligned}|V_Q(J) \cap r\Omega \cap \mathbb{Z}^n| &= |V_0(J) \cap r(g\Omega) \cap g\mathbb{Z}^n| \\ &\geqslant (1-\theta)\lambda(V_0(J) \cap r(g\Omega)) = (1-\theta)\lambda(V_Q(J) \cap r\Omega).\end{aligned}$$

As noted in the beginning, proving this assertion for the interval J completes the proof of the first assertion in Corollary 5.

§10. Values of quadratic forms at integral points. II

It now remains to prove the second part of Corollary 5, for quadratic forms in $n \geqslant 5$ variables. An essentially new element, using the condition that $n \geqslant 5$, involved in this is the fact that when $n \geqslant 5$ any closed orbit of H^0 on G/Γ is noncompact. The latter is a consequence of Meyer's theorem (cf. [35]) that any nondegenerate indefinite rational quadratic form in $n \geqslant 5$ variables has a nontrivial rational zero; if $g \in G$ is such that $H^0 g\Gamma/\Gamma$ is compact then Q_0^g would be a scalar multiple of a rational quadratic form having no nontrivial rational zero (cf. [17, Proposition 9] and [3, Section 8.6]), contradicting the above. Before proceeding to the proof we note the following observation, through which the above fact will come in to play.

10.1. PROPOSITION. *Let $\{y_i\}$ be a sequence of points contained in a compact subset of G/Γ and let $\{T_i\}$ be a sequence in $\mathbb{R}^+$ such that $T_i \to \infty$. Let $\{\pi_i\}$ be the sequence of probability measures on G/Γ such that*

$$\int_{G/\Gamma} f\,d\pi_i = \frac{1}{T_i}\int_{T_i}^{2T_i}\int_M f(u_t m y_i)\,d\sigma\,dt$$

for all bounded continuous functions f on G/Γ. Let π be a limit point of $\{\pi_i\}$ viewed as a sequence of probability measures on the one-point compactification of G/Γ. Then π is a U-invariant probability measure supported on G/Γ and has an ergodic decomposition $\pi = \int \pi_\xi\,d\overline{\pi}(\xi)$, where $\overline{\pi}$ is the quotient measure of the decomposition, such that each π_ξ is either the G-invariant probability measure on G/Γ or is of the form $g\nu$, where $g \in X(H^0, U)$ and ν is a H^0-invariant probability measure supported on a closed H^0-orbit on G/Γ.

PROOF. That π is supported on G/Γ (namely that the point at infinity has zero measure) can be deduced from Theorem 6.1 by an argument as in of the proof of Theorem 3. It is also straightforward to verify that π is U-invariant.

Now let $\delta > 0$ be arbitrary. By Theorem 6.1 there exists a compact subset K_1 of G/Γ such that for any $i \geqslant 1$ and $T \geqslant 0$,

$$l(\{t \in [0, T] \mid u_t y_i \in K_1\}) \geqslant (1-\delta)T.$$

Let Φ be an open neighbourhood of K_1 in G/Γ and let $K = \overline{\Phi}$. As in earlier sections let $\mathscr{H}$ be the class of all proper closed connected subgroups H of G such that $H \cap \Gamma$ is a lattice in H and $\mathrm{Ad}(H\cap\Gamma)$ is Zariski-dense in $\mathrm{Ad}\,H$. Let $\mathscr{H}'$ be the subclass of $\mathscr{H}$ consisting of those subgroups which are not conjugate to H^0. Let $\{C_j\}$ be a sequence of compact subsets such that $\bigcup C_j = \bigcup_{H' \in \mathscr{H}'} X(H', U)$. Let $\mathscr{E}'$ be the class of subsets which are finite unions of the form $\eta_{H_1}^{-1}(D_1) \cup \cdots \cup \eta_{H_k}^{-1}(D_k)$, where $H_1, \dots, H_k \in \mathscr{H}'$ and each D_r, $1 \leqslant r \leqslant k$, is a compact subset of A_{H_r}. Clearly $\mathscr{H}'$ satisfies the condition in Remark 8.3 and hence the general version of Corollary 8.2 indicated in the remark implies that for each j there exist a neighbourhood Ω_j of $C_j\Gamma/\Gamma$ and a $E_j \in \mathscr{E}'$ such that for any compact subset F of $K - E_j\Gamma/\Gamma$ there exists a $T_0 \geqslant 0$ such that for all $y \in F$ and $T \geqslant T_0$,

$$l(\{t \in [0, T] \mid u_t y \in \Omega_j \cap K\}) \leqslant \delta T/2^j. \tag{10.2}$$

For each j let $C_j' = K \cap E_j\Gamma/\Gamma$. Let $j \geqslant 1$ be arbitrary. We claim that for any $y \in G/\Gamma$, $\sigma(\{m \in M \mid my \in C_j'\}) = 0$. Suppose that this is not true. Then for some $y \in G/\Gamma$ there exists an $H' \in \mathscr{H}'$ such that $\sigma(\{m \in M \mid my \in X(H', U)\Gamma/\Gamma\}) > 0$. Since $X(H', U)\Gamma/\Gamma$ is a real analytic submanifold, it follows that there exists an $m_0 \in M$ such that

$M^0 m_0 g \subseteq X(H', U)$, for some $g \in G$ such that $g\Gamma = y$. Then H' contains $(mm_0 g)^{-1} U(mm_0 g)$ for all $m \in M^0$ and hence the closed subgroup generated by them. But the latter is nothing but $g^{-1} H^0 g$ and since H^0 is a maximal connected subgroup of G it follows that $H' = g^{-1} H^0 g$. But this contradicts the assumption that $H' \in \mathscr{H}'$. Hence $\sigma(\{m \in M \mid my \in C_j'\}) = 0$ for all $y \in G/\Gamma$. Since C_j' is compact and $\{y_i\}$ is contained in a compact subset of G/Γ this further implies that there exists a neighbourhood Ω_j' of C_j' such that for all i, $\sigma(\{m \in M \mid my_i \in \Omega_j'\}) < \delta/2^j$. Now applying (10.2) with $K - \Omega_j'$ in the place of F and using the preceding relation we see from the definition of π_i that $\pi_i(\Omega_j \cap \Phi) \leqslant \delta/2^{j-1}$, for all i such that $T_i \geqslant T_0$. This implies that $\pi(K_1 \cap C_j\Gamma/\Gamma) \leqslant \pi(\Omega_j \cap \Phi) \leqslant \delta/2^{j-1}$. Since this is true for all j and since $\pi(G/\Gamma - K_1)$ is at most δ we get that $\pi(\bigcup C_j\Gamma/\Gamma) \leqslant 3\delta$. Since δ was arbitrary this means that $\pi(\bigcup C_j\Gamma/\Gamma) = \pi(\bigcup_{H' \in \mathscr{H}'} X(H', U)\Gamma/\Gamma) = 0$.

By Ratner's classification of invariant measures of unipotent flows (cf. [32, Theorem 1]) any ergodic U-invariant probability measure is a homogeneous measure, namely an L-invariant measure on a closed L-orbit for some closed connected subgroup L of G. Further, by Theorem 2.3 and Corollary 2.13 of [36] the subgroup L is such that the isotropy subgroup of any point in the closed orbit is Zariski-dense in L; this means that L is a conjugate of some $H \in \mathscr{H}$. By the well-known ergodic decomposition theorem (cf. [34]) any U-invariant probability measure has an ergodic decomposition in terms of such measures. In view of the above observation it follows that for the probability measure π the measures supported on closed orbits of subgroups conjugate to those in $\mathscr{H}'$ are redundant in the ergodic decomposition as the corresponding total measure is zero. It is easy to see that all other ergodic measures as above, except for the G-invariant probability measure on G/Γ, are of the form $g\nu$ for some $g \in X(H, U)$ and an H^0-invariant probability measure ν supported on a closed orbit of H^0. Therefore π has an ergodic decomposition as in the statement of the proposition.

Proof of Corollary 5, (ii). Since any Ω as in the hypothesis is intermediate to two balls centered at 0, considering the asymptotics of the volumes it is easy to see that without loss of generality we may assume that Ω is a ball centered at 0; this would somewhat simplify understanding the following proof, but it can be easily avoided if one so wishes.

Let $\mathscr{K}$ and $\varepsilon > 0$, as in the statement of the corollary, be given. Let K_0 be a compact subset of G such that any $Q \in \mathscr{K}$ is of the form Q_0^g for some $g \in K_0$. Let $0 < \delta < 1$. By Theorem 6.1 there exists a compact subset K of G/Γ such that for any $g \in K_0$ and $T \geqslant 0$,

$$l(\{t \in [0, T] \mid u_t g\Gamma \in K\}) \geqslant (1-\delta)T.$$

Let P be the subspace of $\mathbb{R}^n$ consisting of all the points which are fixed by the action of U. Let Θ_1 be a nonempty open subset of S^{n-1}, whose closure is disjoint from $P \cap S^{n-1}$. Let $\Delta_1 = \{\rho v \mid 0 \leqslant \rho < \varepsilon, v \in \Theta_1\}$, the part of the cone over Θ_1 in the ball of radius ε at 0. Let χ_1 be the characteristic function of Δ_1 and let $\tilde{\chi}_1$ be the function on G/Γ associated to χ_1, through the relation (9.1). Since $\int \tilde{\chi}_1 d\mu = \lambda(\Delta_1)$, there exists a bounded continuous function φ_1 on G/Γ such that $0 \leqslant \varphi_1 \leqslant \tilde{\chi}_1$ and $\int \varphi_1 d\mu > \lambda(\Delta_1)/2$. By Theorem 3 there exist proper closed subgroups $H_1, \dots, H_k$ such that $H_j \cap \Gamma$ is a lattice in H_j for all j and compact subsets $D_1, \dots, D_k$ of $X(H_1, U), \dots, X(H_k, U)$ respectively such that for any compact subset F of $K - \bigcup D_j\Gamma/\Gamma$ there exists a $T_0 \geqslant 0$ such that for all $x \in F$ and $T > T_0$,

$$\left|\frac{1}{T}\int_0^T \varphi_1(u_t x)\,dt - \int_{G/\Gamma} \varphi_1\,d\mu\right| < \frac{\lambda(\Delta_1)}{3}. \tag{10.3}$$

By reindexing we may assume that $H_1, \dots, H_r$, where $0 \leqslant r \leqslant k$, are conjugate to H^0 and H_j, $j \geqslant r$, are not conjugate to H^0. For $j = 1, \dots, r$ let $g_j \in G$ be such that $H_j = g_j H^0 g_j^{-1}$ and let $D_j' = D_j g_j$. Then for any $x \in D_j$, $(xg_j)^{-1}U(xg_j) \subseteq g_j^{-1}H_j g_j = H_0$ and hence $D_j' \subseteq X(H^0, U)$ for all $j = 1, \dots, r$. Put $D = \bigcup D_j' \subseteq X(H^0, U)$.

We note that P does not contain any nonzero orbit of gM^0g^{-1} for any $g \in X(H^0, U)$. Suppose that this is not true. Then there exists a $g \in X(H^0, U)$ such that $g^{-1}P$ contains a nonzero M^0-orbit. Since each $v \in g^{-1}P$ is fixed by $g^{-1}Ug$ this implies that $g^{-1}P$ contains a nonzero subspace invariant to the subgroup generated by M^0 and $g^{-1}Ug$. Since $g \in X(H^0, U)$, $g^{-1}Ug \subseteq H^0$. But M^0 is a maximal subgroup of H^0 and therefore M^0 and $g^{-1}Ug$ generate H^0. Thus we get that $g^{-1}P$ contains an H^0-invariant nonzero subspace, but this is a contradiction since in fact the action of H^0 on $\mathbb{R}^n$ is irreducible. Hence P does not contain any orbit of gM^0g^{-1} for any $g \in X(H^0, U)$. Then for such a g and any $v \in S^{n-1}$, $\sigma(\{m \in M^0 \mid gmv \in P\}) = 0$. We conclude from this that for the compact set D of $X(H^0, U)$ as above there exists a compact neighbourhood Φ of $P \cap S^{n-1}$ in S^{n-1}, such that for any $g \in D$ and $v \in S^{n-1}$,

$$\sigma(\{m \in M^0 \mid gmv/\|gmv\| \in \Phi\}) < \delta;$$

without loss of generality we may assume that Φ is disjoint from Θ_1. Now let $\Theta = S^{n-1} - \Phi$, $\Delta = \{\rho v \mid 0 < \rho < \varepsilon, v \in \Theta\}$, and let χ be the characteristic function of Δ. Let B be the open ball with center at 0 and radius $\varepsilon/\sup\{\|g\| \mid g \in D\}$. Then it is easy to see that for any $v \in B$ and $g \in D$, $\int \chi(gmv)\,dm \geqslant (1-\delta)$. Let $\tilde{\chi}$ be the function on G/Γ associated to χ as before, via (9.1).

We next show that there exist $c_0 > 0$ and $T' \geqslant 0$ such that for any $x \in K_0$ and $T > T'$,

$$\frac{1}{T}\int_T^{2T}\int_M \tilde{\chi}(u_t m x\Gamma)\,d\sigma(m)\,dt \geqslant c_0. \tag{10.4}$$

Suppose this is not true. Then there exist sequences $\{x_i\}$ in K_0 and $\{T_i\}$ in $\mathbb{R}^+$ respectively such that $T_i \to \infty$ and

$$\frac{1}{T_i}\int_{T_i}^{2T_i}\int_M \tilde{\chi}(u_t m x_i\Gamma)\, d\sigma(m)\, dt \to 0. \tag{10.5}$$

Let π_i be the probability measure on G/Γ such that

$$\int_{G/\Gamma} f\, d\pi_i = \frac{1}{T_i}\int_{T_i}^{2T_i}\int_M f(u_t m x_i\Gamma)\, d\sigma(m)\, dt$$

for any bounded continuous function f on G/Γ. Let π be a limit point of $\{\pi_i\}$, viewed as a sequence of probability measures on the one-point compactification of G/Γ. Then by Proposition 10.1 π is a U-invariant probability measure on G/Γ with an ergodic decomposition $\int \pi_\xi\, d\overline{\pi}(\xi)$, where $\overline{\pi}$ is the quotient measure for the decomposition, such that any π_ξ is either the G-invariant probability measure on G/Γ or is of the form $g\nu$ where ν is an H^0-invariant probability measure supported on a closed orbit of H^0 and $g \in X(H^0, U)$. We claim that $\int \tilde{\chi} d\pi > 0$. From the choice of K it is clear that $\pi(K) \geqslant (1-\delta) > 0$ and hence to prove the claim it is enough to prove that $\int \tilde{\chi} d\pi_\xi > 0$ for all ergodic components π_ξ such that $\pi_\xi(K) > 0$. If π_ξ is the G-invariant probability measure then clearly $\int \tilde{\chi} d\pi_\xi \geqslant \int \tilde{\chi}_1 d\pi_\xi = \lambda(\Delta_1) > 0$. Now let π_ξ be an ergodic component of the form $g\nu$ as above. If $\pi_\xi(K - \bigcup D_j\Gamma/\Gamma) > 0$ then applying (10.3) to a generic point for $g\nu$ with respect to $\{u_t\}$ we see that $|\int \varphi_1\, d\pi_\xi - \int \varphi_1 d\mu| < \lambda(\Delta_1)/3$ and hence $\int \tilde{\chi} d\pi_\xi \geqslant \int \varphi_1 d\mu - \lambda(\Delta_1)/3 > 0$. Now let π_ξ be of the form $g\nu$ as above, such that $\pi_\xi(K - \bigcup D_j\Gamma/\Gamma) = 0$ and $\pi_\xi(K) > 0$. Then comparing the closed orbits we see that there exists a $g \in D$ such that $\pi_\xi = g\nu$, where ν is the H^0-invariant probability measure on $H^0 g_j^{-1}\Gamma/\Gamma$, for some $j = 1, \ldots, r$. Then

$$\int_{G/\Gamma} \tilde{\chi}\, d\pi_\xi = \int_{G/\Gamma} \tilde{\chi}(gx\Gamma)\, d\nu(x\Gamma) = \int_{G/\Gamma}\int_M \tilde{\chi}(gmx\Gamma)\, d\sigma(m)\, d\nu(x\Gamma).$$

Clearly $x \mapsto \int_M \tilde{\chi}(gmx\Gamma)\, d\sigma(m)$ is associated to $\int_M \chi(gmv)\, d\sigma(m)$ under the correspondence set up by (9.1). Recall that there exists a ball B centered at 0 such that $\int_M \chi(gmv)\, d\sigma(m) \geqslant (1-\delta)$ for all $v \in B$ and $g \in D$. Let ψ be the characteristic function of B and let $\tilde{\psi}$ be the function on G/Γ associated to ψ. Then we have $\int_M \tilde{\chi}(gmx\Gamma)\, d\sigma(m) \geqslant (1-\delta)\tilde{\psi}(x\Gamma)$ and hence

$$\int_{G/\Gamma}\int_M \tilde{\chi}(gmx\Gamma)\, d\sigma(m) d\nu(x\Gamma) \geqslant (1-\delta)\nu(Y),$$

where Y is the set of all $x\Gamma$ such that $B \cap x\mathbb{Z}^n \neq \{0\}$. By Mahler's criterion $G/\Gamma - Y$ is compact. On the other hand since $n \geqslant 5$, as noted earlier by Meyer's theorem any closed H^0-orbit on G/Γ is noncompact and hence, in particular, so is the support of ν. Therefore $\nu(Y) > 0$. Thus we get that $\int_{G/\Gamma} \tilde{\chi}\, d\pi_\xi \geqslant (1-\delta)\nu(Y) > 0$ in this case as well. This completes the proof that $\int_{G/\Gamma} \tilde{\chi}\, d\pi > 0$. Hence there exists a bounded continuous function φ on G/Γ such that $0 \leqslant \varphi \leqslant \tilde{\chi}$ and $\int_{G/\Gamma} \varphi\, d\pi > 0$. However by (10.5) $\int_{G/\Gamma} \varphi\, d\pi_i \to 0$ and hence $\int_{G/\Gamma} \varphi\, d\pi = 0$, which is a contradiction. Hence there exists a $c_0 > 0$ such that (10.4) holds.

In view of (9.2) and (10.4) we now have

$$\frac{1}{T}\int_T^{2T}\int_M \sum_{v\in g\mathbb{Z}^n} \chi(u_t m v)\, d\sigma(m)\, dt \geqslant c_0$$

for all $g \in K_0$ and $T > T'$. From the form of the set Δ it follows that there exists a constant τ such that for any $v \in \mathbb{R}^n$, $l(\{t \in \mathbb{R} \mid u_t v \in \Delta\}) \leqslant \tau$. Now if for $T > 0$ we put $E(T) = \{v \in \mathbb{R}^n \mid u_t v \in \Delta \text{ for some } T \leqslant t \leqslant 2T\}$ and $S(T) = \{v \in \mathbb{R}^n \mid mv \in E(T) \text{ for some } m \in M\}$ then for any $g \in K_0$ and $T > T'$ we get

$$|S(T) \cap g\mathbb{Z}^n|(\tau/T) \sup_{v\in S(T)} \sigma(\{m \in M \mid mv \in E(T)\}) \geqslant c_0.$$

Arguing as in the proof of Corollary 5, (i), one can deduce from this that there exist $c > 0$ and $r_0 \geqslant 0$ such that for any $g \in K_0$ and $r \geqslant r_0$

$$|V_0((-\varepsilon, \varepsilon)) \cap rg\Omega \cap g\mathbb{Z}^n| \geqslant c\lambda(V_0((-\varepsilon, \varepsilon)) \cap rg\Omega).$$

Since there exist constants $c_1, c_2 > 0$ such that $c_1\Omega \subseteq g\Omega \subseteq c_2\Omega$ for any $g \in K_0$, and since $V_0((-\varepsilon, \varepsilon)) = gV_Q((-\varepsilon, \varepsilon))$, this implies the corollary.

REFERENCES

1. E. Artin, *Geometric algebra*, Interscience, New York, 1957.
2. A. Borel, *Linear algebraic groups*, Benjamin, New York, 1969.
3. ———, *Introduction aux groupes arithmetiques*, Publ. de l'Inst. Math. de l'Univ. de Strassbourg XV, Hermann, Paris, 1969.
4. A. Borel and Harish-Chandra, *Arithmetic subgroups of algebraic groups*, Ann. Math. **75** (1962), 485–535.
5. A. Borel and J. Tits, *Groupes réductifs*, Inst. Hautes Etudes Sci. Publ. Math. **27** (1965), 55–150.
6. J. Brezin and C. C. Moore, *Flows on homogeneous spaces: a new look*, Amer. J. Math. **103** (1981), 571–613.
7. M. Burger and P. Sarnak, *Ramanujan duals.* II, Invent. Math. **106** (1991), 1–11.
8. C. Chevalley, *Theorie des groupes de Lie.* II. *Groupes algebraiques*, Hermann, Paris, 1951.
9. S. G. Dani, *Invariant measures of horospherical flows on noncompact homogeneous spaces*, Invent. Math. **47** (1978), 101–138.
10. ———, *A simple proof of Borel's density theorem*, Math. Z. **174** (1980), 81–94.
11. ———, *Invariant measures of horospherical flows on homogeneous spaces*, Invent. Math. **64** (1981), 357–385.
12. ———, *On orbits of unipotent flows on homogeneous spaces*, Ergodic Theory Dynamical Systems **4** (1984), 25–34.

13. ______, *Dynamics of flows on homogeneous spaces — a survey*, Proceedings of the Colloquio de Systemas Dynamicos (Guanajuato, 1983), Aportacione Mat., vol. 1, Soc. Mat. Mexicana, Mexico City, 1985, pp. 1–30.

14. ______, *On orbits of unipotent flows on homogeneous spaces*. II, Ergodic Theory Dynamical Systems **6** (1986), 167–182.

15. S. G. Dani and G. A. Margulis, *Values of quadratic forms at primitive integral points*, Invent. Math. **98** (1989), 405–424.

16. ______, *Orbit closures of generic unipotent flows on homogeneous spaces of* $SL(3, \mathbb{R})$, Math. Ann. **286** (1990), 101–128.

17. ______, *Values of quadratic forms at integral points; an elementary approach*, Enseign. Math. **36** (1990), 143–174.

18. ______, *Asymptotic behaviour of trajectories of unipotent flows on homogeneous spaces*, Proc. Indian Acad. Sci. Math. Sci. **101** (1991), 1–17.

19. ______, *On the limit distributions of orbits of unipotent flows and integral solutions of quadratic inequalities*, C. R. Acad. Sci. Paris Ser. I **314** (1992), 698–704.

20. S. G. Dani and M. McCrudden, *On the factor sets of measures and local tightness of convolution semigroups over Lie groups*, J. Theoret. Probab. **1** (1988), 357–370.

21. S. Helgason, *Differential geometry, Lie groups and symmetric spaces*, Academic Press, New York, 1978.

22. G. A. Margulis, *Arithmetic properties of discrete subgroups*, Uspekhi Mat. Nauk **29** (1974), no. 1, 49–98; English transl. in Russian Math. Surveys **29** (1974), 107–156.

23. ______, *Lie groups and ergodic theory*, Algebra — Some Current Trends (Proceedings, Varna, 1986), Lecture Notes in Math., vol. 1352, Springer-Verlag, Berlin and New York, 1988, pp. 130–146.

24. ______, *Formes quadratiques indéfinies et flots unipotents sur les espaces homogènes*, C. R. Acad. Sci. Paris Ser. I **304** (1987), 247–253.

25. ______, *Indefinite quadratic forms and unipotent flows on homogeneous spaces*, Dynamical Systems and Ergodic Theory, vol. 23, Banach Center Publ., PWN — Polish Scientific Publ., Warsaw, 1989, pp. 399–409.

26. ______, *Number theory, trace formulas and discrete subgroups*, Academic Press, Boston, MA, 1989, pp. 377–398.

27. ______, *Dynamical and ergodic properties of subgroup actions on homogeneous spaces with applications to number theory*, Proceedings of the International Congress of Mathematicians (Kyoto, 1990), The Mathematical Society of Japan, Tokyo, pp. 193–215; Springer-Verlag, Berlin and New York, 1991.

28. G. A. Margulis and G. M. Tomanov, *Measure rigidity for algebraic groups over local fields*, Preprint.

29. M. S. Raghunathan, *Discrete subgroups of Lie groups*, Springer-Verlag, Berlin, 1972.

30. Marina Ratner, *Strict measure rigidity for unipotent subgroups of solvable groups*, Invent. Math. **101** (1990), 449–482.

31. ______, *On measure rigidity of unipotent subgroups of semisimple groups*, Acta. Math. **165** (1990), 229–309.

32. ______, *On Raghunathan's measure conjecture*, Ann. of Math. (2) **134** (1991), 545–607.

33. ______, *Raghunathan's topological conjecture and distributions of unipotent flows*, Duke Math. J. **63** (1991), 235–280.

34. V. A. Rohlin, *Selected topics from metric theory of dynamical systems*, Uspekhi Mat. Nauk **4** (1949), no. 2, 57–128; English transl. in Russian Math. Surveys **49** (1966), 171–239.

35. J. P. Serre, *A course in arithmetic*, Springer-Verlag, Berlin, 1973.

36. Nimish A. Shah, *Uniformly distributed orbits of certain flows on homogeneous spaces*, Math. Ann. **289** (1991), 315–334.

37. C. L. Siegel, *A mean value theorem in geometry of numbers*, Ann. of Math. (2) **46** (1945), 340–347.

38. A. Weil, *L'integration dans les groupes topologiques et ses applications*, Hermann, Paris, 1940.

39. R. J. Zimmer, *Ergodic theory and semisimple groups*, Birkhäuser, Basel, 1984.

SCHOOL OF MATHEMATICS, TATA INSTITUTE OF FUNDAMENTAL RESEARCH, HOMI BHABHA ROAD, BOMBAY 400 005, INDIA

DEPARTMENT OF MATHEMATICS, YALE UNIVERSITY, BOX 2155, NEW HAVEN, CONNECTICUT 06520

ADVANCES IN SOVIET MATHEMATICS
Volume 16, Part 1, 1993

Coinvariants of Nilpotent Subalgebras of the Virasoro Algebra and Partition Identities

BORIS FEIGIN AND EDWARD FRENKEL

Dedicated to I. M. Gelfand on his 80th birthday

§1. Introduction

Let $V_{p,q}^{m,n}$ be the irreducible representation of the Virasoro algebra $\mathscr{L}$ with central charge $c_{p,q} = 1 - 6(p-q)^2/pq$ and highest weight $h_{m,n} = [(np-mq)^2 - (p-q)^2]/4pq$, where $p, q > 1$ are relatively prime integers and m, n are integers, such that $0 < m < p$ and $0 < n < q$. For fixed p and q the representations $V_{p,q}^{m,n}$ form the (p, q) minimal model of the Virasoro algebra [1].

For $N > 0$ let $\mathscr{L}_N$ be the Lie subalgebra of the Virasoro algebra, generated by L_i, $i < -N$. There is a map from the Virasoro algebra to the Lie algebra of polynomial vector fields on $\mathbb{C}^*$ that takes L_i to $z^{-i+1}\frac{\partial}{\partial z}$, where z is a coordinate. This map identifies $\mathscr{L}_N$ with the Lie algebra of vector fields on $\mathbb{C}$ that vanish at the origin along with the first $N+1$ derivatives. The Lie algebra $\mathscr{L}_{2N}$ has a family of deformations $\mathscr{L}(p_1, \dots, p_{N+1})$ that consist of vector fields, vanishing at the points $p_1, \dots, p_{N+1} \in \mathbb{C}$ along with the first derivative.

For a Lie algebra g and a g-module M we denote by $H(g, M)$ the space of coinvariants (or 0th homology) of g in M, which is the quotient $M/g \cdot M$, where $g \cdot M$ is the subspace of M, linearly spanned by vectors $a \cdot x$, $a \in g$, $x \in M$. We will prove the following result.

THEOREM 1. *For any irreducible representation V of the $(2, 2r+1)$ minimal model,*

$$\dim H(\mathscr{L}_{2N}, V) = \dim H(\mathscr{L}(p_1, \dots, p_{N+1}), V).$$

This statement was proved in [2] for $N = 1$ and irreducible representations of general minimal models.

1991 *Mathematics Subject Classification.* Primary 17B68, 81R10; Secondary 05A20.
Research of the second author was supported by a Junior Fellowship from the Harvard Society of Fellows and by NSF grant DMS-9205303.

As an application, we will give a new proof of the Gordon identities, which relies on two different ways to compute characters of the irreducible representations of the $(2, 2r+1)$ minimal models.

Let us note that a priori $\dim H(\mathscr{L}_{2N}, V) \geqslant \dim H(\mathscr{L}(p_1, \dots, p_{N+1}), V)$. Indeed, if we have a family of Lie algebras, then the dimension of homology is the same for generic points of the space of parameters of the family, but it may increase at special points.

Each minimal model of the Virasoro algebra associates a linear space to a punctured complex curve with a representation inserted at each puncture. This space is called the space of conformal blocks. It can be defined as the space of coinvariants of the Lie algebra of meromorphic vector fields on this curve, which are allowed to have poles only at the punctures, in the tensor product of these representations (cf. [1], mathematical aspects of this correspondence are treated in detail in [2, 3]).

In this language, the space $H(\mathscr{L}(p_1, \dots, p_{N+1}), V_{p,q}^{m,n})$ is isomorphic to the direct sum of the spaces of conformal blocks associated to the projective line with punctures $p_1, \dots, p_{N+1}$ with all possible insertions from the (p, q) minimal model, and ∞ with the insertion of $V_{p,q}^{m,n}$. The dimension of this space can be calculated by a Verlinde type argument. Namely, consider a quadratic degeneration of our projective line at $N+1$ different points, such that each of the $N+2$ components of the degenerated curve contains exactly one of the $N+2$ punctures. Under this deformation the dimension of the homology does not change [2, 3], and can therefore be reduced to the well-known result on the dimensions of the spaces of conformal blocks associated to the projective line with three insertions, the so-called fusion coefficients.

This gives an explicit formula for $\dim H(\mathscr{L}(p_1, \dots, p_{N+1}), V_{p,q}^{m,n})$, which represents a lower bound for $\dim H(\mathscr{L}_{2N}, V_{p,q}^{m,n})$. We can also obtain an upper bound by a different method.

This method is based on the calculation of the annihilating ideal of a minimal model [4]. Let us recall the definition of the annihilating ideal. For $c \in \mathbb{C}$ let M_c be the vacuum Verma module with central charge c. It is generated by a vector v_c, such that $L_i v_c = 0$, $i \geqslant -1$. This representation has the structure of a vertex operator algebra (VOA) [5], which acts on any irreducible module M of the Virasoro algebra with the same central charge. It means that each vector of M_c defines a local current (or field), which is a formal power series in z and z^{-1} whose Fourier coefficients are linear operators acting on M. These local currents can be constructed as follows. Consider the projective line $\mathbb{CP}^1$ with three punctures: 0 and ∞ with the insertions of M, and z with the insertion of M_c. The space of conformal blocks is one-dimensional in this case. In other words, denote by $\mathscr{A}(0, z)$ the Lie algebra of meromorphic vector fields on $\mathbb{CP}^1$ that are allowed to have poles only at $0, \infty$, and z. Then the space of coinvariants of $\mathscr{A}(0, z)$ in $M \otimes M \otimes M_c$ is one-dimensional. It has a canonical generator: the projection of the tensor product of the highest weight vectors. The dual to this generator defines for each vector of M_c a linear operator from M to M, depending on z, which is our local current.

These local currents can be constructed explicitly. The monomials

$$\{L_{-m_1} \cdots L_{-m_l} v_c \mid m_1 \geqslant m_2 \geqslant \cdots \geqslant m_l > 1\}$$

linearly span M_c. The corresponding local currents are equal to

$$\frac{1}{(m_1-2)!} \cdots \frac{1}{(m_l-2)!} : \partial_z^{m_1-2} T(z) \cdots \partial_z^{m_l-2} T(z) :, \tag{1}$$

where $T(z) = \sum_{i \in \mathbb{Z}} L_i z^{-i-2}$. The Fourier components of these local currents form a Lie algebra $U_c(\mathscr{L})_{\mathrm{loc}}$, which is called the local completion of the universal enveloping algebra of the Virasoro algebra with central charge c [4, 6]. They act on any representation of $\mathscr{L}$ that has the same central charge.

The module M_c is irreducible if and only if c is not equal to $c_{p,q}$. However, if $c = c_{p,q}$, then M_c contains a (unique) singular vector. The quotient of $M_{c_{p,q}}$ by the submodule generated by this singular vector is isomorphic to $V_{p,q}^{1,1}$. This representation, called the vacuum representation of the (p, q) minimal model, defines another VOA, which is the quotient of the VOA of $M_{c_{p,q}}$ in the appropriate sense [7]. An irreducible representation M of the Virasoro algebra with central charge $c_{p,q}$ is a module over this VOA [7] if and only if the space of coinvariants of the Lie algebra $\mathscr{A}(0, z)$ in $M \otimes M \otimes V_{p,q}^{1,1}$ is one-dimensional. An explicit computation of the dimension of this space, which was made in [2], tells us that it is this case if and only if M is an irreducible representation of the (p, q) minimal model.

The Fourier components of the local current corresponding to the singular vector of $M_{c_{p,q}}$ generate an ideal in $U_{c_{p,q}}(\mathscr{L})_{\mathrm{loc}}$, which is called the annihilating ideal of the (p, q) minimal model. Any element of this ideal acts trivially on any representation of the (p, q) minimal model.

Imposing this condition with respect to the generators of the annihilating ideal immediately leads to certain linear relations among the monomial elements of $V_{p,q}^{m,n}$. This allows one to estimate the dimension of the coinvariants $H_0(\mathscr{L}_{2N}, V_{p,q}^{m,n})$ from above.

The peculiarity of the $(2, 2r+1)$ models is that this upper bound coincides with the lower bound, and is therefore exact. This proves Theorem 1 and leads to a nice combinatorial description of the minimal representations of these models. Namely, as a $\mathbb{Z}$-graded linear space, such a representation is isomorphic to the quotient of the space of polynomials in infinitely many variables by a certain monomial ideal. Thus we obtain an expression for the character of this module, which coincides with the right-hand side of one of the Gordon identities. But it is known that this character is equal to the left-hand side of the identity. Hence, we obtain a new proof of the Gordon identities.

Implementing this program for other minimal models would lead to a result similar to Theorem 1, as well as to nice character formulas for irreducible representations. However, to obtain the exact upper bound for general minimal models, for which the representations are "smaller" and the structure of the annihilating ideal is more complicated, one should impose extra conditions corresponding to other elements of the annihilating ideal.

§2. Lower bound: genus 0 conformal blocks

In this section we will calculate the dimension of $H(\mathscr{L}(p_1, \dots, p_{N+1}), V)$ for an irreducible representation V of a (p, q) minimal model.

Denote by $\mathscr{A}(p_1, \dots, p_{N+1})$ the Lie algebra of meromorphic vector fields on $\mathbb{CP}^1$ that are allowed to have poles only at the distinct points $p_1, \dots, p_{N+1} \in \mathbb{C}$ and ∞. There is an embedding of this Lie algebra into the direct sum of $N+2$ Virasoro algebras, and thus it acts on tensor products $V_1 \otimes \cdots \otimes V_{N+2}$ of $N+2$ representations of the Virasoro algebra with the same central charge.

Let us fix a central charge $c_{p,q}$ and an irreducible representation $V = V_{N+2}$ of the (p, q) minimal model. The following statement follows from the general results on computation of conformal blocks, outlined in [2, §5] (cf. [3] for details).

PROPOSITION 2. (1)

$$\begin{aligned}\dim H(\mathscr{L}(p_1, \dots, p_{N+1}), V) \\ = \sum \dim H(\mathscr{A}(p_1, \dots, p_{N+1}), V_1 \otimes \cdots \otimes V_{N+1} \otimes V),\end{aligned}$$

where the sum is taken over all $(N+1)$-tuples $V_1, \dots, V_{N+1}$ of irreducible representations of the (p, q) minimal model.

(2)

$$\begin{aligned}\dim H(\mathscr{A}(p_1, \dots, p_{N+1}), V_1 \otimes \cdots \otimes V_{N+1} \otimes V) \\ = \sum \dim H(\mathscr{A}(p_1, \dots, p_N), V_1 \otimes \cdots \otimes V_N \otimes W) \\ \cdot \dim H(\mathscr{A}(0, 1), W \otimes V_{N+1} \otimes V),\end{aligned}$$

where the sum is taken over all irreducible representations W of the (p, q) minimal model.

Using this result, we can calculate $\dim H(\mathscr{L}(p_1, \dots, p_{N+1}), V)$ by induction.

Let $V_1, \dots, V_s$ be the set of all irreducible representations of the (p, q) minimal model ($s = (p-1)(q-1)/2$), and let $\mathbf{u}_N$ be the s-vector whose components are $u_N^i = \dim H(\mathscr{L}(p_1, \dots, p_{N+1}), V_i)$, $i = 1, \dots, s$. Put $c_{ijk} = \dim H(\mathscr{A}(0, 1), V_i \otimes V_j \otimes V_k)$. The numbers c_{ijk} are usually called the fusion coefficients. They define the fusion algebra of this minimal model, which has generators g_i, $1 \leqslant i \leqslant s$, and relations $g_i \cdot g_j = \sum_k c_{ijk} g_k$.

Let us introduce the $s \times s$ matrix $\mathscr{M} = \mathscr{M}_{p,q}$ whose (i, j)th entry is equal to $\sum_k c_{ijk}$. This matrix can be interpreted as the matrix of action of the sum $\sum_k g_k$ on the fusion algebra.

According to Proposition 2, $u_N^i = \sum_{j,k} c_{ijk} u_{N-1}^j$. This gives us the recursion relation for $\mathbf{u}_N$:

$$\mathbf{u}_N = \mathscr{M}\mathbf{u}_{N-1}.$$

For any i, the representation V_i is generated from the highest weight vector v_i by the action of the Lie algebra $\mathscr{L}_0$, hence $u_0^i = 1$. Therefore, we obtain the formula

$$\mathbf{u}_N = \mathscr{M}^N \mathbf{u}_0,$$

where $\mathbf{u}_0^t = [1, 1, \ldots, 1]$.

This formula enables us to calculate the dimensions u_N^i explicitly by diagonalizing the matrix $\mathscr{M}$.

In the rest of the paper we will focus on the $(2, 2r+1)$ models. In such a model we have r irreducible representations $V_i = V_{2,2r+1}^{1,i}$, $i = 1, \ldots, r$. The (i, j)th entry of the corresponding $r \times r$ matrix $\mathscr{M}$ is equal to $\min\{i, j\}$. This matrix is equal to the square of the matrix

$$M_r = \begin{pmatrix} 0 & 0 & \cdots & 0 & 1 \\ 0 & 0 & \cdots & 1 & 1 \\ \multicolumn{5}{c}{\dotfill} \\ 0 & 1 & \cdots & 1 & 1 \\ 1 & 1 & \cdots & 1 & 1 \end{pmatrix}. \tag{2}$$

This fact has the following interpretation: in the fusion algebra of this model $\sum_k g_k$ is equal to the square of g_r, and the action of g_r is given by the matrix M_r.

Let us introduce vectors $\mathbf{x}_i$, whose ith component is equal to 1 and all other components are equal to 0. Since $\mathbf{u}_0 = M_r^2 \mathbf{x}_1$, we have $u_N^i = \mathbf{x}_i^t M_r^{2N+2} \mathbf{x}_1$. This gives us a lower bound for $v_N^i = \dim H(\mathscr{L}_{2N}, V_i)$.

PROPOSITION 3. $v_N^i \geqslant \mathbf{x}_i^t M_r^{2N+2} \mathbf{x}_1$.

In the next section we will show that this bound is exact.

§3. Upper bound: the annihilating ideal

The annihilating ideal of the $(2, 2r+1)$ minimal model is the ideal of the Lie algebra $U_{c_{2,2r+1}}(\mathscr{L})_{\text{loc}}$, generated by the Fourier components of the local current of the singular vector of the vacuum Verma module $M_{c_{2,2r+1}}$. All elements of this ideal act by 0 on any irreducible representation of the $(2, 2r+1)$ minimal model. In particular, the generators of the ideal act by

0. This leads to certain relations between vectors in the irreducible minimal representations, as explained in [4].

Introduce a filtration on the irreducible representation V_i of the $(2, 2r+1)$ models as follows: $0 \subset V_i^0 \subset V_i^1 \subset \cdots \subset V_i^\infty = V_i$, where V_i^k is linearly spanned by the monomials $L_{-m_1} \cdots L_{-m_l} v_i$, such that $m_1 \geqslant m_2 \geqslant \cdots \geqslant m_l \geqslant 1$ and $l \leqslant k$. Let Ω_i be the adjoint graded space:

$$\Omega_i = \bigoplus_{k \geqslant 0} \Omega_i^k, \qquad \Omega_i^k = V_i^k / V_i^{k-1}.$$

The space Ω_i is a polynomial algebra in variables a_j, $j > 0$, where a_j is the image of $L_{-j} v_i$.

The symbol of the singular vector of $M_{c_{2,2r+1}}$ is equal to $L_{-2}^r v_0$. According to formula (1), the symbol of the corresponding local current is equal to $: T(z)^r :$. Hence the symbols of the Fourier components of this current are given by the formula

$$\sum_{j_1 + \cdots + j_r = n} : L_{-j_1} \cdots L_{-j_r} : .$$

Their action on Ω_i is given by the multiplication by

$$S_n = \sum_{\substack{j_k \geqslant 0; \\ j_1 + \cdots + j_r = n}} a_{j_1} \cdots a_{j_r} \tag{3}$$

for $n \geqslant r$, and 0 for $n < r$.

It is known that V_i is the quotient of the Verma module with highest weight $h_{1,i}$ by the maximal submodule generated by two singular vectors. The symbol of one of them is equal to $L_{-1}^i v_i$. This gives us a surjective map from the quotient Ω_i' of $\mathbb{C}[a_j]_{j>0}$ by the ideal I_i generated by S_n, $n > r$, and a_1^i, to Ω_i.

Let Ω_i^{mon} be the quotient of $\mathbb{C}[a_j]_{j>0}$ by the ideal I_i^{mon}, generated by the monomials $a_j^u a_{j+1}^v$, $0 \leqslant u < r$, $u + v = r$, $j > 0$, and a_1^i.

We can introduce a $\mathbb{Z}$-grading on the module V_i by putting $\deg v_i = 0$, $\deg L_{-j} = j$. The space Ω_i inherits this grading. We can also introduce a compatible $\mathbb{Z}$-grading on the space $\mathbb{C}[a_j]_{j>0}$ by putting $\deg a_{-j} = j$. The spaces Ω_i' and Ω_i^{mon} inherit this grading.

For any $\mathbb{Z}$-graded linear space $V = \bigoplus_{n \geqslant 0} V(n)$ such that $\dim V(n) \leqslant \infty$, let $\operatorname{ch} V = \sum_{n \geqslant 0} \dim V(n) q^n$ be its character. We will write $\operatorname{ch} V \leqslant \operatorname{ch} V'$ if $\dim V(n) \leqslant \dim V'(n)$ for any $n \geqslant 0$.

We have

$$\operatorname{ch} V_i = \operatorname{ch} \Omega_i \leqslant \operatorname{ch} \Omega_i' \leqslant \operatorname{ch} \Omega_i^{\text{mon}}.$$

The last inequality follows from the fact that each of the generators S_n of the ideal I_i has as a summand one and only one generator of the ideal I_i^{mon}, namely, $a_j^u a_{j+1}^v$ is a summand of S_{rj+v}.

By the construction of our filtration, the character of the space of coinvariants $H(\mathscr{L}_{2N}, V_i)$ is equal to the character of the quotient $\Omega_{i,N}$ of the space Ω_i by the ideal generated by a_j with $j > 2N$. Again, we have:

$$\operatorname{ch} H(\mathscr{L}_{2N}, V_i) = \operatorname{ch} \Omega_{i,N} \leqslant \operatorname{ch} \Omega_{i,N}^{\mathrm{mon}}, \tag{4}$$

where $\Omega_{i,N}^{\mathrm{mon}}$ is the quotient of Ω_i^{mon} by the ideal generated by a_j, $j > 2N$. Therefore,

$$v_N^i = \dim H(\mathscr{L}_{2N}, V_i) \leqslant \omega_N^i = \dim \Omega_{i,2N}^{\mathrm{mon}}.$$

We can easily calculate ω_N^i by induction [8].

Let $C_i^r = \{(m_1, \dots, m_l) \mid m_1 \geqslant \cdots \geqslant m_l \geqslant 1,\ m_i \geqslant m_{i+r-1} + 2,\ m_{l-i+1} > 1\}$. The monomials

$$\{a_{m_1} \cdots a_{m_l} \mid (m_1, \dots, m_l) \in C_i^r\}$$

constitute a linear basis in Ω_i^{mon}.

For every integer $N > 0$, introduce the subspaces $W_{k,N}^i$, $1 \leqslant k \leqslant r$, of Ω_i^{mon} that are linearly spanned by the monomials

$$\{a_{m_1} \cdots a_{m_l} \mid (m_1, \dots, m_l) \in C_i^r,\ \ m_1 \leqslant N,\ \ m_k \leqslant N-1\}.$$

Clearly, $W_{1,N}^i$ is isomorphic to $\Omega_{i,N-1}^{\mathrm{mon}}$.

Due to the relations $a_{N-1}^u a_N^v = 0$ for $u + v = r$, the dimension of the space $W_{k,N}^i$ that is spanned by the monomials of degree less than k in a_N is equal to the sum of dimensions of the spaces $W_{l,N-1}^i$ with $l = 1, \dots, r-k+1$. Let us introduce the r-vector $\mathbf{w}_N^i$, whose components are $w_{k,N}^i = \dim W_{k,N}^i$, $k = 1, \dots, r$. We obtain the formula [8]:

$$\mathbf{w}_N^i = M_r \mathbf{w}_{N-1}^i,$$

where M_r is given by (2), which shows that $\mathbf{w}_{2N+1}^i = M_r^{2N} \mathbf{w}_1^i$.

REMARK 1. As shown in [8], there is a q-deformation of this formula, which gives an expression for the characters of $W_{k,N}^i$.

By definition, $w_{k,1}^i = \min\{i, k\}$. One can check that $\mathbf{w}_1^i = M_r^2 \mathbf{x}_i$, and so $\mathbf{w}_{2N+1}^i = M_r^{2N+2} \mathbf{x}_i$. Therefore, $\omega_N^i = w_{1,2N+1}^i = \mathbf{x}_1^t M_r^{2N+2} \mathbf{x}_i$, and this gives us an upper bound for v_N^i.

PROPOSITION 4. $v_N^i \leqslant \mathbf{x}_1^t M_r^{2N+2} \mathbf{x}_i$.

Since $M_r^t = M_r$, this upper bound coincides with the lower bound from Proposition 3. Therefore, we have the equality

$$u_N^i = v_N^i = \omega_N^i = \mathbf{x}_1^t M_r^{2N+2} \mathbf{x}_i.$$

This completes the proof of Theorem 1.

Note that we have also proved that Ω_i is isomorphic to $\Omega_i' = \mathbb{C}[a_j]_{j>0}/I_i$.

REMARK 2. For general minimal models the graded space Ω of an irreducible representation is also isomorphic to the quotient of $\mathbb{C}[a_j]_{j>0}$ by a certain ideal I. This ideal contains the operators S_n given by formula (3) with $r=(p-1)(q-1)/2$. They correspond to the action of the symbols of generators of the annihilating ideal of the minimal model. However, in general these elements do not generate the ideal I. There are other generators corresponding to the action of the symbols of other elements of the annihilating ideal on Ω. It is an interesting problem to find explicit formulas for them. This will hopefully lead to a nice combinatorial description of general irreducible representations.

§4. Application: Gordon identities

According to formula (4), $\dim H(\mathscr{L}_{2N}, V_i)(n) \leqslant \dim \Omega^{\mathrm{mon}}_{i,2N}(n)$ for any n. In the previous section we proved that $\dim H(\mathscr{L}_{2N}, V_i) = \dim \Omega^{\mathrm{mon}}_{i,2N}$; therefore $\dim H(\mathscr{L}_{2N}, V_i)(n) = \dim \Omega^{\mathrm{mon}}_{i,2N}(n)$ for any n and N. Clearly, $\dim H(\mathscr{L}_{2N}, V_i)(n) = \dim V_i(n)$ and $\dim \Omega^{\mathrm{mon}}_{i,2N}(n) = \dim \Omega^{\mathrm{mon}}_i(n)$ for N large enough. Hence $\dim V_i(n) = \dim \Omega^{\mathrm{mon}}_i(n)$ for any n, and $\mathrm{ch}\, V_i = \mathrm{ch}\, \Omega^{\mathrm{mon}}_i$. This can be interpreted as follows.

PROPOSITION 5. *The monomials*

$$\{L_{-m_1} \cdots L_{-m_l} v_i \mid (m_1, \dots, m_l) \in C^r_i\}$$

constitute a linear basis in the irreducible representation V_i of the $(2, 2r+1)$ minimal model.

This was conjectured in [11].

Thus, we obtain the following formula for the character of the module V_i:

$$\mathrm{ch}\, V_i = \sum_{n\geqslant 0} |C^r_i(n)| q^n, \tag{5}$$

where $C^r_i(n)$ is the subset of C^r_i consisting of the elements $(m_1, \dots, m_l) \in C^r_i$ with $\sum m_j = n$.

The right-hand side of formula (5) is known to be equal to

$$\sum_{n_1, \dots, n_{r-1} \geqslant 0} \frac{q^{N_1^2+\cdots+N_{r-1}^2+N_i+\cdots+N_{r-1}}}{(q)_{n_1} \cdots (q)_{n_{r-1}}},$$

where $N_j = n_j + \cdots + n_{r-1}$ and $(q)_n = \prod_{l=1}^{l=n}(1-q^l)$ (cf. [9]).

On the other hand, it is known [10, 11] that

$$\mathrm{ch}\, V_i = \prod_{\substack{n>0,\\ n\neq 0, \pm i \bmod (2r+1)}} (1-q^n)^{-1}.$$

This formula follows from the Weyl-Kac type character formula for irreducible minimal representations of the Virasoro algebra.

Thus, we have obtained a new proof of the Gordon identities:

$$\prod_{\substack{n>0,\\ n\neq 0,\pm i \bmod (2r+1)}} (1-q^n)^{-1} = \sum_{n_1,\dots,n_{r-1}\geqslant 0} \frac{q^{N_1^2+\dots+N_{r-1}^2+N_i+\dots+N_{r-1}}}{(q)_{n_1}\cdots(q)_{n_{r-1}}}.$$

For $r=2$ these are the Rogers-Ramanujan identities:

$$\prod_{\substack{n>0,\\ n\neq 0,\pm 1 \bmod 5}} (1-q^n)^{-1} = \sum_{n\geqslant 0} \frac{q^{n(n+1)}}{(q)_n}$$

and

$$\prod_{\substack{n>0,\\ n\neq 0,\pm 2 \bmod 5}} (1-q^n)^{-1} = \sum_{n\geqslant 0} \frac{q^{n^2}}{(q)_n}.$$

They correspond to two irreducible representations of the $(2,5)$ minimal model.

Acknowledgments

The main part of this work was done while the first author was visiting the Isaac Newton Institute of the University of Cambridge. He would like to thank the Institute for hospitality. The second author thanks A. Szenes for valuable discussions.

References

1. A. Belavin, A. Polyakov, and A. Zamolodchikov, *Infinite conformal symmetry in two-dimensional quantum field theory*, Nuclear Phys. B **241** (1984), 333–380.
2. B. L. Feigin and D. B. Fuchs, *Cohomology of some nilpotent subalgebras of the Virasoro and Kac-Moody algebras*, Geometry and Physics (S. Gindikin and I. M. Singer, eds.), Essays in Honor of I. M. Gelfand on the Occasion of his 75th Birthday, North-Holland, Amsterdam, 1991, pp. 209–235.
3. A. Beilinson, B. Feigin, and B. Mazur, *Introduction to algebraic field theory on curves* (to appear).
4. B. Feigin, T. Nakanishi, and H. Ooguri, *The annihilating ideals of minimal models*, Internat. J. Modern Phys. A **1** (1992), 217–238.
5. I. Frenkel and Y. Zhu, *Vertex operator algebras associated to representations of affine and Virasoro algebras*, Duke Math. J. **66** (1992), 123–168.
6. B. Feigin and E. Frenkel, *Affine Kac-Moody algebras at the critical level and Gelfand-Dikii algebras*, Internat. J. Modern Phys. Suppl. A **1** (1992), 197–215.
7. I. Frenkel, Y.-Z. Huang, and J. Lepowsky, *On axiomatic approaches to vertex operator algebras and modules*, Preprint, 1990.
8. E. Frenkel and A. Szenes, *Dilogarithm identities, q-difference equations, and the Virasoro algebra*, Preprint hep-th/9212094; Internat. Math. Res. Notices, Duke Math. J. (to appear).

9. G. E. Andrews, *The theory of partitions*, Addison-Wesley, Reading, MA, 1976.

10. V. Kac and M. Wakimoto, *Modular invariant representations of infinite-dimensional Lie algebras and superalgebras*, Proc. Nat. Acad. Sci. U.S.A. **85** (1988), 4956–4960.

11. V. Kac, *Modular invariance in mathematics and physics*, Mathematics into the Twenty-first Century, Proc. Centennial of the AMS Centennial Symp. (Aug. 8–12, 1988), Amer. Math. Soc., Providence, RI, 1992, pp. 337–350.

Landau Institute for Theoretical Physics, Moscow 117334, Russia

Department of Mathematics, Harvard University, Cambridge, Massachusetts 02138

ADVANCES IN SOVIET MATHEMATICS
Volume 16, Part 1, 1993

Induction and Restriction of Character Sheaves

VICTOR GINZBURG

To I. M. Gelfand on his 80th birthday

Introduction

This paper may be viewed as a continuation of [Gi]. In [Gi], we gave various equivalent definitions of character sheaves that, we believe, are more transparent than the original definitions of Lusztig [Lu1, Lu2]. The purpose of the present paper is quite similar: we provide simple conceptual proofs to a number of results concerning restriction and induction, proved in [Lu2] by rather long arguments involving case by case analysis. We do not claim however to be able to obtain all of the results of [Lu2] in an elementary way. The part of [Lu2] concerning classification of character sheaves remains as difficult (and as mysterious) as it was. There are actually certain reasons preventing that part being simplified. We will try to explain those reasons in the rest of the introduction below.

Let G be a connected complex semisimple Lie group with Lie algebra $\mathfrak{g}$. Let N^* denote the nilpotent cone in $\mathfrak{g}^*$, the dual of $\mathfrak{g}$. The character sheaves on G can be defined as irreducible $\operatorname{Ad} G$-equivariant perverse sheaves on G with nilpotent characteristic variety (see [Gi]). There are analogous objects on $\mathfrak{g}$ (see [Lu3] and §9 below), the irreducible $\operatorname{Ad} G$-equivariant perverse sheaves on $\mathfrak{g}$ whose characteristic variety is contained in $\mathfrak{g} \times N^*$. Hence, a perverse sheaf M on $\mathfrak{g}$ is a character sheaf if and only if the Fourier transform of M is the intersection cohomology complex associated to an irreducible G-equivariant local system on a G-orbit in N^*. Thus, character sheaves on $\mathfrak{g}$ are parametrized by all irreducible G-equivariant local systems on nilpotent G-orbits in $\mathfrak{g}^*$.

Now, let M be a character sheaf on G. Pulling M back to $\mathfrak{g}$ via the exponential map gives a character sheaf on $\mathfrak{g}$. Let $u(M)$ denote the irreducible local system on a nilpotent G-orbit in $\mathfrak{g}^*$ attached to the latter sheaf via the above parametrization. The assignment $M \rightsquigarrow u(M)$ does not lead, however, to a parametrization of character sheaves on G, because there is

1991 *Mathematics Subject Classification.* Primary 20G05, 22E47.

an additional invariant of character sheaves on G, called a *central character* (see [Gi, §1] for an algebraic definition of a central character), that does not exist for character sheaves on $\mathfrak{g}$. The central character of a character sheaf M is a semisimple conjugacy class, $h(M) \subset G^\vee$, where $G^\vee$ stands for the complex semisimple Lie group dual to G in the sense of Langlands. To define $h(M)$ geometrically, choose a Levi subgroup $L \subset G$ of minimal dimension such that $M_{|L} \neq 0$ (cf. §7). Let C be the connected center of L. It turns out that the complex $M_{|L}$ is locally-constant along C-orbits (of the multiplication action) in L. The monodromy of M along the C-orbits gives rise to a finite set of points in the dual torus $C^\vee \subset G^\vee$ that belong, in fact, to the same conjugacy class in $G^\vee$. This conjugacy class is $h(M)$, by definition. Clearly, there is no Lie algebra analogue of such a definition, for all orbits of the additive group Lie(C) are simply connected.

One would like to view the invariants $h(M)$ and $u(M)$ attached to a character sheaf M on G as the "semisimple" and the "unipotent" parts of a single invariant, in analogy with the Jordan decomposition of an element of an algebraic group. There is a difference, spoiling that analogy, however. In the case of the Jordan decomposition of an element a, the semisimple part $h(a)$ dominates over the unipotent part $u(a)$ in the sense that, usually, the semisimple part is found first, and the unipotent part can then be looked for in the centralizer of $h(a)$, a smaller group. In the character sheaves case, the invariants $h(M)$ and $u(M)$ have been defined in a totally different way so that $h(M)$ by no means plays a dominant role. In particular, there is no counterpart to the key property of the Jordan decomposition, saying that the semisimple and the unipotent parts commute. For that (and other) reasons Lusztig used another invariant $v(M)$ instead of $u(M)$, which fits better into the Jordan decomposition philosophy and which is dominated, in the above sense, by $h(M)$. The replacement of $u(M)$ by $v(M)$ was made however at the expense of simplicity; there is no conceptual definition of the invariant $v(M)$ as yet. Lusztig defined the assignment $M \rightsquigarrow v(M)$ by hand, on a case by case basis (see [Lu2]). We will not reproduce his definition, just saying instead that $v(M)$ is an Ad-equivariant local system on a special unipotent orbit in the Langlands dual group $G^\vee$ (rather than on G). Thus, character sheaves on G are parametrized by the set of all pairs $(h(M), v(M))$, which is similar to the Jordan decomposition of conjugacy classes in $G^\vee$, in conjunction with Langlands's philosophy. The sophisticated nature of the invariant $v(M)$ is due to the absence of a procedure relating objects on G to those on $G^\vee$ (see, however, [Gi4]).

We indicate an approach to the definition of the invariant $v(M)$ in the framework of [Gi] (due to Ian Grojnowski; see his MIT Thesis, 1992), which is different from the original definition of Lusztig. Let $D(G)$ denote the algebra of regular algebraic differential operators on G, $U(\mathfrak{g})$ the enveloping algebra of the Lie algebra $\mathfrak{g}$, and $Z(\mathfrak{g})$ the center of $U(\mathfrak{g})$. The action of G on itself by left and right translations gives rise to an injective algebra homomorphism: $U(\mathfrak{g}) \otimes_{Z(\mathfrak{g})} U(\mathfrak{g}) \hookrightarrow D(G)$. Recall [Gi] that a finitely generated $D(G)$-module M is said to be admissible if the action on M of the subalgebra $Z(\mathfrak{g}) \subset D(G)$ and the action of the Lie subalgebra $\mathfrak{g}_{\mathrm{ad}} = \{x \otimes 1 - 1 \otimes x \in U(\mathfrak{g}) \otimes_{Z(\mathfrak{g})} U(\mathfrak{g}) \subset D(G), x \in \mathfrak{g}\}$ are both locally finite. By a result of [Gi], character sheaves are precisely the perverse sheaves corresponding to simple admissible $D(G)$-modules via the Riemann-Hilbert correspondence. Given such a simple admissible module M, choose a simple $(U(\mathfrak{g}) \otimes_{Z(\mathfrak{g})} U(\mathfrak{g}))$-submodule M_0. Then M_0 is an irreducible Harish-Chandra $(\mathfrak{g} \times \mathfrak{g}, G_{\triangle})$-module. Hence, $SS(M_0)$, the characteristic variety of M_0, is the closure of a single special nilpotent orbit in $\mathfrak{g}^*$. The orbit does not depend on the choice of submodule $M_0 \subset M$ due to [Gi, Theorem 4.3.3] and the equality $M = D(G) \cdot M_0$. Let $\mathscr{O}$ be the unipotent orbit in G corresponding to that nilpotent orbit in $\mathfrak{g}^* \simeq \mathfrak{g}$ via the exponential map. It is known further that there is a (one-to-one) correspondence between special unipotent orbits in G and those in $G^\vee$. The special orbit in $G^\vee$ corresponding to $\mathscr{O}$ turns out to be the support of the local system $v(M)$.

Admissible modules form an abelian category Admiss(G). One would like to have not only a parametrization of the simple objects of Admiss(G), i.e., of character sheaves, but also an explicit description of the category itself. This goes as follows. Let L be a Levi subgroup of G and M a cuspidal character sheaf on L (see [Lu2] and §7 below). To each G-conjugacy class of pairs (L, M) one associates a full abelian subcategory $\mathrm{Admiss}_L^G(M) \subset \mathrm{Admiss}(G)$ (see §8), called a "block". It turns out that the category Admiss(G) breaks up into a direct sum of subcategories:

$$\mathrm{Admiss}(G) = \bigoplus_{\text{all blocks}} \mathrm{Admiss}_L^G(M).$$

Thus, it suffices to describe each of the blocks separately. We prove that the category $\mathrm{Admiss}_L^G(M)$ is equivalent to $A(L, M)$-Mod, the category of finite-dimensional modules over an explicitly constructed algebra $A(L, M)$ (see §8). The algebra $A(L, M)$ has a natural grading by nonnegative integers and its zero-degree component is a group algebra introduced by Lusztig in [Lu1, §9.2]. Moreover, $A(L, M)$ turns out to be a Koszul algebra (see [BGS]).

The algebra $A(L, M)$ also has a "dual" geometric interpretation in terms of the group $G^\vee$. To that end, recall that the character sheaf M is the intersection cohomology complex associated with a cuspidal local system $\mathscr{M}$ on L. Write the "Jordan decomposition" of $\mathscr{M}$ (see [Lu2]) as $\mathscr{M} = \mathscr{M}_C \boxtimes \mathscr{M}_{\mathscr{O}}$, where $\mathscr{M}_C$ is a one-dimensional local system on a connected component of C, the center of L, and $\mathscr{M}_{\mathscr{O}}$ is (essentially) a cuspidal local system on a special unipotent orbit $\mathscr{O} \subset L$. Let $L^\vee$ be the Levi subgroup in the Langlands dual of G corresponding to L. Let $C^\vee$ be the center of $L^\vee$. Thus, $C^\vee$ is the Langlands dual of C, that is, the group of all one-dimensional local systems on C. Further, let $\mathscr{O}^\vee$ be the special unipotent orbit in $L^\vee$

corresponding to the orbit $\mathscr{O}$ via the bijection between special orbits in L and $L^\vee$. Let $c \in C^\vee$ denote the point corresponding to the local system $\mathscr{M}_C$. Form the set $c \cdot \mathscr{O}^\vee \subset L^\vee$, and let $\operatorname{Ad} G^\vee(c \cdot \mathscr{O}^\vee)$ denote the set of elements of $G^\vee$ that are conjugate to $c \cdot \mathscr{O}^\vee$.

Now, Lusztig has defined a bijective correspondence from the set of cuspidal local systems on $\mathscr{O}$ to a set of $L^\vee$-equivariant local systems on $\mathscr{O}^\vee$. Let $\mathscr{M}_{\mathscr{O}}^\vee$ denote the local system on $\mathscr{O}^\vee$ corresponding to $\mathscr{M}_{\mathscr{O}}$ and let $IC(\mathscr{M}_{\mathscr{O}}^\vee)$ be the intersection cohomology complex on $L^\vee$ associated with $\mathscr{M}_{\mathscr{O}}^\vee$. The element c is central in $L^\vee$ and it acts on $L^\vee$ by multiplication. Let $c_* IC(\mathscr{M}_{\mathscr{O}}^\vee)$ denote the push-forward of $IC(\mathscr{M}_{\mathscr{O}}^\vee)$ with respect to that action on $L^\vee$. Form the induced sheaf $M^\vee = \operatorname{Ind}_{L^\vee}^{G^\vee}(c_* IC(\mathscr{M}_{\mathscr{O}}^\vee))$ (see §5). The complex $M^\vee$ is supported on the closure of the set $\operatorname{Ad} G^\vee(c \cdot \mathscr{O}^\vee)$ and, moreover, it is an object of the $\operatorname{Ad} G^\vee$-equivariant derived category of constructible complexes on $G^\vee$ (see, e.g., [Gi4]). Let $\operatorname{Ext}^\bullet_{G^\vee}$ stand for the Ext-groups in that category.

THEOREM 0.1. *There is a natural graded algebra isomorphism*

$$\operatorname{Ext}^\bullet_{G^\vee}(M^\vee, M^\vee) \cong A(L, M).$$

In the case when L is a maximal torus in G the theorem was first proved in [Gi3] using the results of [Gi5] (for $q = 1$). In the case of arbitrary Levi subgroup L and $\mathscr{M}_C$ = trivial local system, the theorem follows from [Lu4, §5] (for $q = \exp(c_i) = 1$). The arguments of [Lu4] can be extended to the general case.

The results of the last paragraphs fit into the framework of a general Koszul-Langlands philosophy, an extension of the Langlands philosophy advocated for some time by Wolfgang Soergel and the author. The Langlands philosophy says that, generally, simple objects of a representation-theoretic category associated with a reductive group are parametrized by certain objects associated with the Langlands dual group. The Koszul-Langlands philosophy says that, moreover, an appropriate mixed version (see [BGS, Chapter 2]) of the representation-theoretic category in question is, under favorable circumstances, a Koszul category [BGS, Chapter 2], and that the Koszul dual category [BGS] has a natural description in terms of the Langlands dual group. In particular, the simple objects of the two categories correspond to each other (whence the ordinary Langlands philosophy); however, it is more appropriate to regard simple objects of one of the categories as corresponding to the indecomposable projectives in the dual category. Now, in the special case studied in this paper, one can define a category $\operatorname{Admiss}_{\mathrm{mix}}(G)$, a mixed version of the category $\operatorname{Admiss}(G)$ formed by suitable mixed Hodge modules on G (in the sense of M. Saito). One then shows, modifying the proof of Theorem 8.1, that the category $\operatorname{Admiss}_{\mathrm{mix}}(G)$ is equivalent to the category of graded modules over a Koszul algebra, hence, is a Koszul category.

Furthermore, the Koszul dual category is equivalent, due to Theorem 0.1, to an appropriate category of $\mathrm{Ad}\, G^\vee$-equivariant perverse sheaves on $G^\vee$.

A detailed treatment of the above ideas, including a complete proof of Theorem 0.1, will be given elsewhere.

§1. Naive restriction

Let G be a complex connected reductive group. Lusztig has defined in [Lu1, §3] a decomposition of G into smooth locally-closed algebraic subvarieties. His construction was extended in [Gi, §3.4] to the more general case of a complex symmetric variety G/K. However, the definition given in [Gi] is incorrect. Below, we give a corrected definition of decomposition of a symmetric variety, although it will be used in this paper in the group case only.

Let θ be an involutive automorphism of G. Let $K = G^\theta$ be the fixed point subgroup of G, and let P be the identity component of the set $\{x \in G,\, \theta(x) = x^{-1}\}$. A torus (resp., a semisimple element of G) contained in P is said to be *split*. The centralizer of a split torus is called a split Levi subgroup of G. This is a θ-stable reductive subgroup of G (in [Gi] the term "relevant" instead of "split" was used). Following [Lu1, Definition 2.6], call a split element $h \in P$ *isolated* if its connected centralizer $Z^0_G(h)$ is not contained in a proper split Levi subgroup of G. Observe further that the set P is stable under the adjoint K-action. A K-orbit in P is said to be isolated if any element of the orbit has an isolated semisimple part. In particular, any unipotent K-orbit is isolated.

Given a split Levi subgroup L, let L^{reg} denote the Zariski open part of L defined by

$$L^{\mathrm{reg}} = \{x = s_x \cdot u_x \in L \mid Z^0_G(s_x) \subset L\}, \tag{1.1}$$

where $x = s_x \cdot u_x$ stands for the Jordan decomposition of x. Let $Z^0(L)$ denote the connected center of L and let $\overline{L} = L/Z^0(L)$. The involution θ on G induces an involution on $\overline{L}$ and we let $K(\overline{L})$ and $P(\overline{L})$ denote the "$\overline{L}$-counterparts" of the objects K and P (for G).

We now define, imitating [Lu1, §3], a decomposition $P = \bigsqcup P_{L,\mathscr{O}}$ indexed by all pairs $(L, \mathscr{O})$, where L is a split Levi subgroup of G and $\mathscr{O}$ is an isolated $K(\overline{L})$-orbit in $P(\overline{L})$. Let $\widetilde{\mathscr{O}}$ be the inverse image of $\mathscr{O}$ under the natural projection $P(L) \to P(\overline{L})$. The piece $P_{L,\mathscr{O}}$ attached to a pair $(L, \mathscr{O})$ is defined by

$$P_{L,\mathscr{O}} = \textit{union of } K(G)\textit{-orbits in } P(G) \textit{ that meet } (\widetilde{\mathscr{O}} \cap L^{\mathrm{reg}}). \tag{1.2}$$

With this definition of the partition of P, all the arguments of [Gi, §3] go through.

From now on, we restrict our attention to the "group case", i.e., assume that $\theta = \mathrm{id}$. Then we have $G = K = P$, and we write $G_{L,\mathscr{O}}$ for the piece of the decomposition of G given by (1.2).

Lemma 1.3. *Let L be a Levi subgroup of G. Then for any piece $G_{M,\mathscr{O}}$ the intersection $L^{\mathrm{reg}} \cap G_{M,\mathscr{O}}$ is transverse.*

Proof. Let $x = s \cdot u$ be the Jordan decomposition of $x \in L^{\mathrm{reg}}$. Let M be the centralizer of the connected center of $Z_G^0(s)$. Then $s, u \in M$ and, moreover, $x \in G_{M,\mathscr{O}}$, where $\mathscr{O}$ is the $\overline{M}$-conjugacy class of the image of x in $\overline{M}$. Let $\mathfrak{g}$, $\mathfrak{m}$, $\mathfrak{l}$, $\mathfrak{r}$ denote the Lie algebras of the groups G, M, L, $Z_G^0(s)$ respectively. Observe that $z(\mathfrak{l}) \subset \mathfrak{r}$, for $s \in L$. Moreover, $s \in L^{\mathrm{reg}}$ yields $\mathfrak{r} \subset \mathfrak{l}$, hence, $z(\mathfrak{l}) \subset z(\mathfrak{r})$. It follows that $\mathfrak{l} = z(z(\mathfrak{l})) \supset z(z(\mathfrak{r})) = \mathfrak{m}$. Whence, $\mathfrak{m} \cap \mathfrak{l}^{\perp} = 0$, where $\mathfrak{l}^{\perp}$ stands for the annihilator of $\mathfrak{l}$ in $\mathfrak{g}$ with respect to an invariant form.

Now let N_1 and N_2 denote the fibers at x of conormal bundles to the subvarieties $L^{\mathrm{reg}} \subset G$ and $G_{M,\mathscr{O}} \subset G$ respectively. We have:

$$N_1 = \mathfrak{l}^{\perp} \quad \text{and} \quad N_2 = [\mathfrak{m}, \mathfrak{m}] \cap z_{\mathfrak{m}}(u) \tag{1.4}$$

(see [Gi, (3.6.6)]). Observe that $[\mathfrak{m}, \mathfrak{m}] \cap z_{\mathfrak{m}}(u) \subset \mathfrak{m}$. Thus, $N_1 \cap N_2 \subset \mathfrak{l}^{\perp} \cap \mathfrak{m} = 0$, and the lemma follows. □

Let $\mathrm{Perv}_{\mathrm{strat}}(G)$ denote the abelian category of $\mathrm{Ad}\, G$-equivariant perverse sheaves on G which are locally constant along the pieces $G_{M,\mathscr{O}}$. By Lemma 1.3, any object of $\mathrm{Perv}_{\mathrm{strat}}(G)$ is noncharacteristic with respect to the subvariety $L^{\mathrm{reg}} \subset G$. Hence, the restriction $V \mapsto V_{|L^{\mathrm{reg}}}$ gives rise to an exact functor from $\mathrm{Perv}_{\mathrm{strat}}(G)$ to $\mathrm{Ad}\, L$-equivariant perverse sheaves on L^{reg}.

Now, let V be an admissible module on G. Then, $V \in \mathrm{Perv}_{\mathrm{strat}}(G)$ and $SSV \subset \mu^{-1}(N_{\mathfrak{g}}^*)$ ([Gi, Theorem 1.4.2]). The restriction to L^{reg} being noncharacteristic, it follows easily that $SS(V_{|L^{\mathrm{reg}}}) \subset N_{\mathfrak{l}}$. In particular, $V_{|L^{\mathrm{reg}}}$ is smooth in the direction of orbits of the group $Z(L)$, the center of L.

In §4 we will prove the following

Proposition 1.5. *For any admissible module V on G the restriction $V_{|L^{\mathrm{reg}}}$ can be uniquely extended to a $Z(L)$-monodromic module on L.*

The uniqueness of a $Z(L)$-monodromic extension of $V_{|L^{\mathrm{reg}}}$ trivially follows from the equality $L = Z(L) \cdot L^{\mathrm{reg}}$, which shows that any $Z(L)$-monodromic module on L is completely determined by its restriction to L^{reg}.

The existence of a $Z(L)$-monodromic extension implies, in particular, that for any $Z(L)$-orbit $C \subset L$, the locally-constant complex $V_{|C \cap L^{\mathrm{reg}}}$ has no monodromy about $C \cap (L \setminus L^{\mathrm{reg}})$. We will not attempt to prove this monodromy vanishes. Instead, we will give an alternative construction in §3 of a module on L that extends $V_{|L^{\mathrm{reg}}}$.

Definition 1.6. For an admissible module V let $\mathrm{Res}_L^G V$ denote the $Z(L)$-monodromic extension of $V_{|L^{\mathrm{reg}}}$.

The uniqueness of the extension yields

COROLLARY 1.7. *The assignment* $V \mapsto \operatorname{Res}_L^G V$ *gives rise to an exact functor commuting with the Verdier duality.*

REMARK 1.8. Let $i: L \hookrightarrow G$ be the inclusion. One might expect naively that $\operatorname{Res}_L^G V = i^* V[\dim G - \dim L]$, for the equality, obviously holds on L^{reg}. However, in general, one has two functors i^* and $i^!$ and there is no way to make a choice between them compatible with the Verdier duality. The correct answer given in §3 below involves, in a sense, a combination of both i^* and $i^!$.

§2. Digression to the Harish-Chandra homomorphism

Let P be a parabolic subgroup of G, U_P the unipotent radical of P, and L a Levi subgroup of P so that $P = L \cdot U_P$. Let $\mathfrak{g}$, $\mathfrak{p}$, $\mathfrak{l}$, $\mathfrak{u}$ denote the respective Lie algebras and let $\bar{\mathfrak{u}}$ be the nilpotent Lie subalgebra of $\mathfrak{g}$ opposite to $\mathfrak{u}$, so that $\mathfrak{g} = \bar{\mathfrak{u}} \oplus \mathfrak{l} \oplus \mathfrak{u}$.

Let $Z(\mathfrak{g})$ and $Z(\mathfrak{l})$ be the centers of the enveloping algebras $U(\mathfrak{g})$ and $U(\mathfrak{l})$ respectively. First, we recall the definition of the Harish-Chandra homomorphism $a: Z(\mathfrak{g}) \to Z(\mathfrak{l})$ and after that explain its geometric meaning.

The decomposition $\mathfrak{g} = (\bar{\mathfrak{u}} \oplus \mathfrak{l}) \oplus \mathfrak{u}$ gives rise to a vector space decomposition:

$$U(\mathfrak{g}) = U(\bar{\mathfrak{u}} \oplus \mathfrak{l}) \oplus U(\mathfrak{g}) \cdot \mathfrak{u}. \tag{2.1}$$

Let a denote the first projection.

PROPOSITION 2.2. (i) *If* $x \in U(\mathfrak{g})^{\mathfrak{l}}$ ($=$*the ring of* $\operatorname{ad}\mathfrak{l}$*-invariants*), *then* $a(x) \in Z(\mathfrak{l})$ ($\subset U(\mathfrak{l}) \subset U(\bar{\mathfrak{u}} \oplus \mathfrak{l})$).

(ii) *The restriction of* a *to* $Z(\mathfrak{g})$ *gives rise to an algebra homomorphism* $Z(\mathfrak{g}) \to Z(\mathfrak{l})$.

PROOF. The direct sum decomposition (2.1) is, clearly, $\operatorname{ad} L$-stable. Hence for $x \in U(\mathfrak{g})^{\mathfrak{l}}$ we have $a(x) \in U(\bar{\mathfrak{u}} \oplus \mathfrak{l})^{\mathfrak{l}}$. Let h be an element in the center of $\mathfrak{l}$ such that all the eigenvalues of $\operatorname{ad} h$ on $\bar{\mathfrak{u}}$ (resp. $\mathfrak{l}$) are > 0 (resp. $= 0$). We see that any $\operatorname{ad} h$-invariant element of $U(\bar{\mathfrak{u}} + \mathfrak{l})$ belongs to $U(\mathfrak{l})$. Hence, $U(\bar{\mathfrak{u}} + \mathfrak{l})^{\mathfrak{l}} = U(\mathfrak{l})^{\mathfrak{l}} = Z(\mathfrak{l})$ and (i) follows.

Verification of (ii) is left to the reader. □

Set $Y_P = G/U_P$. The group L normalizes U_P, so that there is a right L-action on Y_P commuting with the natural left G-action.

Let $D(Y_P)$ denote the ring of globally defined differential operators on Y_P. The left G-action and the right L-action on Y_P give rise to algebra homomorphisms

$$e_L: U(\mathfrak{g}) \to D(Y_P) \quad \text{and} \quad e_R: U(\mathfrak{l}) \to D(Y_P)$$

(the subscript L stands for "left" and R for "right") with commuting images. The intersection $\operatorname{im} e_L \cap \operatorname{im} e_R$ turns out to be equal to $e_L(Z(\mathfrak{g}))$ and we have

PROPOSITION 2.3. $e_L(z) = e_R(a(z))$, $z \in Z(\mathfrak{g})$.

To prove the proposition one regards $e_L(z)$ and $e_R(a(z))$ as sections of the vector bundle $\underline{\mathfrak{u}} \cdot U(\underline{\mathfrak{g}})$ (see [Gi, Proposition 5.6.2]). The equality is then verified fiber by fiber, using Proposition 2.2 (i). The details are left to the reader. □

§3

Given a parabolic subgroup $P = L \cdot U_P$, let $i_P\colon P \hookrightarrow G$ denote the inclusion and $q_P\colon P \to P/U_P = L$ the natural projection. Following Lusztig, for a D-module V on G, set

$$\mathrm{res}_P V = (q_P)_*(i_P^! V)[\dim U_P].$$

This is a complex of D-modules on L.

LEMMA 3.1. *If V is an admissible module on G, then $\mathrm{res}_P V$ is an admissible complex on L, i.e., a complex with admissible cohomology.*

PROOF. Let $q\colon G \to G/U_P = Y_P$ be the natural projection and $i\colon L = P/U_P \hookrightarrow G/U_P = Y_P$ the inclusion. The base change theorem yields $\mathrm{res}_P V = (q_P)_*(i_P^! V)[\dim U_P] = i^!(q_* V)[\dim U_P]$. We show that the $Z(\mathfrak{l})$-action on cohomology of the complex $i^!(q_* V)$ is locally-finite.

Consider the direct image $q_* V$ first. We construct a Zariski-open affine covering of Y_P by right L-stable subsets W, so that the action of $e_R(Z(\mathfrak{l}))$ (see §2) on $\Gamma(W, H^i q_* V)$ turns out to be locally-finite. To that end, note that the map q is smooth and affine. Hence the complex $q_* V$ can be computed by means of the relative de Rham complex, which effectively reduces to the Koszul complex:

$$0 \to q.V \to q.V \otimes \mathfrak{u}^* \to \cdots \to q.V \otimes \Lambda^{\dim \mathfrak{u}} \mathfrak{u}^* \to 0 \tag{3.2}$$

for the Lie algebra cohomology $H^*(\mathfrak{u}, q.V)$, where $\mathfrak{u}$ is viewed as the Lie algebra of vector fields on G generated by the right U_P-action.

Now, let W be a Zariski-open affine part of Y_P which is stable under the right L-action (i.e., W is the inverse image of a subset in G/P via the natural projection $G/U_P \to G/P$). We choose W in such a way that $q^{-1}(W)$ is affinely embedded into G and such that the hypersurface $G \setminus q^{-1}(W)$ is defined by an equation $f = 0$, where the function f is an eigenfunction with respect to the right P-action on G. A subset W of that type may be obtained, for instance, as the inverse image of the open B-orbit in G/P, where B is an appropriate Borel subgroup of G, so that $q^{-1}(W) = B \cdot P$.

Let $(\cdot)_{(f)}$ denote the localization with respect to the function f on G, associated to W. The localization functor being exact and the variety G being affine, we get (using 3.2):

$$\Gamma(W, H^j q_* V) = H^j(\mathfrak{u}, \Gamma(G, V))_{(f)}. \tag{3.3}$$

We now examine the $Z(\mathfrak{g})$-action on the right-hand side of (3.3). The $Z(\mathfrak{g})$-action on $\Gamma(G, V)$ is locally-finite since V is admissible. The $Z(\mathfrak{g})$-action on $\Gamma(G, V) \otimes \Lambda^j \mathfrak{u}^*$ is also locally-finite, for the Lie algebra $\mathfrak{g}$ acts trivially on the factor $\Lambda^i \mathfrak{u}^*$. Thus, the $Z(\mathfrak{g})$-action on $H^j(\mathfrak{u}, \Gamma(G, V))$, and on its image in $\Gamma(W, H^j q_* V)$, is locally-finite. But, the left-hand side of (3.3) may be viewed as a D-module on W. So, the $Z(\mathfrak{g})$-action on the left-hand side is induced by the action of $e_R(Z(\mathfrak{l}))$ via the Harish-Chandra homomorphism 2.3. Hence, the $e_R(Z(\mathfrak{l}))$-action on the image of $H^j(\mathfrak{u}, \Gamma(G, V))$ is locally-finite (we have used here that $Z(\mathfrak{l})$ is a finitely-generated module over $a(Z(\mathfrak{g}))$, the image of the Harish-Chandra homomorphism $a\colon Z(\mathfrak{g}) \to Z(\mathfrak{l})$).

Further, the equality (3.3) shows that $\Gamma(W, H^j q_* V)$ is a quotient of

$$H^j(\mathfrak{u}, \Gamma(G, V)) \otimes_{\mathbb{C}} \mathbb{C}[f^{-1}]. \tag{3.4}$$

We have just established that the first factor here is a locally-finite $Z(\mathfrak{l})$-module. The factor $\mathbb{C}[f^{-1}]$, on the other hand, is a direct sum of one-dimensional $\mathfrak{l}$-submodules spanned by f^{-k}, for f was chosen to be an L-eigenfunction. Hence, (3.4) is a locally-finite $Z(\mathfrak{l})$-module by [Gi, Proposition 2.2] applied to the Lie algebra $\mathfrak{l}$. Therefore, the same holds for $\Gamma(W, H^j q_* V)$.

Next, we study the restriction of $q_* V$ to the submanifold $L = P/U_P \subset Y_P$. Let W be an open neighborhood of L in Y_P of the type considered above, so that its image in G/P is a neighborhood of the base point $l \in G/P$. Choose a local coordinate system on G/P with the origin at the base point l and pull the coordinate functions back on G/U_P. The resulting functions $t_1, \dots, t_r$ are right L-invariant functions defined on a neighborhood of L in Y_P, so that $L = \{y \in Y_P \mid t_i(y) = 0, i = 1, \dots, r\}$. Hence, we may compute the restriction functor $i^!$ by means of the standard Koszul complex $K(t_1, \dots, t_r)$ associated to the functions $t_1, \dots, t_r$. Thus, the object $i^!(q_* V)$ is represented by the simple complex associated with the double complex (cf. (3.2)–(3.3))

$$\cdots \to K(t_1, \dots, t_r)\Gamma(G, V)_{(f)} \otimes \Lambda^j \mathfrak{u}^*$$
$$\to K(t_1, \dots, t_r)\Gamma(G, V)_{(f)} \otimes \Lambda^{j+1} \mathfrak{u}^* \to \cdots.$$

Let

$$H^n i^!(H^m q_* V) = E_2^{n,m} \Rightarrow \operatorname{Gr} H^{n+m}(i^! q_* V)$$

be the standard spectral sequence associated with a double complex. As we know, the $e_R(Z(\mathfrak{l}))$-action on $H^m q_* V$ is locally finite. Furthermore, the multiplication by t_j commutes with the $e_R(Z(\mathfrak{l}))$-action, for t_j is a right-invariant function on Y_P. Hence, the induced $e_R(Z(\mathfrak{l}))$-action on a Koszul-complex cohomology $H^n i^!(H^m q_* V)$ is locally-finite. The spectral sequence yields that the same holds for the $Z(\mathfrak{l})$-action on $H^{n+m}(i^! q_* V)$. This completes the proof. □

We now indicate the proof of [Gi, Proposition 8.7.1]. Consider the following commutative diagram:

$$\begin{array}{ccc} G \times (Y/T) & \xrightarrow{q} & (Y \times Y)/T \\ & {\scriptstyle \mathrm{pr}_2}\searrow \quad \swarrow{\scriptstyle \mathrm{pr}} & \\ & Y/T & \end{array}$$

where both pr_2 and pr denote the second projection (see [Gi] for notation). Given $y \in Y/T$, let $U_y = y \cdot U \cdot y^{-1}$ be the unipotent radical of a Borel subgroup of G. The fiber of the map q over a point of $pr^{-1}(y)$ is naturally isomorphic to U_y. We see that the map q may be viewed as a relative version (with respect to the base Y/T) of the projection $G \to G/U$. Hence, the direct image functor q_* can be computed via the "relative" Koszul complex with respect to the family of Lie algebras $\{\mathrm{Lie}\, U_y, y \in Y/T\}$. Repeating the argument of the first part of the proof of Lemma 3.1 (applied to $P = T \cdot U =$ Borel subgroup), one shows that the right $U(\mathfrak{t})$-action on ${}^pH^i(q_* p^! V)$ is locally-finite, provided V is an admissible $D(G)$-module.

§4

Let $\mathrm{res}^i_P V$ denote the ith cohomology D-module of the complex $\mathrm{res}_P V$, introduced at the beginning of §3.

Observe that any admissible module on L is a $Z(L)$-monodromic module, since the Lie algebra of the group $Z(L)$ is obviously contained in $Z(\mathfrak{l})$, the center of the enveloping algebra. With that understood we can state

THEOREM 4.1. *Let V be an admissible module on G. Then:*

(i) *The complex $\mathrm{res}_P V$ is concentrated in degree 0, i.e., $\mathrm{res}^i_P V = 0$ for $i \neq 0$.*

(ii) *$\mathrm{res}^0_P V$ is the $Z(L)$-monodromic extension of $V_{|L^{\mathrm{reg}}}$ so that*

$$\mathrm{res}^0_P V = \mathrm{Res}^G_L V$$

(see §1).

(iii) *The assignment $V \mapsto \mathrm{Res}^G_L V$ gives rise to an exact functor from the category of admissible $D(G)$-modules to the category of admissible $D(L)$-modules.*

(iv) *The functor Res^G_L commutes with the Verdier duality and preserves the weights (of mixed modules).*

COROLLARY 4.2. *The functor res_P does not depend on the choice of parabolic subgroup P that has L as a Levi factor.*

COROLLARY 4.3. *If V is a semisimiple admissible module, then so is $\mathrm{Res}^G_L V$.*

Corollary 4.2 follows from parts (i), (ii) of the theorem and Corollary 4.3 from part (iv), since Res^G_L maps pure objects into pure ones.

PROOF OF THEOREM 4.1. Choose a one-dimensional torus C ($\cong \mathbb{C}^*$) in the center of L in such a way that all the weights of the adjoint C-action on the Lie algebra $\mathfrak{u}$ are positive integers. Since the group U_P is unipotent, it is algebraically and C-equivariantly isomorphic to $\mathfrak{u}$ via the exponential map. So, the projection $q_P: P \to P/U_P = L$ may (and will) be viewed as a vector bundle on L with C-action along the fibers.

Now, $i_P^! V$ is obviously an $\operatorname{Ad} L$-equivariant, hence C-equivariant, complex on P. But for a C-equivariant complex on a vector bundle, one knows that $(q_P)_* = i_L^*$ where $i_L: L \hookrightarrow P$ denotes the zero-section embedding (see, e.g., [Gi2]; it is used here that all the weights of the C-action on the vector bundle are of the same sign). Hence,

$$\operatorname{res}_P V = i_L^* i_P^! V.$$

We see that the complex $\operatorname{res}_P V$ depends only on the restriction of V on a small open neighborhood of L.

Let $x \in L^{\text{reg}}$. It follows from Lemma 1.3 that both the restriction from G to P and the restriction from P to L are noncharacteristic at x. Hence, on L^{reg} we have $i_L^* = i_L^![\dim P - \dim L]$, so that

$$(\operatorname{res}_P V)_{|L^{\text{reg}}} = (i_L \cdot i_P)^! V[\dim P - \dim L]_{|L^{\text{reg}}} = V_{|L^{\text{reg}}}.$$

This shows that $(\operatorname{res}_P^i V)_{|L^{\text{reg}}} = 0$ for $i \neq 0$ and $(\operatorname{res}_P^0 V)_{|L^{\text{reg}}} = V_{|L^{\text{reg}}}$. But since the complex $\operatorname{res}_P V = 0$ is $Z(L)$-monodromic (Lemma 3.1), we conclude that $\operatorname{res}_P^i = 0$ (for $i \neq 0$) everywhere on L and that $\operatorname{res}_P^0 V = \operatorname{Res}_L^G V$. This proves (i) and (ii). Part (iii) follows from Lemma 3.1 and from the exactness of Res_L^G.

Finally, it is well known that the functors $i_P^!$ and $(q_P)_*$ increase the weights of mixed complexes. Hence, so does res_P. But since res_P commutes with the Verdier duality (Corollary 1.7), it must also decrease the weights. Part (iv) follows. □

§5. Induction

(Cf. [Lu2, Chapter 4].) Given a parabolic subgroup P of G, set

$$G_P = \{(h, y) \in G \times (G/P) \mid h \in y \cdot P \cdot y^{-1}\}. \tag{5.1}$$

Clearly, G_P is a G-equivariant fiber bundle on G/P with fiber P. The restriction to the fiber P over the base point $l \in G/P$ sets up a bijective correspondence between G-equivariant complexes on G_P and $\operatorname{Ad} P$-equivariant complexes on P. Let $\widetilde{V}_P$ denote the complex on G_P corresponding to an equivariant complex V_P on P.

Let U_P be the unipotent radical of P and $L = P/U_P$, the Levi factor. Given an $\operatorname{Ad} L$-equivariant complex V on L, let V_P be its pull-back on P via the natural projection $P \to P/U_P = L$. It is clear that V_P is an $\operatorname{Ad} P$-equivariant complex on P. We set $\operatorname{ind}_P V = f_* \widetilde{V}_P$ where $f: G_P \to G$

is defined by $f(h, y) = h$ (see (5.1)). The complex $\mathrm{ind}_P V$ is said to be *induced* from V.

Viewing L as a Levi subgroup of G, we have the following simple

LEMMA 5.2 [Lu2, (4.6.2)]. *The functor* ind_P *is the right adjoint of* res_P.

Next, we shall prove the following

PROPOSITION 5.3. *Let V be an L-admissible $D(L)$-module. Then, any cohomology group of* $\mathrm{ind}_P V$ *is a G-admissible $D(G)$-module.*

PROOF. Let $\mathfrak{g}$, $\mathfrak{p}$, $\mathfrak{u}$, $\mathfrak{l}$ denote the Lie algebras of the groups G, P, U_P, L. The algebra $\mathfrak{l}$ may be viewed as a Levi subalgebra of $\mathfrak{p}$, so that $\mathfrak{p} = \mathfrak{l} \oplus \mathfrak{u}$.

First, it follows from the construction that $\mathrm{ind}_P V$ is an $\mathrm{Ad}\, G$-equivariant complex on G. By [Gi, Theorem 1.4.2], it suffices to show that the characteristic variety $SS(\mathrm{ind}_P V)$ is contained in $\mu^{-1}(N^*_\mathfrak{g})$, where $N^*_\mathfrak{g}$ is the nilpotent variety of $\mathfrak{g}^*$ and $\mu\colon T^*G \to N^*_\mathfrak{g}$ is the moment map.

We identify T^*G with the trivial bundle $\mathfrak{g}^* \times G$, T^*L with $\mathfrak{l}^* \times L$, etc., and write

$$T^*(G \times G/P) = \{(\lambda, x, \xi_y) \in \mathfrak{g}^* \times G \times T^*(G/P)\}$$

where ξ_y denotes a covector at a point $y \in G/P$. Now, let the module $\widetilde{V}_P$ be viewed as a D-module on $G \times (G/P)$ supported on G_P. The module being G-equivariant, one obtains

$$SS\widetilde{V}_P = \{(\lambda, h, \xi_y) \in g^* \times G \times T^*(G/P) \mid (y^{-1} \cdot \lambda \cdot y_{|\mathfrak{p}}, \; y^{-1} \cdot h \cdot y) \in SSV_P\}. \tag{5.4}$$

Let $\bar{\mathfrak{u}}$ denote the nilpotent subalgebra of $\mathfrak{g}$ opposite to $\mathfrak{u}$, so that $\mathfrak{g} = \bar{\mathfrak{u}} + \mathfrak{l} + \mathfrak{u}$. We identify $\mathfrak{g}^*$ with $\mathfrak{g}$ and $\mathfrak{p}^*$ with $\bar{\mathfrak{u}} + \mathfrak{l}$ via the Killing form. Then we can write (notation of (5.4)) $y^{-1} \cdot \lambda \cdot y = \bar{v} + x + v$, $\bar{v} \in \bar{\mathfrak{u}}$, $x \in \mathfrak{l}$, $v \in \mathfrak{u}$. But we have $\bar{v} = 0$ and $x \in N^*_\mathfrak{l}$, since $\widetilde{V}_P$ is the pull-back to P of an admissible $D(L)$-module. Thus

$$SS\widetilde{V}_P \in \{(\lambda, h, \xi_y) \mid y^{-1} \cdot \lambda \cdot y = x + v \in N_\mathfrak{l} + \mathfrak{u}\}. \tag{5.5}$$

Now, $\mathrm{ind}_P V$ is the direct image of $\widetilde{V}_P$ with respect to the first projection $G \times (G/P) \to G$. The projection being proper, one knows (see [Ka]) that $SS(\mathrm{Ind}_P V) \subset \mathrm{pr}(SS\widetilde{V}_P)$, where pr denotes the projection $T^*(G \times G/P) = T^*G \times T^*(G/P) \to T^*G$. Hence, the estimate (5.5) yields

$$SS(\mathrm{ind}_P V) \subset \{(\lambda, h) \in \mathfrak{g} \times G \mid y^{-1} \cdot \lambda \cdot y \in N^*_\mathfrak{l} + \mathfrak{u}\} \subset N^*_\mathfrak{g} \times G.$$

An equality $N^*_\mathfrak{g} \times G = \mu^{-1}(N^*_\mathfrak{g})$ completes the proof.

§6

Recall the category $\mathrm{Perv}_{\mathrm{strat}}(G)$ introduced in §1.

PROPOSITION 6.1. (i) *The functor* ind_P *yields an exact functor* $\mathrm{Perv}_{\mathrm{strat}}(L) \to \mathrm{Perv}_{\mathrm{strat}}(G)$.

(ii) *If* V_L *is the intersection cohomology complex associated to an* $\mathrm{Ad}\,L$-*equivariant local system on a stratum* $L_{M,\mathscr{O}}$, *then* $\mathrm{ind}_P(V_L)$ *is the intersection cohomology complex associated to an* $\mathrm{Ad}\,G$-*equivariant local system on the stratum* $G_{M,\mathscr{O}}$.

The proof of part (ii) is essentially due to Lusztig. More specifically, given a stratum $L_{M,\mathscr{O}}$, let

$$\widetilde{G}_{M,\mathscr{O}} = G \times_{\mathrm{Ad}\,P} (L_{M,\mathscr{O}} \cdot U_P)$$

be the corresponding "stratum" of G_P (see §5). Clearly, $f(\widetilde{G}_{M,\mathscr{O}}) = \overline{G}_{M,\mathscr{O}}$, where the bar stands for the closure. One can derive from dimension estimates of [Lu1, §1] that the map $f\colon \widetilde{G}_{M,\mathscr{O}} \to \overline{G}_{M,\mathscr{O}}$ is small (cf. [BM]). Statement (ii) now follows from the decomposition theorem [BBD] (this was observed in [Lu1, Proposition 4.5]).

To prove (i), observe that for any simple object $V \in \mathrm{Perv}_{\mathrm{strat}}(L)$ we have just obtained that

$${}^pH^i(\mathrm{ind}_P V) = 0 \quad \text{for } i \neq 0.$$

Hence, ind_P is an exact functor from $\mathrm{Perv}_{\mathrm{strat}}(L)$ to $\mathrm{Perv}_{\mathrm{strat}}(G)$. □

Given an admissible module V on L, set $\mathrm{Ind}_L^G V = \mathrm{ind}_P V$.

THEOREM 6.2. (i) *The functor* Ind_L^G *is an exact functor from* $\mathrm{Admiss}(L)$ *to* $\mathrm{Admiss}(G)$; *it is the right adjoint of* Res_L^G.

(ii) *The functor* Ind_L^G *does not depend on the choice of parabolic subgroup* P *containing* L *as a Levi factor.*

(iii) Ind_L^G *commutes with the Verdier duality and takes semisimple (pure) modules on* L *into similar ones on* G.

PROOF. (i) follows from Proposition 5.3, Proposition 6.1 (i), and Lemma 5.2. Part (ii) follows from the adjointness of Ind_L^G and Res_L^G and the analogue of (ii) for Res (see Theorem 4.1 (ii)). Part (iii) is clear since the projection $P \to P/U_P$ is a smooth morphism and the map $G_P \to G$ is proper. □

§7. Cuspidal modules

DEFINITION 7.1 [Lu1, Lu2]. An irreducible module $V \in \mathrm{Perv}_{\mathrm{strat}}(G)$ is said to be cuspidal if, for any proper parabolic subgroup $P \subset G$, we have $\mathrm{res}_P^0 V = 0$.

Cuspidal modules are the most basic ones in the following sense.

PROPOSITION 7.2 [Lu2, Theorem 4.4 (a)]. *Let* $V \in \mathrm{Admiss}(G)$ *be an irreducible module. If* V *is not cuspidal, then there exists a Levi subgroup* $L \neq G$ *and an irreducible cuspidal module* $V_L \in \mathrm{Admiss}(L)$ *such that* V *is a direct summand of* $\mathrm{Ind}_L^G(V_L)$.

PROOF. If V is an admissible module, then by Theorem 4.1 (ii) we have $\mathrm{res}_P^0 V = \mathrm{Res}_L^G V$.

Let L be a Levi subgroup of minimal dimension such that $\mathrm{Res}_L^G V \neq 0$. Then $\mathrm{Res}_L^G V$ is a sum of cuspidal modules on L by the transitivity of restriction and Theorem 4.1 (iv). Observe next that $L \neq G$, for V is not cuspidal. So, if V_L is a simple constituent of $\mathrm{Res}_L^G V$ then:

$$\mathrm{Hom}(V, \mathrm{Ind}_L^G(V_L)) = \mathrm{Hom}(\mathrm{Res}_L^G V, V_L) \neq 0.$$

Hence, V is a direct summand of $\mathrm{Ind}_L^G(V_L)$ (Theorem 6.2 (iii)). □

We have the following result that explains the relation between cuspidal and admissible modules.

THEOREM 7.3. *Let* $V \in \mathrm{Perv}_{\mathrm{strat}}(G)$ *be an irreducible perverse sheaf with support* $\overline{G}_{M,\mathscr{O}}$. *Then the following conditions are equivalent*:

(i) V *is an admissible module and* $M = G$;

(ii) V *is cuspidal.*

PROOF OF (i) $\Longrightarrow$ (ii). Suppose V is admissible but not cuspidal. Then V is a direct summand of $\mathrm{Ind}_L^G(V_L)$ for a certain cuspidal module V_L on a Levi subgroup $L \neq G$ (Proposition 7.2). Since V_L is irreducible, we have $\mathrm{supp}\, V_L = L_{M',\mathscr{O}'}$, where $M' \subset L$. Hence, $\mathrm{supp}\, V = \overline{G}_{M',\mathscr{O}'}$ by Proposition 4.1 (ii). But the assumption of the theorem implies $\mathrm{supp}\, V = G_{G,\mathscr{O}} \neq G_{M',\mathscr{O}'}$, since $M' \subset L \neq G$. The contradiction completes the proof.

(ii) $\Longrightarrow$ (i). Let V be a cuspidal module with $\mathrm{supp}\, V = G_{M,\mathscr{O}}$. Choose a parabolic subgroup $P = M \cdot U_P$ with Levi component M. Obviously, $V_{|M^{\mathrm{reg}}} \neq 0$. Hence, $(\mathrm{res}_P^0 V)_{|M^{\mathrm{reg}}} \neq 0$. It follows that $M = G$. Admissibility of V will be proved later in §9. □

COROLLARY 7.4. *The support of a cuspidal module is the closure of a distinguished stratum* (*see* [Gi, §3]).

PROOF. Let V be a cuspidal module and let $S = G_{G,\mathscr{O}}$ be the open stratum in $\mathrm{supp}\, V$. Obviously, we have $T_S^*G \subset SSV$. On the other hand, [Gi, Theorem 1.4.2] and 7.3 yield $SSV \subset \mu^{-1}(N_{\mathfrak{g}}^*)$, so that $T_S^*G \subset \mu^{-1}(N_{\mathfrak{g}}^*)$. [Gi, Proposition 3.5.1] completes the proof. □

The following result was verified by Lusztig using case by case analysis.

COROLLARY 7.5. *Any cuspidal sheaf is a characteristic sheaf.*

The proof follows from Theorem 7.3 and [Gi, Theorem 1.6.1]. □

COROLLARY 7.6. *Any cuspidal sheaf* V *is strongly cuspidal in the sense of* [Lu2, *Chapter* 7], *that is,* $\mathrm{res}_P V = 0$ *for any proper parabolic subgroup* $P \subsetneq G$.

The proof follows from Theorem 7.3 and Theorem 4.1 (i), (ii). □

§8

We now fix a Levi subgroup $L \subset G$ and a cuspidal module M on L. As we know, $\mathrm{Ind}_L^G M$ is a semisimple admissible module on G (Theorem 7.3 and Theorem 6.2). Let $\mathrm{Admiss}_L^G(M)$ be the full subcategory of Admiss(G) consisting of all those modules whose irreducible subquotients are direct summands of $\mathrm{Ind}_L^G M$.

To study the category $\mathrm{Admiss}_L^G(M)$ more closely we introduce some notation. Let $\mathrm{supp}\, M = \overline{L_{L,\mathscr{O}}}$, so that M is the intersection cohomology complex associated to an equivariant irreducible local system on $Z^{\mathrm{reg}} \cdot \mathscr{O}$, where Z^{reg} is the regular part of a component of $Z(L)$. The local system can be written as $\mathscr{M} = \mathscr{M}_Z \times \mathscr{M}_{\mathscr{O}}$ where $\mathscr{M}_Z$ and $\mathscr{M}_{\mathscr{O}}$ are local systems on Z^{reg} and $\mathscr{O}$ respectively. Observe that $\mathscr{M}_Z$ extends to a local system on the whole of Z, since M is $Z(L)$-monodromic. Hence, $\mathscr{M}_Z$ is a one-dimensional local system associated with a character $\lambda \in \mathrm{Hom}(\pi_1(Z(L)), \mathbb{C}^*) = Z(L)^\vee$, where $Z(L)^\vee$ denotes the dual torus.

Let $N(L)$ denote the normalizer of L in G. Following [Lu1, §3.4], form the group $N(L, M)$ of all $n \in N(L)$ such that $n \cdot Z \cdot n^{-1} = Z$, $n \cdot \mathscr{O} \cdot n^{-1} = \mathscr{O}$, and such that the automorphism $x \mapsto n \cdot x \cdot n^{-1}$ of $Z^{\mathrm{reg}} \cdot \mathscr{O}$ can be lifted to a morphism $\mathscr{M} \to (\mathrm{Ad}\, n)^* \mathscr{M}$ of local systems. Then $L \subset N(L, M)$ and we set $W(L, M) = N(L, M)/L$ (this was shown by Lusztig to be a Coxeter group).

For $n \in N(L, M)$, the space $\mathrm{Hom}(\mathscr{M}, (\mathrm{Ad}\, n)^*\mathscr{M})$ is one-dimensional ($\mathscr{M}$ irreducible) and can be canonically identified with a similar space for $n' \in N(L, M)$, provided $n' \equiv n \bmod L$. So, if w is the image of n in $W(L, M)$, we write $A_w := \mathrm{Hom}(\mathscr{M}, (\mathrm{Ad}\, n)^*\mathscr{M})$. The space $A := \bigoplus_{w \in W(L,M)} A_w$ has a natural algebra structure (see [Lu1, §3.4]); the algebra A is isomorphic to the group algebra of the group $W(L, M)$. In particular, A is a semisimple algebra.

Let $\mathfrak{z} = \mathrm{Lie}\, Z(L)$, and let $S(\mathfrak{z})$ be the symmetric algebra on $\mathfrak{z}$. The adjoint action of $N(L, M)$ on L preserves $Z(L)$ and induces an action of the group $W(L, M)$ on $\mathfrak{z}$, hence, on $S(\mathfrak{z})$. Let $A(L, M) := A \# S(\mathfrak{z})$ be the smash-product of algebras arising from this action.

THEOREM 8.1. *The category* $\mathrm{Admiss}_L^G(M)$ *is equivalent to the category of finite-dimensional* $A(L, M)$*-modules.*

This theorem strengthens the following result of Lusztig.

THEOREM 8.2 [Lu1]. $\mathrm{Hom}(\mathrm{Ind}_L^G M, \mathrm{Ind}_L^G M) \cong A$.

SKETCH OF PROOF OF THEOREM 8.1 (cf. [Gi4]). Let S_+ denote the augmentation ideal of $S(\mathfrak{z})$ and $S_n := S(\mathfrak{z})/S_+^n$, $n = 1, 2, \ldots$. We view S_n as a $\mathfrak{z}$-module by restricting the natural $S(\mathfrak{z})$-action on S_n to $\mathfrak{z}$. Clearly, the $\mathfrak{z}$-action is nilpotent.

Next, we define a unipotent representation e_n of the group $\pi_1(Z(L))$ in the space S_n, by identifying $\pi_1(Z(L))$ with a lattice in $\mathfrak{z}$ and setting:

$$e_n(z) \cdot s = (\exp z) \cdot s, \qquad s \in S_n, \ z \in \mathfrak{z}.$$

Let $\mathscr{E}_n$ denote the local system on $Z^0(L)$ corresponding to the representation e_n of $\pi_1(Z(L))$. The local systems $\mathscr{E}_n$ form a projective system:

$$\mathscr{E}_1 \leftarrow \mathscr{E}_2 \leftarrow \cdots.$$

Now, let $\mathscr{M} = \mathscr{M}_Z \times \mathscr{M}_{\mathscr{O}}$ be the local system on $Z^{\mathrm{reg}} \cdot \mathscr{O}$ associated with the cuspidal module M. We pull the systems $\mathscr{E}_n$ on Z via a $Z^0(L)$-equivariant isomorphism $Z^0(L) \xrightarrow{\sim} Z$ and form the tensor product $\mathscr{E}_n \otimes \mathscr{M}_Z$. So, the collection $\{\mathscr{E}_n \otimes \mathscr{M}\}$ may be regarded as a projective system of $\mathrm{Ad}\, L$-equivariant local systems on $Z^{\mathrm{reg}} \cdot \mathscr{O}$. We have $\mathscr{E}_1 \otimes \mathscr{M} \cong \mathscr{M}$ and, moreover, all irreducible subquotients of $\mathscr{E}_n \otimes \mathscr{M}$ are isomorphic to $\mathscr{M}$.

Let M_n be the intersection cohomology complex on L associated with the local system $\mathscr{E}_n \otimes \mathscr{M}$, $E_n = \mathrm{Ind}_L^G M_n$, and let $E = \varprojlim E_n$. By Proposition 4.1 (ii), we have $E_n = \overline{G_{L,\mathscr{O}}}$. To prove the theorem it suffices to show the following statements:

(A) E_n is an object of $\mathrm{Admiss}_L^G(M)$, for each $n = 1, 2, \ldots$;
(B) E is a projective generator of the category $\mathrm{Admiss}_L^G(M)$;
(C) $\mathrm{Hom}(E_{|G_{L,\mathscr{O}}}, E_{|G_{L,\mathscr{O}}}) \cong A(L, M)$;
(D) the natural restriction morphism $\mathrm{Hom}(E, E) \to \mathrm{Hom}(E_{|G_{L,\mathscr{O}}}, E_{|G_{L,\mathscr{O}}})$ is an isomorphism.

We will comment on the proofs of (A)–(D). One first notes that all the simple subquotients of M_n are isomorphic to M because of a similar property of the local systems $\mathscr{E}_n \otimes \mathscr{M}$. Hence, the exactness of the induction functor yields (A).

Next, observe (cf. [Lu1]) that the restriction of $\mathrm{Ind}_L^G M$ to $G_{L,\mathscr{O}}$ is an equivariant local system. Hence, the same is true for E. The proof of isomorphism (C) is straightforward. This is similar to an isomorphism

$$\mathrm{Hom}(\mathrm{Ind}_L^G M_{G_{L,\mathscr{O}}}, \mathrm{Ind}_L^G M_{G_{L,\mathscr{O}}}) = A$$

established in [Lu1].

Let $j_{!*}$ denote the DGM-extension functor of a local system on $G_{L,\mathscr{O}}$. We have $E = j_{!*}(E_{G_{L,\mathscr{O}}})$, by Proposition 6.1 (ii). This yields (D). Finally, proving that E is projective amounts to showing that $\mathrm{Ext}^1(E, \mathrm{Ind}_L^G M) = 0$. Let

$$0 \to \mathrm{Ind}_L^G M \to N \to E \to 0$$

be a nontrivial extension. Since both $\mathrm{Ind}_L^G M$ and E are obtained by applying $j_{!*}$ they have neither quotients nor submodules supported on $\overline{G_{L,\mathscr{O}}} \setminus G_{L,\mathscr{O}}$. Hence, the same holds for N, so that N is also obtained by applying $j_{!*}$. Thus, it suffices to show the sequence

$$0 \to \mathrm{Ind}_L^G M_{|G_{L,\mathscr{O}}} \to N_{|G_{L,\mathscr{O}}} \to E_{|G_{L,\mathscr{O}}} \to 0$$

is split. This is easy. □

§9. Admissible modules on a semisimple Lie algebra

Let G be a connected complex semisimple Lie group with Lie algebra $\mathfrak{g}$, $D(\mathfrak{g})$ the ring of differential operators on $\mathfrak{g}$ with polynomial coefficients, and $Z(\mathfrak{g}) \subset D(G)$ the subring of $\mathrm{Ad}\, G$-invariant differential operators with constant coefficients. We have $T^*\mathfrak{g} = \mathfrak{g} \times \mathfrak{g}^*$. Recall that N denotes the nilpotent variety of $\mathfrak{g}$ and N^* the "nilpotent" variety of $\mathfrak{g}^*$, i.e., the variety arising from N via the Killing form isomorphism $\mathfrak{g} \xrightarrow{\sim} \mathfrak{g}^*$.

Recall that given a $D(\mathfrak{g})$-module V, one can define a $D(\mathfrak{g}^*)$-module FV, the Fourier transform of V. It is defined via the natural algebra isomorphism $D(\mathfrak{g}) = D(\mathfrak{g}^*)$ which is an algebraic counterpart of the isomorphism

$$T^*(\mathfrak{g}) \cong \mathfrak{g} \times \mathfrak{g}^* \cong T^*(\mathfrak{g}^*). \tag{9.1}$$

It is known [Br] that $SS(FV) = SSV$, provided V is a $\mathbb{C}^*$-monodromic module, with respect to the natural $\mathbb{C}^*$-action on $\mathfrak{g}$ by multiplication. In that case FV is regular holonomic iff so is V.

Proposition 9.2. *Let V be an $\mathrm{Ad}\, G$-equivariant $D(\mathfrak{g})$-module. Then the following conditions are equivalent*:

(i) *V is a locally-finite $Z(\mathfrak{g})$-module*;

(ii) *V is regular holonomic such that $SSV \subset \mathfrak{g} \times N^*$*;

(iii) *FV is supported on N^**.

Proof. The equivalence (i) $\Longleftrightarrow$ (iii) is straightforward. Let us prove that (iii) $\Longrightarrow$ (ii). Recall that any equivariant D-module on a G-variety that has finitely-many G-orbits is regular holonomic. The module FV is, clearly, $\mathrm{Ad}\, G$-equivariant. Hence, it is regular holonomic. Furthermore FV is a $\mathbb{C}^*$-monodromic module, for it is known that any $\mathrm{Ad}\, G$-orbit in N^* is a $\mathbb{C}^*$-stable variety. It follows that $F(FV)$ is a regular holonomic module such that $SSF(FV) \subset \mathfrak{g} \times N^*$. But the module V is obtained from $F(FV)$ by the sign-involution of the space $\mathfrak{g}$. Thus, (ii) is proved.

To prove that (ii) $\Longrightarrow$ (iii), we introduce a stratification $\mathfrak{g} = \bigsqcup \mathfrak{g}_{\mathfrak{l},\mathscr{O}}$ similar to the stratification of G, considered in §1. The strata are indexed by pairs $(\mathfrak{l}, \mathscr{O})$ where $\mathfrak{l}$ is a Levi subalgebra of $\mathfrak{g}$ and $\mathscr{O}$ is a nilpotent conjugacy class in $\mathfrak{l}$. The stratum $\mathfrak{g}_{\mathfrak{l},\mathscr{O}}$ is defined by

$$\mathfrak{g}_{\mathfrak{l},\mathscr{O}} = \mathrm{Ad}\, G\text{-saturation of } \mathfrak{z}^{\mathrm{reg}} + \mathscr{O} \tag{9.3}$$

where $\mathfrak{z}^{\mathrm{reg}} = \{x \in Z(\mathfrak{l}) \mid Z_{\mathfrak{g}}(x) = \mathfrak{l}\}$.

Now let V be a regular holonomic module such that $SSV \subset \mathfrak{g} \times N^*$. One can verify, as in [Gi, §3], that V is smooth along the strata $\mathfrak{g}_{\mathfrak{l},\mathcal{O}}$. Further, it follows from (9.3) that each stratum of the stratification is a $\mathbb{C}^*$-stable variety. Hence, V is a $\mathbb{C}^*$-monodromic module, and the statement (iii) follows. □

REMARK 9.4. Proposition 9.2 is a Lie algebra counterpart of [Gi, Theorem 1.4.2]. Strangely enough, the Lie algebra case turns out to be much easier to prove.

DEFINITION 9.5. An $\mathrm{Ad}\,G$-equivariant $D(\mathfrak{g})$-module V is called admissible if it satisfies the equivalent conditions (i)–(iii) of Proposition 9.2.

Given a parabolic subalgebra $\mathfrak{q}$ of $\mathfrak{g}$ let $\mathfrak{u}_{\mathfrak{q}}$ denote the nilradical of $\mathfrak{q}$, $i_{\mathfrak{q}}: \mathfrak{q} \hookrightarrow \mathfrak{g}$ the inclusion, and $p_{\mathfrak{q}}: \mathfrak{q} \to \mathfrak{q}/\mathfrak{u}_{\mathfrak{q}}$ the natural projection. Let V be an irreducible regular holonomic $\mathrm{Ad}\,G$-equivariant $D(\mathfrak{g})$-module supported on N. Set $\mathrm{res}_{\mathfrak{q}} V = (p_{\mathfrak{q}})_*(i_{\mathfrak{q}}^! V)[\dim \mathfrak{u}_{\mathfrak{p}}]$. The module V is said to be *cuspidal* (cf. [Lu3]) if, for any proper parabolic subalgebra $\mathfrak{l} \subset \mathfrak{g}$, we have

$$^{p}H^0(\mathrm{res}_{\mathfrak{q}} V) = 0.$$

THEOREM 9.6 (cf. [Lu3]). *Let V be an irreducible* $\mathrm{Ad}\,G$*-equivariant $D(G)$-module. The following properties are equivalent*:

(i) *V is cuspidal*;
(ii) *V is an admissible $D(\mathfrak{g})$-module supported on N*;
(iii) *Both V and FV are supported on nilpotent varieties (of $\mathfrak{g}$ and $\mathfrak{g}^*$, respectively)*.

PROOF. The equivalence of (ii) and (iii) follows from Proposition 9.2.

(i) $\Longrightarrow$ (iii). We identify $\mathfrak{g}^*$ with $\mathfrak{g}$ via the Killing form. Since V is supported on N, we see that FV is an admissible irreducible $D(\mathfrak{g})$-module. Let $\mathfrak{g}_{\mathfrak{l},\mathcal{O}}$ be the open stratum in $\mathrm{supp}\,FV$ and $\mathfrak{l} \neq \mathfrak{g}$. The restriction of FV to $\mathfrak{g}_{\mathfrak{l},\mathcal{O}}$ is a smooth local system. Let $\mathfrak{q} = \mathfrak{l}+\mathfrak{u}$ be a parabolic subalgebra of $\mathfrak{g}$ with Levi component $\mathfrak{l}$. For any $s \in \mathfrak{z}^{\mathrm{reg}}$ and $n \in \mathcal{O}$ the affine space $s+n+\mathfrak{u}$ is contained in $\mathfrak{g}_{\mathfrak{l},\mathcal{O}}$. Hence, the restriction to $s+n+\mathfrak{u}$ of the complex $i_{\mathfrak{q}}^* FV$ is the constant sheaf. It follows that $^{p}H^m(p_{\mathfrak{q}})_!(i_{\mathfrak{q}}^* FV) \neq 0$. The identity $F((p_{\mathfrak{q}})_*(i_{\mathfrak{q}}^! V)) = (p_{\mathfrak{q}})_!(i_{\mathfrak{q}}^* FV)$ yields $^{p}H^m(p_{\mathfrak{q}})_*(i_{\mathfrak{q}}^* FV) \neq 0$. This contradicts the assumption that V is cuspidal. Hence, $\mathfrak{l} = \mathfrak{g}$ so that $\mathrm{supp}\,FV \subset N$ and (iii) is proved.

The proof of (ii) $\Longrightarrow$ (i) is similar to the proof of the implication (i) $\Longrightarrow$ (ii) of Theorem 7.3. We omit the details. □

We are now able to complete the proof of Theorem 7.3 by proving implication (ii) $\Longrightarrow$ (i) of that theorem. Let V be a cuspidal module on G. An elementary reduction argument shows that we may assume that G is semisimple and that $\mathrm{supp}\,V$ is the closure of a unipotent orbit. Let $\exp: \mathfrak{g} \to G$ be the exponential map and $\exp^* V$ the pull-back of V on $\mathfrak{g}$. It is not hard to verify that $\exp^* V$ is a cuspidal module on $\mathfrak{g}$. So, it is admissible by Theorem 9.6. Hence, the characteristic variety of $\exp^* V$ is contained in $\mathfrak{g} \times N^*$. It follows that $SSV \subset G \times N^* \subset T^*G$. Thus, V is admissible by [Gi, Theorem 1.4.2].

REFERENCES

[BBD] A. Beilinson, J. Bernstein, and P. Deligne, *Faisceaux pervers*, Astérisque **100** (1982).

[BM] W. Borho and R. MacPherson, *Représentations des groupes de Weyl et homologie d'intersection pour les variétiés nilpotents*, C. R. Acad. Sci. Paris Sér. I Math. **292** (1981), 707–710.

[BGS] A. Beilinson, V. Ginzburg, and W. Soergel, *Koszul duality patterns in representation theory*, J. Amer. Math. Soc. (to appear).

[Br] J.-L. Brylinski, *Transformations canoniques, dualité projective, theorie de Lefschetz, transformation de Fourier et sommes trigonometriques*, Astérisque **140** (1986), 3–134.

[Gi] V. Ginzburg, *Admissible modules on a symmetric space*, Astérisque **173–174** (1989), 199–255.

[Gi2] ———, *Characteristic varieties and vanishing cycles*, Invent. Math. **84** (1986), 327–402.

[Gi3] ———, *The Fourier-Langlands transform on reductive groups*, Funktsional. Anal. i Prilozhen. **22** (1988), no. 2, 71–72; English transl., Functional Anal. Appl. **22** (1988), no. 2, 143–144.

[Gi4] ———, *Perverse sheaves on a loop group and Langlands' duality*, Inst. Hautes Études Sci. Publ. Math. (to appear).

[Gi5] ———, *Deline-Langlands conjecture and representations of affine Hecke algebras*, Preprint, Moscow, 1985.

[Ka] M. Kashiwara, *B-function and holonomic systems*, Invent. Math. **38** (1976), 33–53.

[Lu1] G. Lusztig, *Intersection cohomology complexes on a reductive group*, Invent. Math. **75** (1984), 205–272.

[Lu2] ———, *Character sheaves*. I–V, Adv. in Math. **56** (1985), 193–237; **57** (1985), 226–315; **59** (1986), 1–63; **61** (1986), 103–155; **62**, 313–314.

[Lu3] ———, *Fourier transforms on a semisimple Lie algebra over* $\mathbb{F}_q$, Lecture Notes in Math., vol. 1271, Springer-Verlag, Berlin, Heidelberg, and New York, 1987, pp. 177–188.

[Lu4] ———, *Cuspidal local systems and graded Hecke algebras*, Inst. Hautes Études Sci. Publ. Math. **67** (1988), 145–202.

ADVANCES IN SOVIET MATHEMATICS
Volume 16, Part 1, 1993

Explicit Construction of Characteristic Classes

A. B. GONCHAROV

To my teacher Israel Moiseevich Gelfand on the occasion of his 80th birthday

ABSTRACT. Let E be a vector bundle over an algebraic manifold X. An explicit local construction is given of characteristic classes $c_n(E)$ with values in *bi-Grassmannian* cohomology that are defined in §1. In the special case $n = \dim E$ it reduces to the construction in [BMS] of $c_n(E)$ with values in the Grassmannian cohomology.

Our construction implies immediately an explicit construction of Chern classes with values in $H^n(X, \underline{\underline{K}}_n^M)$, where $\underline{\underline{K}}_n^M$ is the sheaf of Milnor's K-groups.

A construction of classes $c_n(E)$ with values in motivic cohomology is given for $n \leqslant 3$. For $n = 2$ it could be considered as a motivic analog of the local combinatorial formula of Gabrielov, Gelfand, and Losik for the first Pontryagin class (see [GGL]). The reason for the restriction $n \leqslant 3$ is the present lack of a good theory of n-logarithms for $n \geqslant 4$. Explicit constructions of the universal Chern classes $c_n \in H^n(BGL_{m^\bullet}, \underline{\underline{K}}_n^M)$ and, for $n \leqslant 3$, of classes $c_n \in H^{2n}_{\mathcal{M}}(BGL_{m^\bullet}, \mathbb{Z}(n))$ ($H^\bullet_{\mathcal{M}}$ is the motivic cohomology) are given.

§1. Introduction

1.1. Chern classes with values in $H^n(X, \underline{\underline{K}}_n^M)$. Let L be a line bundle over X. There is the following classical construction of $c_1(L) \in H^1(X, \mathcal{O}^*)$. Choose a Zariski covering $\{U_i\}$ of X such that $L|_{U_i}$ is trivial. Choose nonzero sections $s_i \in \Gamma(U_i, L)$. Then $s_i/s_j \in \Gamma(U_i \cap U_j, \mathcal{O}^*)$ satisfies the cocycle relation and hence defines a cohomology class $c_1(L) \in H^1(X, \mathcal{O}^*)$.

Let us define the presheaf of Milnor's K-groups on X as follows: its section over an open set U is the quotient group of $\underbrace{\mathcal{O}^*(U) \otimes \cdots \otimes \mathcal{O}^*(U)}_{n \text{ times}}$ by the subgroup generated by elements

$$g_1 \otimes \cdots \otimes g_k \otimes f \otimes (1-f) \otimes g_{k+3} \otimes \cdots \otimes g_n, \qquad g_i, f, 1-f \in \mathcal{O}^*(U).$$

1991 *Mathematics Subject Classification.* Primary 57R20, 55R40; Secondary 19D45, 55R50.

Let us denote by $\underline{\underline{K}}_n^M$ the sheaf associated with this presheaf. We will denote by $\{f_1, \dots, f_n\}$ the image of $f_1 \otimes \cdots \otimes f_n \in \mathcal{O}^*(U)^{\otimes n}$ in $\underline{\underline{K}}_n^M(U)$.

In § 3 for any vector bundle E over X an explicit construction of the Chern classes $c_n(E) \in H^n(X, \underline{\underline{K}}_n^M)$ will be given.

The construction of $c_n(E^n)$ for an n-dimensional vector bundle E^n follows from the results of [S1] and [BMS, Chapter 1]. More precisely, let U_i be a Zariski covering such that $E^n|_{U_i}$ is trivial. Choose a section $s_i \in \Gamma(U_i, E^n)$ such that $s_{i_1}(x), \dots, s_{i_{n+1}}(x)$ are in general position on $U_{i_1 \dots i_{n+1}} := U_{i_1} \cap \cdots \cap U_{i_{n+1}}$. Then $s_{i_{n+1}} = \sum_{k=1}^n a_{i_k}(x) \cdot s_{i_k}(x)$ and

$$\{a_{i_1}(x), \dots, a_{i_n}(x)\} \in K_n^M(U_{i_1 \dots i_{n+1}})$$

is a cocycle in the Čech complex.

I will generalize this construction to vector bundles of arbitrary dimension and show that, being applied to $c_1(E)$, it gives the above cocycle for $c_1(\det E)$.

1.2. An application. There is a canonical map of sheaves

$$\underline{\underline{K}}_n^M \to \Omega^n_{\log} \hookrightarrow \Omega^n_{\mathrm{cl}} \hookrightarrow \Omega^n,$$
$$\{f_1, \dots, f_n\} \mapsto d\log f_1 \wedge \cdots \wedge d\log f_n.$$

Here $\Omega^n_{\log}$ (respectively Ω^n_{cl}) is the sheaf of n-forms with logarithmic singularities at infinity (respectively closed n-forms). Therefore we get a construction of characteristic classes with values in $H^n(X, \Omega^n_{\log})$ and $H^n(X, \Omega^n_{\mathrm{cl}})$. Note that Atiyah's construction provides us with characteristic classes in $H^n(X, \Omega^n)$ ([A], see also [Har]).

1.3. The Grassmannian bicomplex and bi-Grassmannian cohomology (see [G1, G2], compare with [GGL, BMS, S3]). Let Y be a set and $\widetilde{C}_n(Y)$ be the free abelian group generated by elements $(y_0, \dots, y_n)$ of $Y^{n+1} := \underbrace{Y \times \cdots \times Y}_{n+1}$. There is a complex $(\widetilde{C}_*(Y), d)$, where

$$d(y_0, \dots, y_n) := \sum_{i=0}^{n} (-1)^i (y_0, \dots, \widehat{y}_i, \dots, y_n). \tag{1.1}$$

This is just the simplicial complex of the simplex whose vertices are labeled by elements of Y. Suppose that a group G acts on Y. Let us call elements of the quotient set $G\backslash Y^{n+1}$ *configurations* of elements of Y. Denote by $C_n(Y)$ the free abelian group generated by configurations of $(n+1)$ elements of Y. There is a complex $(C_*(Y), d)$, where d is defined by the same formula (1.1) and $C_*(Y) = \widetilde{C}_*(Y)_G$. We will also apply this construction to subsets of $G \backslash Y^{n+1}$ of "configurations in general position".

Now let us denote by $C_n(m)$ the free abelian group generated by configurations of $n+1$ vectors in general position in an m-dimensional vector space V^m over F (i.e., any m vectors of the configuration are linearly independent). It does not depend on the choice of vector space V_m. In this case there is another map:

$$d': C_n(m) \to C_{n-1}(m-1)$$
$$d': (v_0, \dots, v_n) \mapsto \sum_{i=0}^{n} (-1)^i (v_i \mid v_0, \dots, \widehat{v}_i, \dots, v_n).$$

Here $(v_i \mid v_0, \dots, \widehat{v}_i, \dots, v_n)$ is a configuration of vectors in $V_m/\langle v_i\rangle$ obtained by projection of vectors $v_j \in V^m$, $j \neq i$. Then there is the following bicomplex:

$$\begin{array}{ccccccc}
 & & \vdots & & \vdots & & \vdots \\
 & & \downarrow & & \downarrow & & \downarrow \\
\cdots \longrightarrow & & C_{n+4}(n+2) & \xrightarrow{d} & C_{n+3}(n+2) & \xrightarrow{d} & C_{n+2}(n+2) \\
 & & \downarrow{\scriptstyle d'} & & \downarrow{\scriptstyle d'} & & \downarrow{\scriptstyle d'} \\
\cdots \longrightarrow & & C_{n+3}(n+1) & \xrightarrow{d} & C_{n+2}(n+1) & \xrightarrow{d} & C_{n+1}(n+1) \\
 & & \downarrow{\scriptstyle d'} & & \downarrow{\scriptstyle d'} & & \downarrow{\scriptstyle d'} \\
\cdots \longrightarrow & & C_{n+2}(n) & \xrightarrow{d} & C_{n+1}(n) & \xrightarrow{d} & C_n(n)
\end{array} \tag{1.2}$$

We will call it the *Grassmannian bicomplex* (over $X = \operatorname{Spec} F$).

There is a subcomplex $(C_*(n), d)$

$$\to C_{n+2}(n) \xrightarrow{d} C_{n+1}(n) \xrightarrow{d} C_n(n), \tag{1.3}$$

of the bicomplex (1.2). This is the *Grassmannian complex* introduced in [S1, BMS], see also [Q2].

Let us denote by $(BC_*(n), \partial)$ the total complex associated with the bicomplex (1.2); in particular, $BC_n(n) := C_n(n)$. We assume that $BC_k(n)$ placed in degree k and ∂ has degree -1.

Note that

$$BC_*(n+1) = BC_*(n)/C_*(n)$$

and there is a sequence of surjective homomorphisms of complexes

$$BC_*(1) \twoheadrightarrow BC_*(2) \twoheadrightarrow BC_*(3) \twoheadrightarrow \cdots .$$

The complex $BC_*(n)$ is "homological", i.e., its differential has degree -1. Let us make from it a "cohomological" complex $BC^*(n)$ with differential ∂ of degree $+1$. By definition $BC^i(n) := BC_{2n-i}(n)$,

$$\cdots \xrightarrow{\partial} BC^0(n) \xrightarrow{\partial} BC^1(n) \xrightarrow{\partial} \cdots \xrightarrow{\partial} BC^n(n).$$

Here $BC^i(n)$ is placed at degree i. There is also a "cohomological" version $C^*(n)$ of the Grassmannian complex $C_*(n)$.

Now let us give a more geometrical interpretation of the Grassmannian bicomplex that also explains the name.

Let $(e_1, \dots, e_{p+q+1})$ be a basis in a vector space V. Let us denote by $\widehat{\mathbf{G}}^p_q$ the open subset of the Grassmannian of q-dimensional subspaces of P^{p+q} consisting of subspaces that are transverse to coordinate hyperplanes. In [M] R. MacPherson constructed an isomorphism

$$m\colon \widehat{\mathbf{G}}^p_q \xrightarrow{\sim} \left\{ \begin{array}{l} \text{configurations of } p+q+1 \text{ vectors in general} \\ \text{position in a } p\text{-dimensional vector space} \end{array} \right\}. \tag{1.4}$$

Namely, $m(\xi)$ is a configuration formed by the images of e_i in V/ξ.

Remark. The set $\widehat{\mathbf{G}}^p_q$ is defined in terms of projective geometry. However, the isomorphism (1.4) depends on the choice of the vectors e_i. This additional data can be also visualized inside P^{p+q}: one must add a generic hyperplane (affinization).

Let

$$\mathbb{Z}\colon \mathrm{Var} \to \mathrm{Ab} \tag{1.5}$$

be a functor from the category of algebraic varieties over F to the category of abelian groups that sends a variety X to the free abelian group generated by F-points of X. Applying it to (1.4) we get an isomorphism

$$\mathbb{Z}[\widehat{\mathbf{G}}^p_q] \xrightarrow{\sim} C_{p+q}(p). \tag{1.6}$$

For each integer i such that $0 \leqslant i \leqslant p+q$, there are intersection maps a_i and projection maps b_i:

$$\begin{array}{ccc} \widehat{\mathbf{G}}^p_q & \xrightarrow{a_i} & \widehat{\mathbf{G}}^p_{q-1} \\ \downarrow{\scriptstyle b_i} & & \\ \widehat{\mathbf{G}}^{p-1}_q & & \end{array}$$

Here the subspace $a_i(\xi)$ is the intersection of ξ with the ith coordinate hyperplane and the subspace $b_i(\xi)$ is the image of ξ under the projection with the center at the ith vertex of the simplex. We get a bi-Grassmannian $\widehat{\mathbf{G}}(n)$:

$$\begin{array}{ccccccc} & & & & & & \downarrow\vdots\downarrow \\ & & & & & \overset{\cdots}{\rightrightarrows} & \widehat{\mathbf{G}}^{n+2}_0 \\ & & & & \downarrow\vdots\downarrow & & {\scriptstyle b_0}\downarrow\vdots\downarrow{\scriptstyle b_{n+1}} \\ \widehat{\mathbf{G}}(n)\colon & & & \overset{\cdots}{\rightrightarrows} & \widehat{\mathbf{G}}^{n+1}_1 & \underset{a_{n+1}}{\overset{a_0}{\rightrightarrows}} & \widehat{\mathbf{G}}^{n+1}_0 \\ & \downarrow\vdots\downarrow & & & {\scriptstyle b_0}\downarrow\vdots\downarrow{\scriptstyle b_{n+1}} & & {\scriptstyle b_0}\downarrow\vdots\downarrow{\scriptstyle b_n} \\ \overset{\cdots}{\rightrightarrows} & \widehat{\mathbf{G}}^n_2 & \underset{a_{n+1}}{\overset{a_0}{\rightrightarrows}} & & \widehat{\mathbf{G}}^n_1 & \underset{a_n}{\overset{a_0}{\rightrightarrows}} & \widehat{\mathbf{G}}^n_0 \end{array} \tag{1.7}$$

Applying functor (1.5) to it, considering differentials $d = \sum(-1)^i a_i$ and $d' = \sum(-1)^i b_i$ and using isomorphism (1.6) we get the Grassmannian bicomplex.

Now let us sheafify these constructions.

A bicomplex $\underline{\underline{\mathbb{Z}}}[\widehat{\mathbf{G}}(n)]$ of sheaves on X, called the Grassmannian bicomplex, is constructed as follows. For a point $x \in X$, the stalk of $\underline{\underline{\mathbb{Z}}}[\widehat{\mathbf{G}}(n)]$ at x is the formal linear combinations of germs at x of maps from X to $\widehat{\mathbf{G}}^p_q$. The corresponding bicomplex looks as follows:

$$\begin{array}{ccccc} & & \vdots & & \vdots \\ & & \downarrow & & \downarrow \\ \underline{\underline{\mathbb{Z}}}[\widehat{\mathbf{G}}(n)]: \cdots & \longrightarrow & \underline{\underline{\mathbb{Z}}}[\widehat{\mathbf{G}}^{n+1}_1] & \xrightarrow{d} & \underline{\underline{\mathbb{Z}}}[\widehat{\mathbf{G}}^{n+1}_0] \\ & & \downarrow{\scriptstyle d'} & & \downarrow{\scriptstyle d'} \\ \cdots & \longrightarrow & \underline{\underline{\mathbb{Z}}}[\widehat{\mathbf{G}}^{n}_1] & \xrightarrow{d} & \underline{\underline{\mathbb{Z}}}[\widehat{\mathbf{G}}^{n}_0] \end{array} \tag{1.8}$$

Here $\underline{\underline{\mathbb{Z}}}[\widehat{\mathbf{G}}^n_0]$ is placed at degree $(n, 0)$ and d (respectively d') has degree $(1, 0)$ (respectively $(0, 1)$). The hypercohomology of the total complex associated with this bicomplex of sheaves is the *bi-Grassmannian cohomology* of X. We will denote it by $H^*(X, \underline{\underline{\mathbb{Z}}}[\widehat{\mathbf{G}}(n)])$ or $H^*(X, \underline{\underline{BC}}^*(n))$. Note that the Grassmannian cohomology of [BMS] maps canonically to the bi-Grassmannian one, but there is no inverse map.

In § 2 we construct characteristic classes $c_n(E) \in H^{2n}(X, \underline{\underline{\mathbb{Z}}}[\widehat{\mathbf{G}}(n)])$. There is a homomorphism of complexes of sheaves

$$\underline{\underline{\mathbb{Z}}}[\widehat{\mathbf{G}}(n)] \to \underline{\underline{K}}^M_n[-n] \tag{1.9}$$

(see § 3) that provides a construction of characteristic classes

$$c_n(E) \in H^n(X, \underline{\underline{K}}^M_n).$$

1.4. Polylogarithms (compare with [GGL, BMS, HM]). Now let $F = \mathbb{C}$. Note that $\widehat{\mathbf{G}}^n_0$ is almost canonically isomorphic to $(\mathbb{C}^*)^n$ (see Remark in 1.3 above). Indeed, according to (1.4) a point $\xi \in \widehat{\mathbf{G}}^n_0$ defines an (ordered) configuration of $n+1$ vectors in general position in $\mathbb{C}^n$: $m(\xi) = (v_0, \dots, v_n)$. So $v_0 = \sum_{i=1}^n z_i v_i$ and the map $\xi \mapsto (z_1, \dots, z_n)$ provides an isomorphism $\widehat{\mathbf{G}}^n_0 \xrightarrow{\sim} (\mathbb{C}^*)^n$. Therefore there is a canonical multivalued holomorphic $n-1$ form

$$w^n_0 := \frac{1}{n} \sum_{i=1}^{n} (-1)^i \log z_i \, d\log z_1 \wedge \cdots \wedge \widehat{d \log z_i} \wedge \cdots \wedge d\log z_n \tag{1.10}$$

on $\widehat{\mathbf{G}}^n_0$.

Consider the multivalued Deligne complex $\widetilde{\mathbb{Q}}(n)_Y$ on a variety Y ($\mathbb{Q}$ is placed at degree 0, d has degree $+1$):

$$\mathbb{Q} \xrightarrow{(2\pi i)^n} \widetilde{\Omega}^0(Y) \xrightarrow{d} \widetilde{\Omega}^1(Y) \xrightarrow{d} \cdots \xrightarrow{d} \widetilde{\Omega}^{p-1}(Y) \to 0.$$

Here $\widetilde{\Omega}^i$ denotes multivalued holomorphic differential forms, i.e., holomorphic differential forms defined on the universal covering space $\widetilde{Y}$ of Y. We want to consider a triple complex $\mathbb{D}$, which is the multivalued complex $\widetilde{\mathbb{Q}}(n)_{\widehat{\mathbf{G}}(n)}$ in the vertical direction and is a double complex constructed from the bi-Grassmannian $\widehat{\mathbf{G}}(n)$ in the horizontal directions. All differentials have degree $+1$.

A $2n$-cocycle in the complex $\mathbb{D}$ is just a collection of $(2n-1-p-q)$-forms $\{\omega_q^p\}$ such that

$$d\omega_q^p = \sum(-1)^i a_i^* \omega_{q-1}^p + \sum(-1)^i b_i^* \omega_q^{p-1}. \tag{1.11}$$

CONJECTURE 1.1. *There exists a $2n$-cocycle $\mathbb{L}_n$ in the triple complex $\mathbb{D}$ such that its ω_0^n-component is given by formula* (1.10).

The collection of forms $\{\omega_q^n\}$ is, of course, the Grassmannian n-logarithm conjectured in [BMS, HM]. However, for an explicit construction of the Chern classes in Deligne cohomology we have to construct the entire bi-Grassmannian n-logarithm and it is *not* sufficient to construct only its Grassmannian part. The main construction of this paper (see § 2) yields a construction of

$$c_n(E) \in H_{\mathscr{D}}^{2n}(X, \mathbb{Q}(n)),$$

using the bi-Grassmannian polylogarithm $\mathbb{L}_n$. The coincidence of this class with the one constructed by A. A. Beilinson [B2] is guaranteed by formula (1.10) (see Theorem 5.11). The problem of construction of a collection of forms $\{\omega_q^p\}$ satisfying (1.11) goes back to [GGL], see also [You], where the real-valued forms on the corresponding manifolds over $\mathbb{R}$ were considered (forms $S^{p,q}$).

The most interesting component of $\mathbb{L}_n$ is a multivalued function $P_n := \omega_{n-1}^n$ on $\widehat{\mathbf{G}}_{n-1}^n$. The cocycle condition means that it should satisfy two " $(2n+1)$-term" functional equations

$$\sum_{i=0}^{2n}(-1)^i a_i^* P_n = (2\pi i)^n q_1, \tag{1.12a}$$

$$\sum_{i=0}^{2n}(-1)^i b_i^* P_n = (2\pi i)^n q_2, \tag{1.12b}$$

where $q_1, q_2 \in \mathbb{Q}$. Note that a_i^*, b_i^* make sense after lifting maps a_i, b_i to the simply connected covering spaces.

For a much more precise "motivic" version of Conjecture 1.1 see Conjecture 6.1. It is formulated for any field F and implies Conjecture 1.1, when $F = \mathbb{C}$.

Instead of the Deligne complex $\underline{\underline{\widetilde{\mathbb{Q}}}}(n)_{\mathcal{D}}$ one could consider the real Deligne complex $\underline{\underline{\mathbb{R}}}(n)_{\mathcal{D}}$ that is the total complex of the bicomplex

$$\underline{\underline{\mathbb{R}}}(n)_{\mathcal{D}}: \quad \begin{array}{ccccccccc} S_X^0 & \xrightarrow{d} & S_X^1 & \xrightarrow{d} & \cdots & \xrightarrow{d} & S_X^n & \xrightarrow{d} & S_X^{n+1} & \xrightarrow{d} & \cdots \\ & & & & & & \uparrow{\scriptstyle \pi_n} & & \uparrow{\scriptstyle \pi_n} & & \\ & & & & & & \Omega_X^n & \xrightarrow{\partial} & \Omega_X^{n+1} & \xrightarrow{\partial} & \cdots \end{array} \tag{1.13}$$

where $(S_X^\bullet, d)$ is the de Rham complex of the real-valued forms, $(\Omega^\bullet, \partial)$ is the de Rham complex of the holomorphic forms with logarithmic singularities at infinity $\pi_n = \mathrm{Re}$ for odd n and $\pi_n = \mathrm{Im}$ for even n, and S_X^0 is placed at degree 1.

One can consider the triple complex $\mathbb{D}$, which is the complex $\underline{\underline{\mathbb{R}}}(n)_{\mathcal{D}}$ in the vertical direction and is the double complex constructed from the bi-Grassmannian $\underline{\underline{\widehat{\mathbf{G}}}}(n)$ in the horizontal directions. Actually it is more natural to consider the complex that computes hypercohomology of the bi-Grassmannian $\widehat{\mathbf{G}}(n)$ with coefficients in $\underline{\underline{\mathbb{R}}}(n)_{\mathcal{D}}$ (for this we should replace the complex $(\Omega_X^{\geqslant n}, \partial)$ in (1.13) by its Dolbeaux resolution $(\mathcal{D}_X^{\geqslant n, q})$ for example), but it is not important for our purposes.

Conjecture 1.1$'$. *There exists a $2n$-cocycle $\mathbb{L}_n'$ in the triple complex $\mathbb{D}'$ such that its component over $\widehat{\mathbf{G}}_0^n$ is given by the following formulas*:

$$\begin{aligned} \omega_0^{n'} &= \pi_n \operatorname{Alt}\Big(\sum_{k=0}^{[(n-1)/2]} \frac{i^{n-2k-1}}{(2k+1)!(n-2k-1)!} \log|z_1|\, d\log|z_2| \wedge \cdots \\ &\qquad \wedge d\log|z_{2k+1}| \wedge d\arg z_{2k+2} \wedge \cdots \wedge d\arg z_n \Big) \in S_{\widehat{\mathbf{G}}_0^n}^{n-1}, \\ \omega_0^{n''} &= d\log z_1 \wedge \cdots \wedge d\log z_n \in \Omega_X^n \end{aligned} \tag{1.14}$$

($d\omega_0^{n'} = \pi_n(\omega_0^{n''})$). *The corresponding component P_n' of $\mathbb{L}_n'$ on $\widehat{\mathbf{G}}_{n-1}^n$ should satisfy the "clean" $(2n+1)$-term equation*

$$\sum_{i=0}^{2n} (-1)^i a_i^* P_n' = 0, \tag{1.14a}$$

$$\sum_{i=0}^{2n} (-1)^i b_i^* P_n' = 0. \tag{1.14b}$$

Remark. $(\omega_0^{n'}, \omega_0^{n''})$ is just the product in the real Deligne cohomology of 1-cocycles $(\log|z_i|, d\log z_i) \in H^1(\widehat{\mathbf{G}}_0^n, \mathbb{R}(1)_{\mathcal{D}})$.

In [GM] I. M. Gelfand and R. MacPherson constructed for even n a function $\widetilde{P}_n$ on the real "middle Grassmannian" $\widehat{G}^n_{n-1}(\mathbb{R})$, that satisfies the conditions (1.12).

On the other hand, there are the *classical* polylogarithms $\mathrm{Li}_n(z)$ that are functions in *one* complex variable z. They were defined by L. Euler as functions on the unit disc $|z| \leqslant 1$ given by absolutely convergent series

$$\mathrm{Li}_n(z) = \sum_{k=1}^{\infty} \frac{z^k}{k^n},$$

and can be continued analytically to a multivalued function on $\mathbb{C}P^1 \setminus \{0, 1, \infty\}$ using the formulas

$$\mathrm{Li}_1(z) = -\log(1-z),$$
$$\mathrm{Li}_n(z) = \int_0^z \mathrm{Li}_{n-1}(t)\frac{dt}{t}.$$

It turns out that $\mathrm{Li}_n(z)$ has a remarkable single-valued version ($B_0 = 1$, $B_1 = -1/2$, $B_2 = 1/6$, ... are Bernoulli numbers)

$$\mathscr{L}_n(z) = \begin{matrix}\mathrm{Re}(n\colon \text{odd}) \\ \mathrm{Im}(n\colon \text{even})\end{matrix} \left(\sum_{k=0}^{n} \frac{B_k \cdot 2^k}{k!} \log^k |z| \cdot \mathrm{Li}_{n-k}(z) \right), \qquad n \geqslant 2,$$

$$\mathscr{L}_1(z) = \log|z|.$$

For example

$$\mathscr{L}_2(z) = \mathrm{Im}(\mathrm{Li}_2(z)) + \arg(1-z)\cdot\log|z|$$

is the Bloch-Wigner function, and

$$\mathscr{L}_3(z) = \mathrm{Re}(\mathrm{Li}_3(z) - \log|z|\cdot\mathrm{Li}_2(z) + 1/3\log^2|z|\cdot\mathrm{Li}_1|z|)$$

was used in [G1]. The functions $\mathscr{L}_n(z)$ for arbitrary n were written by D. Zagier [Z]. Explicit formulas expressing the bi-Grassmannian polylogarithms $\mathbb{L}_n$, $\mathbb{L}'_n$ in terms of the classical polylogarithms for $n \leqslant 3$ were given in [G1] (see also [G2] and §§ 5, 6 of this paper). For example $\mathbb{L}'_3$, that is a function on the 9-dimensional manifolds $\widehat{G}^3_2$, is expressed in terms of $\mathscr{L}_3(z)$. However for $n \geqslant 4$ the "natural" cocycle $\mathbb{L}_n$ can*not* be expressed by the classical polylogarithms (the reason was explained in § 1 of [G1]).

An interesting geometrical construction of the Grassmannian 2- and 3-logarithms was suggested by M. Hanamura and R. MacPherson [HaM]. The existence of the Grassmannian n-logarithms for $n \leqslant 3$ was proved in [HM].

It is interesting that in formulas for $\mathbb{L}'_n$, $n \leqslant 3$, from § 9 of [G1] all forms ω_q^{n+i} vanish for $i > 0$. This means that the bi-Grassmannian n-logarithms for $n \leqslant 3$ reduces essentially to its Grassmannian part $\{\omega_q^n\}$. This is a nontrivial fact about the Grassmannian n-logarithms, $n \leqslant 3$. But this is *not true* for $n \geqslant 4$. For example, already forms ω_1^{n+1} cannot be chosen equal to zero for $n \geqslant 4$. This is another important difference between cases $n \leqslant 3$

and $n > 4$. It shows why we must enlarge the Grassmannian polylogarithms to the bi-Grassmannian ones.

1.5. The universal Chern classes $c_n \in H^n(BGL_{m\bullet}, \underline{\underline{K}}_n^M)$. Recall that the classifying space for a group G can be represented by the simplicial scheme

$$BG_\bullet : * \Leftarrow G \Lleftarrow G^2 \cdots .$$

In § 4 an explicit construction will be given for the universal Chern classes $c_n \in H^n(BGL_m(F)_\bullet, \underline{\underline{K}}_n^M)$, $m \geqslant n$. It is a refinement of the construction from § 2 and, of course, implies it immediately. More precisely, a Zariski covering $\{U_i\}_{i\in I}$ defines a simplicial scheme $U_\bullet$:

$$\coprod_{i\in I} U_i \Leftarrow \coprod_{i_0<i_1\in I} U_{i_0 i_1} \Lleftarrow \coprod_{i_0<i_1<i_2\in I} U_{i_0 i_1 i_2} \cdots .$$

A G-bundle E over X given by its transition functions $g_{ij} \in \Gamma(U_{ij}, G)$ defines a canonical map of simplicial schemes $u: U_\bullet \to BG_\bullet$. Our G-bundle is the inverse image of the canonical G-bundle $EG_\bullet \xrightarrow{G} BG_\bullet$ over $BG_\bullet$ and $c_n(E) = u^* c_n$.

As a byproduct, an explicit algebraic construction of cohomology classes generating the ring $H^*(GL_m)$ is obtained. The existence of such a description of the usual topological cohomology of GL_m was conjectured by A. A. Beilinson [B3].

1.6. The universal motivic Chern classes. In § 4 an explicit construction of such Chern classes

$$c_n \in H^{2n}_{\mathscr{M}}(BGL_{m^\bullet}, \mathbb{Z}(n)), \qquad n \leqslant 3,$$

will be given. It implies, in particular, an explicit construction of the Chern classes $c_n(E)$ with values in Deligne cohomology $H^{2n}_{\mathscr{D}}(X, \mathbb{Z}(n))$ in terms of classical n-logarithms ($n \leqslant 3$). A cocycle representing the usual topological characteristic class $c_n(E) \in H^{2n}(X, \mathbb{Z})$ in the Čech complex was constructed by J.-L. Brylinski and D. MacLaughlin [BM2].

A local combinatorial formula for all Pontryagin classes was suggested by I. M. Gelfand and R. MacPherson [GM2].

Let $H^*_{\text{cts}}(G, R)$ be the ring of continuous cohomology of a Lie group G. It is known that

$$H^*_{\text{cts}}(GL_m(\mathbb{C}), R) = \Lambda^*_R(b_1^{(m)}, b_3^{(m)}, \dots, b_{2m-1}^{(m)}),$$
$$b_{2k-1}^{(m)} \in H^{2k-1}_{\text{cts}}(GL_m(\mathbb{C}), \mathbb{R}).$$

As a byproduct of the construction of the universal Chern classes $c_n \in H^{2n}_{\mathscr{D}}(BGL_m(\mathbb{C}), \mathbb{R}(n))$, we get an explicit formula for (measurable) cocycles representing classes $b_{2n-1}^{(m)}$ for $n \leqslant 3$ and arbitrary $m \geqslant 2n-1$ using the classical n-logarithm. The formula for b_1 is well known: $b_1(g) := \log|\det g|$,

$g \in GL_m(\mathbb{C})$, is a 1-cocycle. The formula for $b_3^{(2)}$ was given by D. Wigner in the middle of the seventies, and the formula for $b_5^{(3)}$ by the author ([G1], see also [G2]). A formula for $b_3^{(m)}$ was written also by Kioshi Igusa (unpublished).

Note that there is a canonical map

$$H_n(X, \underline{\underline{K}}_n^M) \to H^n(X, \underline{\underline{K}}_n),$$

and it was shown by Soulé [So] and by Nesterenko and Suslin [NS] that this map is an isomorphism modulo torsion. This together with characteristic classes $c_n(E) \in H^n(X, \underline{\underline{K}}_n)$ of Gillet [Gil] proves the existence of $c_n(E) \in H^n(X, \underline{\underline{K}}_n^M)$ but does not give any precise construction.

This work was initiated by A. A. Beilinson who explained to me that there is no explicit construction of the Chern classes with values in $H^n(X, \underline{\underline{K}}_n^M)$ as well as in $H^n(X, \Omega^n_{\log})$ or $H^n(X, \Omega^n_{\log})$ and emphasized importance of such a construction.

I hope it is clear from the introduction how much I benefited from the paper [GGL] of A. M. Gabrielov, I. M. Gelfand, and M. V. Losik.

The final draft of this paper was prepared during my stay at MIT, Max-Planck-Institut für Mathematik, and MSRI, and was supported by NSF Grant DMS-9022140. I am grateful to these institutions for their hospitality and to S. Bloch, J.-L. Brylinski, D. Kazhdan, R. MacPherson, V. V. Schechtman, and especially A. A. Beilinson for useful conversations and encouragement to publish this paper.

§2. Affine flags and Chern classes in bi-Grassmannian cohomology

2.1. Affine flags. Let V be a vector space over a field F. By definition a *p-flag* in V is a sequence of subspaces

$$0 \subset L^1 \subset L^2 \subset \cdots \subset L^p, \qquad \dim L^i = i.$$

An *affine p-flag* $L^\bullet$ is a p-flag together with choice of vectors $l^i \in L^i/L^{i-1}$, $i = 1, \dots, p$ ($L^0 = 0$). We will denote affine p-flags as $(l^1, \dots, l^p)$. Subspaces L^i can be recovered as the ones generated by $l^1, \dots, l^i$: $L^i = \langle l^1, \dots, l^i \rangle$.

We say that an $(n+1)$-tuple of affine flags

$$L_0^\bullet = (l_0^1, \dots, l_0^p), \dots, L_n^\bullet = (l_n^1, \dots, l_n^p) \tag{2.1}$$

is generic (or in the general position)

$$\dim(L_0^{i_0} + \cdots + L_n^{i_n}) = i_0 + \cdots + i_n \quad \text{whenever} \quad i_0 + \cdots + i_n \leqslant \dim V. \tag{2.2}$$

Let $A^p(m)$ be the manifold of all affine p-flags in an m-dimensional vector space V_m. It is a $GL(V_m)$-set, so as usual (see 1.3 of the Introduction) one

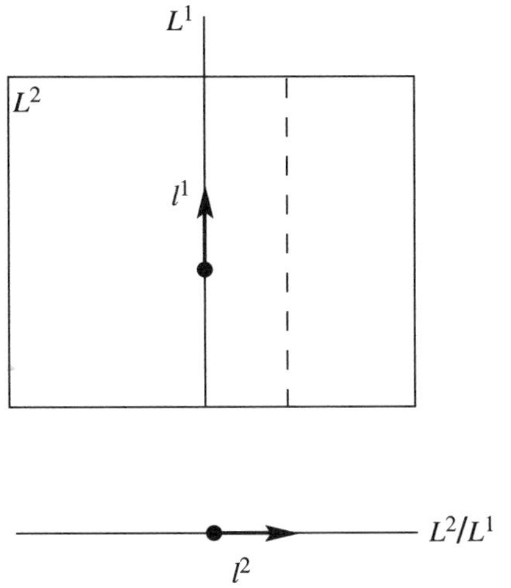

FIGURE 1. An affine 2-flag

can consider free abelian groups $C_n(A^p(m))$ of configurations of $(n+1)$-tuples of affine p-flags in general position in V_m. Further, there is a complex of affine p-flags $C_*(A^p(m))$:

$$\begin{gathered}\cdots \xrightarrow{d} C_{n+1}(A^p(m)) \xrightarrow{d} C_n(A^p(m)) \xrightarrow{d} C_{n-1}(A^p(m)) \xrightarrow{d} \cdots, \\ d\colon (L_0^\bullet, \ldots, L_n^\bullet) \mapsto \sum_{i=0}^{n} (-1)^i (L_0^\bullet, \ldots, \widehat{L_i^\bullet}, \ldots, L_n^\bullet).\end{gathered} \tag{2.3}$$

In particular, $C_*(A^1(m)) \equiv C_*(m)$. Let us define a map of complexes

$$T\colon C_*(A^{p+1}(n+p)) \to BC_*(n) \tag{2.4}$$

as follows: for

$$a_k^{p+1} = (v_0^1, \ldots, v_0^{p+1}; \ldots; v_k^1, \ldots, v_k^{p+1}) \in C_k(A^{p+1}(n+p)) \qquad (k \geqslant n),$$

set

$$\begin{aligned} T(a_k^{p+1}) &:= \bigoplus_{q=0}^{k-n} \sum_{\substack{i_0+\cdots+i_k=p-q \\ i_k \geqslant 0}} (L_0^{i_0} \oplus \cdots \oplus L_k^{i_k} \mid v_0^{i_0+1}, \ldots, v_k^{i_k+1}) \\ &\in \bigoplus_{q=0}^{k-n} C_k(n+q) =: BC_k(n).\end{aligned} \tag{2.5}$$

Here $(L_0^{i_0} \oplus \cdots \oplus L_k^{i_k} \mid v_0^{i_0+1}, \ldots, v_k^{i_k+1})$ is the configuration of vectors in the space $V_I := V_m / \bigoplus_{\alpha=0}^{k} L_\alpha^{i_\alpha}$ equal to projections of vectors $v_0^{i_0+1}, \ldots, v_k^{i_k+1}$. (Since $v_\alpha^{i_\alpha+1}$ is a vector in $V_m/L_\alpha^{i_\alpha}$, we can project it to V_I.)

KEY LEMMA 2.1. *T is a morphism of complexes.*

PROOF. Let $T_k(n+q)\colon C_k(A^{p+1}(n+p)) \to C_k(n+q)$ be the $C_k(n+q)$-component of the map T. We have to prove that (see (2.6))

$$d \circ T_k(n+q) = T_{k-1}(n+q) - d' \circ T_k(n+q+1),$$

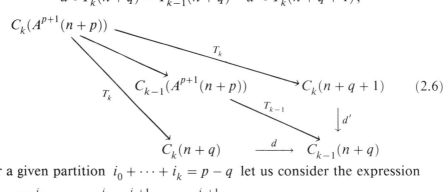

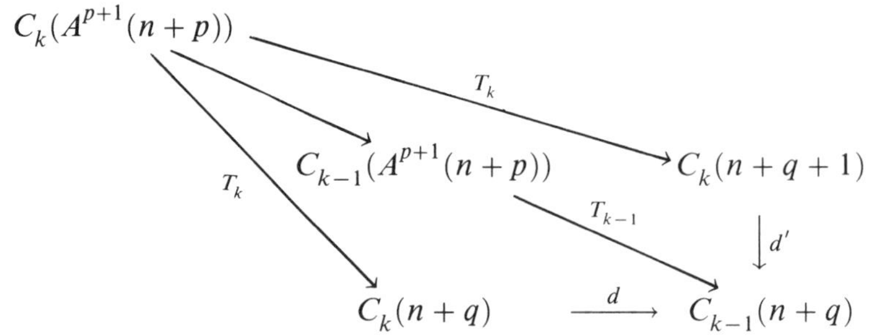

(2.6)

For a given partition $i_0 + \cdots + i_k = p - q$ let us consider the expression

$$\begin{aligned} &d(L_0^{i_0} \oplus \cdots \oplus L_k^{i_k} \mid v_0^{i_0+1}, \dots, v_k^{i_k+1}) \\ &= \sum_{j=0}^{k} (-1)^j (L_0^{i_0} \oplus \cdots \oplus L_k^{i_k} \mid v_0^{i_0+1}, \dots, \widehat{v_j^{i_j+1}}, \dots, v_k^{i_k+1}). \end{aligned} \tag{2.7}$$

For $i_j = 1$ the corresponding term in (2.6) will appear in the formula for $T_{k-1}(n+q)(a_k^{p+1})$. For $i_j > 1$ such a term appears in the formula for

$$d'(L_0^{i_0} \oplus \cdots \oplus L_j^{i_j-1} \oplus L_k^{i_k} \mid v_0^{i_0+1}, \dots, v_j^{i_j}, \dots, v_k^{i_k+1}).$$

2.2. A construction of Chern classes in bi-Grassmannian cohomology. Let us denote by $\mathscr{A}_E^p(X)$ the bundle of affine p-flags in fibers of a vector bundle E over X. Choose a Zariski covering $\{U_i\}$ of X such that $E\,|_{U_i}$ is trivial. Choose sections

$$L_i^{\bullet}(x) \in \Gamma(U_i, \mathscr{A}_E^p(x))$$

such that for any $i_0 < \cdots < i_k$ affine p-flags $L_{i_0}^{\bullet}(x), \dots, L_{i_k}^{\bullet}(x)$ are in general position for every $x \in U_{i_0 \dots i_k}$.

Let $(\mathscr{F}(n)_X^*, \partial)$ be a complex of sheaves in the Zariski topology on X obtained by sheafification from a complex $(\mathscr{F}(n)^*, \partial)$, where ∂ has degree $+1$. Denote by $\mathscr{F}(n)_*$ the corresponding homological complex:

$$\cdots \xrightarrow{\partial} \mathscr{F}(n)_i \xrightarrow{\partial} \mathscr{F}(n)_{i-1} \xrightarrow{\partial} \mathscr{F}(n)_{i-2} \xrightarrow{\partial} \cdots,$$

where, by definition, $\mathscr{F}(n)_i := \mathscr{F}(n)^{2n-i}$.

Suppose that we are given a morphism of complexes

$$T_n\colon C_*(A^{p+1}(n+p)) \to \mathscr{F}(n)_*. \tag{2.8}$$

Then

$$T_n(L_{i_0}(x), \dots, L_{i_k}(x)) \in \mathscr{F}(n)^{2n-k}\,|_{U_{i_0 \dots i_k}} \tag{2.9}$$

is a $2n$-cocycle in the Čech complex for the covering $\{U_i\}$ with values in the complex of sheaves $\mathcal{F}(n)^*_X$. (In fact, the chain (2.9) is a cocycle if and only if the map (2.8) is a morphism of complexes.)

A different choice of sections $L^\bullet_i(x)$ gives a cocycle that is canonically cohomologous to the previous one. Therefore the cohomology class

$$c_n(E) \in H^{2n}(\check{C}(U_\bullet, \mathcal{F}(n)^*))$$

is well defined.

Applying this construction to the bi-Grassmanian cohomology and using the Key Lemma 2.1 we immediately obtain the following theorem.

THEOREM 2.2. $T(L^\bullet_{i_0}(x), \dots, L^\bullet_{i_k}(x)) \in \underline{\underline{BC}}^{2n-k}(n)\,|_{U_{i_0 \dots i_k}}$ *is a cocycle in the Čech complex for the covering* $\{U_i\}$ *with values in the bi-Grassmannian complex.*

The homomorphism (2.8) for other cohomology theories $\mathcal{F}(n)^*$, including the motivic one (hence Deligne, l-adic, K^M, etc. cohomology), should be obtained as a composition

$$C_*(A^{p+1}(n+p)) \xrightarrow{T} BC_*(n) \longrightarrow \mathcal{F}(n)_*.$$

This will be proved for K^M_n-cohomology in § 3 and for motivic cohomology of weight $\leqslant 3$ in § 5.

However, I want to emphasize that the bi-Grassmannian cohomology is certainly different (larger) than the motivic one.

§3. Chern classes with values in $H^n(X, \underline{\underline{K}}^M_n)$

3.1. In § 2 we constructed Chern classes with values in $H^{2n}(\underline{\underline{BC}}^*(n))$. To obtain Chern classes with values in $H^n(X, \underline{\underline{K}}^M_n)$ it is sufficient to define a morphism of complexes

$$BC_*(n) \to K^M_n(F)[-n], \qquad (3.1)$$

i.e., a homomorphism $\overline{f}_n(n)\colon C_n(n) \to K^M_n(F)$ such that $f_n(n) \circ d = f_n(n) \circ d' = 0$:

$$\begin{array}{ccccc}
 & & & & \downarrow d' \\
 & & C_{n+2}(n+1) & \xrightarrow{d} & C_{n+1}(n+1) \\
 & & \downarrow d' & & \downarrow d' \\
\xrightarrow{d} & & C_{n+1}(n) & \xrightarrow{d} & C_n(n) \\
 & & & & \downarrow f_n(n) \\
 & & & & K^M_n(F)
\end{array}$$

Now let us define a homomorphism

$$f_n(n)\colon C_n(n) \to \Lambda^n F^*,$$

as follows (compare with § 3.2 in [G2]). Choose a volume form $w \in \det(V_n)^* \equiv \Lambda^n(V_n)^*$ (where $\dim V^n = n$). Set

$$\Delta(v_1, \dots, v_n) := \langle w, v_1 \wedge \dots \wedge v_n \rangle \in F^*, \qquad v_i \in V_n,$$
$$f_n(n)(v_0, \dots, v_n) := \mathrm{Alt} \bigwedge_{1 \leqslant i \leqslant n} \Delta(v_0, \dots, \widehat{v_i}, \dots, v_n) \in \Lambda^n F^*. \tag{3.2}$$

Here $\mathrm{Alt}\, g(v_0, \dots, v_n) := \sum_{\sigma \in S_{n+1}} (-1)^{|\sigma|} g(v_{\sigma(0)}, \dots, v_{\sigma(n)})$. For example, up to a 2-torsion we have

$$f_2(2)(v_0, v_1, v_2) := 2(\Delta(v_0, v_2) \wedge \Delta(v_0, v_1) - \Delta(v_1, v_2) \wedge \Delta(v_0, v_1) + \Delta(v_0, v_2) \wedge \Delta(v_1, v_2)).$$

LEMMA 3.1. *$f_n(n)(v_0, \dots, v_n)$ does not depend on w.*

PROOF. Let $f'_n(n)$ be a homomorphism defined using another volume form $w' = \lambda w$. Then

$$(f_n(n) - f'_n(n))(v_0, \dots, v_n) = \lambda \wedge \sum \Lambda_{i,j},$$

where $\Lambda_{i,j} \in \Lambda^{n-1} F^*$ depends on $v_0, \dots, \widehat{v_i}, \dots, \widehat{v_j}, \dots, v_n$. So $\Lambda_{i,j}$ is symmetric on v_i, v_j. But the left-hand side is antisymmetric by definition. Therefore $\Lambda_{i,j} = 0$.

LEMMA 3.2. *The composition*

$$C_{n+1}(n+1) \xrightarrow{d'} C_n(n) \xrightarrow{f_n(n)} \Lambda^n F^* \tag{3.3}$$

is equal to zero modulo 2-torsion.

PROOF (Compare with the proof of Lemma 3.4 in [G1]).

$$f_n(n) \circ d'(v_0, \dots, v_{n+1}) = \mathrm{Alt} \bigwedge_{j=2}^{n+1} \Delta(v_0, v_1, \dots, \widehat{v_j}, \dots, v_{n+1}) = 0,$$

because $\Delta(v_0, v_1, \dots, \widehat{v_j}, \dots, v_{n+1})$ is invariant under the transposition of v_0 and v_1 modulo 2-torsion.

PROPOSITION 3.3. *The composition*

$$C_{n+1}(n) \xrightarrow{d} C_n(n) \xrightarrow{\overline{f_n(n)}} K_n^M(F) \tag{3.4}$$

is equal to zero.

PROOF (Compare with the proof of Proposition 2.4 in [S1]). There exists a duality $*\colon C_{m+n-1}(m) \to C_{m+n-1}(n)$, $*^2 = \mathrm{id}$, that satisfies the following properties (see § 3.8 in [G2]).

1. $*$ commutes with the action of the permutation group S_{m+n}.

2. If $*(l_1, \dots, l_{m+n}) = (l'_1, \dots, l'_{m+n})$ then

$$*(l_1, \dots, \widehat{l_i}, \dots, l_{m+n}) = (l'_i \mid l'_1, \dots, \widehat{l_i}, \dots, \widehat{l}_{m+n}).$$

3. Choose volume forms in V_m and V_n and consider the partition

$$\{1, \dots, m+n\} = \{i_1 < \dots < i_m\} \cup \{j_1 < \dots < j_n\}.$$

Then $\dfrac{\Delta(l_{i_1}, \dots, l_{i_m})}{\Delta(l_{j_1}, \dots, l_{j_n})}$ does not depend on a partition.

This duality can be defined as follows. A configuration of $(m+n)$ vectors in an m-dimensional coordinate vector space can be represented as columns of the $m \times (m+n)$ matrix (I_m, A). The dual configuration is represented by the $n \times (m+n)$ matrix $(-A^t, I_n)$. Using the duality we can reformulate Proposition 3.3 as follows: the composition

$$C_{n+1}(2) \xrightarrow{d'} C_n(1) \xrightarrow{\widetilde{f}_n(n)} K_n^M(F)$$

is equal to 0. Here

$$\widetilde{f}_n(n)(v_0, \dots, v_n) := \operatorname{Alt} \Delta(v_0) \wedge \Delta(v_1) \wedge \dots \wedge \Delta(v_{n-1}) \in \Lambda^n F^*.$$

Consider the following diagram:

$$\begin{array}{ccc} C_{n+1}(2) & \xrightarrow{d'} & C_n(1) \\ \big\downarrow \widetilde{f}_{n+1}(n) & & \big\downarrow \widetilde{f}_n(n) \\ \mathbb{Z}[P_F^1 \setminus \{0, 1, \infty\}] \otimes \Lambda^{n-2} F^* & \xrightarrow{\delta} & \Lambda F^* \end{array}$$

Here $\mathbb{Z}[P_F^1 \setminus \{0, 1, \infty\}]$ is a free abelian group generated by symbols $\{x\}$, $x \in P_F^1 \setminus \{0, 1, \infty\}$, $\delta\colon \{x\} \otimes y_1 \wedge \dots \wedge y_{n-2} \mapsto (1-x) \wedge x \wedge y_1 \wedge \dots \wedge y_{n-2}$. Note that by definition $\operatorname{Coker} \delta = K_n^M(F)$. The homomorphism $\widetilde{f}_{n+1}(n)$ is defined as follows:

$$\widetilde{f}_{n+1}(n)(v_0, \dots, v_{n+1}) := n!\,[v_0, \dots, v_{n+1}],$$

where $[v_0, \dots, v_{n+1}]$ is defined by induction:

$$[v_0, v_1, v_2, v_3] := \{r(v_0, v_1, v_2, v_3)\} \in \mathbb{Z}[P_F^1 \setminus \{0, 1, \infty\}],$$

$$[v_0, \dots, v_{n+1}] := \gamma_n^{-1} \cdot \operatorname{Alt}\left(\varepsilon_1 \cdot \binom{n+1}{1} [v_1, \dots, v_{n+1}] \otimes \Delta(v_0, v_1)\right.$$

$$\left. + \sum_{k=2}^{n-2} \varepsilon_k \binom{n+1}{k} [v_0, v_{k+1}, \dots, v_{n+1}] \otimes \Delta(v_0, v_1) \wedge \dots \wedge \Delta(v_0, v_k)\right).$$

Here $\varepsilon_i = \pm 1$. More precisely, $\gamma_n = 2^{n+1} - (2 + \binom{n+1}{n-1} + \binom{n+1}{n} + \binom{n+1}{n+1})$, $\varepsilon_1 = +1$, and $\varepsilon_i = (-1)^i$, $i > 1$, for even n and $\gamma_n = 2^{n+1} - (\binom{n+1}{n-1} + \binom{n+1}{n} + \binom{n+1}{n+1})$,

$\varepsilon_1 = -1$, $\varepsilon_i = +1$, $i > 1$, for odd n. To prove the last formula one can write

$$[v_0, \dots, v_{n+1}] := \mathrm{Alt}\Big(\alpha_1 \cdot [v_1, \dots, v_{n+1}] \otimes \Delta(v_0, v_1) + \sum_{k=2}^{n-3} \alpha_k [v_0, v_k, \dots, v_{n+1}] \otimes \Delta(v_0, v_1) \wedge \dots \wedge \Delta(v_0, v_k)\Big)$$

with some unknown α_i. Then the condition

$$\delta[v_0, \dots, v_{n+1}] = (1/n!)\,\mathrm{Alt}\,\Delta(v_0, v_1) \wedge \dots \wedge \Delta(v_0, v_n)$$

gives exactly $n-3$ simple linear equations on α_i.

3.2. We get the following construction of the Chern classes $c_n(E) \in H^n(X, \underline{\underline{K}}^M_n)$. Choose a Zariski covering $\{U_i\}$ of X such that $E|_{U_i}$ is trivial. Choose sections $L^\bullet_i(x) \in \Gamma(U_i, \mathscr{A}^p_E(x))$ such that for any $i_0 < \dots < i_n$ affine flags $L^\bullet_{i_0}(x), \dots, L^\bullet_{i_n}(x)$ are in general position for every $x \in U_{i_0 \dots i_n}$.

THEOREM 3.4. *The chain*

$$\overline{f}_n(n)(T(L^\bullet_{i_0}(x), \dots, L^\bullet_{i_n})) \in \underline{\underline{K}}^M_n(\mathscr{O}^*(U_{i_0 \dots i_n})) \tag{3.5}$$

is a cocycle in the Čech complex for the covering $\{U_i\}$.

PROOF. Follows from Lemmas 3.2, 3.3 and Theorem 2.2.

By definition, $c_n(E)$ is the cohomology class of the cocycle from Theorem 3.4. It does not depend on the choice of sections $L^\bullet_i(x)$.

EXAMPLE 3.5. Recall that $c_1(E) = c_1(\det E)$. So $c_1(E)$ can be computed as follows: choose $m = \dim E$ linearly independent sections $l^\alpha_i(X)$ ($1 \leqslant \alpha \leqslant m$) of $E|_{U_i}$. Then $(l^\alpha_i(x)) = g_{ij}(x) \cdot (l^\beta_i(x))$, where $g_{ij}(x) \in GL_n(F)$ is the transition matrix and $\det g_{ij}(x)$ is a 1-cocycle representing $c_1(E)$.

Now let $(l^1_i, \dots, l^m_i)$ be the affine flag corresponding to the m-tuple of vectors $(l^1_i; \dots; l^m_i)$. Let us prove that the cocycle (3.5) computed for these flags cooincides with $\det g_{ij}$.

PROPOSITION 3.6. *We have* $f_1(1)(c((l^1_i, \dots, l^m_i), (l^1_j, \dots, l^m_j))) = \det g_{ij}$.

PROOF. Let us say that a frame $(f^1; \dots; f^m)$ is associated with an affine m-flag $(l^1, \dots, l^m)$ if

$$\langle f^1, \dots, f^k \rangle = \langle l^1, \dots, l^k \rangle \equiv L^k,$$

and the images of f^{k+1} and l^{k+1} in L^{k+1}/L^k coincide.

The set of all frames associated with a given m-flag is a principal homogeneous space over the group of unipotent matrices.

LEMMA-CONSTRUCTION 3.7. *For two affine m-flags*

$$L_1^\bullet = (v_1, \ldots, v_m) \quad \text{and} \quad L_2^\bullet = (w_1, \ldots, w_m),$$

in general position in V^m there exist exactly two frames associated with both of them.

PROOF. We have the following isomorphisms of one-dimensional vector spaces:

$$s_1\colon L_1^k/L_1^{k-1} \xrightarrow{\sim} L_1^k \cap L_2^{m-k+1},$$
$$s_2\colon L_2^{m-k+1}/L_2^{m-k} \xrightarrow{\sim} L_1^k \cap L_2^{m-k+1}.$$

Put $f_1^k := s_1(v_k)$, $f_2^{m-k+1} := s_2(w_{m-k+1})$. Then the frames $(f_1^1; \ldots; f_1^m)$ and $(f_2^1; \ldots; f_2^m)$ are associated with both $L_1^\bullet$ and $L_2^\bullet$.

Let $f_1^k = \lambda_k \cdot f_2^k$, $\lambda_k \in F^*$, and

$$(v_1, \ldots, v_m) = g \cdot (w_1, \ldots, w_m), \qquad g \in GL_m(F).$$

Then $\det g = \prod_{k=1}^m \lambda_k$ since $g = n_+ \cdot \lambda \cdot n_-$:

$$(w_i) \xrightarrow{n_-} (f_2^k) \xrightarrow{\lambda=(\lambda_k)} (f_1^k) \xrightarrow{n_+} (v_j),$$

where n_- (n_+) is a lower (upper) unipotent matrix and λ is the diagonal matrix with entries λ_k (the Gauss decomposition).

On the other hand, the left-hand side of the formula in Proposition 3.6 is equal to

$$f_1(1)\Big(\sum_{k=1}^m (L_1^k \oplus L_2^{m-k} \mid l_1^k, l_2^{m-k+1})\Big) = f_1(1)(f_1^k, f_2^k)\Big) = \prod_{k=1}^m \lambda_k.$$

§4. The universal Chern class $c_n \in H^n(BGL(m)_\bullet, \underline{\underline{K}}_n^M)$

4.1. The Gersten resolution in Milnor's K-theory ([Ka]). Let F be a field with a discrete valuation v and the residue class field $\overline{F}_v (= \overline{F})$. The group of units U has a natural homomorphism $U \to \overline{F}^*$, $u \mapsto \overline{u}$. An element $\pi \in F^*$ is prime if $\operatorname{ord}_v(\pi) = 1$. There is a canonical homomorphism (see [M1]):

$$\partial\colon K_{n+1}^M(F) \to K_n^M(\overline{K}_v) \qquad (n \geqslant 0),$$

uniquely determined by the properties ($u_i \in U$)

1. $\partial(\{\pi, u_1, \ldots, u_n\}) = \{\overline{u}_1, \ldots, \overline{u}_n\}$;
2. $\partial(\{\pi, u_1, \ldots, u_{n+1}\}) = 0$.

Let X be an excellent scheme [EGA, 3, IV, §7], $X_{(i)}$ the set of all codimension i points x, $F(x)$ the field of functions corresponding to a point $x \in X_{(i)}$.

There is a sequence of groups $\mathscr{K}(n)_\bullet$; here $K_n^M(x) := K_n^M(F(x))$:

$$K_n^M(F(X)) \xrightarrow{\partial} \bigoplus_{x\in X_{(1)}} K_{n-1}^M(x) \xrightarrow{\partial} \bigoplus_{x\in X_{(2)}} K_{n-2}^M(x) \to \cdots \to \bigoplus_{x\in X_{(n)}} \mathbb{Z}. \quad (4.1)$$

We follow [Ka] in the definition of ∂. Let us define for $y \in X_{(i)}$ and $x \in X_{(i+1)}$ a homomorphism $\partial_x^y: K_{*+1}^M(y) \to K_*^M(x)$ as follows. Let Y be the normalization of the reduced scheme $\{\overline{y}\}$. Set $\partial_x^y := \sum_{x'} N_{F(x')/F(x)} \circ \partial_{x'}$, where x' ranges over all points of Y lying over x, $\partial_{x'}: K_{*+1}^M(y) \to K_*(x)$ is the tame symbol associated with the discrete valuation ring $\mathscr{O}_{Y,x'}$, and $N_{F(x')/F(x)}$ is the norm map $K_*^M(x') \to K_*^M(x)$ (see [BT, Chapter I, § 5] and [Ka, § 1.7]). The coboundary ∂ is by definition the sum of these homomorphism ∂_x^y.

Proposition 4.1. $\partial^2 = 0$.

Proof. See the proof of Proposition 1 in [Ka].

Theorem 4.2. *The complex $\mathscr{K}(n)_\bullet$ is exact.*

Proof. See [So, Chapter 6] or [NS].

4.2. Explicit formula for a class $c \in H^n(BGL(m)_\bullet, \underline{\underline{K}}_n^M)$. Set $G^n := \underbrace{G \times \cdots \times G}_{n \text{ times}}$. Recall that

$$BG_\bullet := \text{pt} \overset{s_0}{\underset{s_1}{\Longleftarrow}} G \overset{s_0}{\underset{s_2}{\Lleftarrow}} G^2 \cdots$$

is the simplicial scheme representing the classifying space for a group G. We will compute $H^n(BG_\bullet, \underline{\underline{K}}_n^M)$ using the Gersten resolution (4.1). Therefore, the cochain we must construct lives in the following bicomplex ($G := GL(m)$):

$$\begin{array}{ccccc}
\vdots & & & & \\
\uparrow \partial & & & & \\
\bigoplus_{x\in G^{n-2}_{(2)}} K_{n-2}^M(F(x)) & \xrightarrow{s^*} & \bigoplus_{x\in G^{n-1}_{(2)}} K_{n-2}^M(F(x)) & & \\
& & \uparrow \partial & & \\
& & \bigoplus_{x\in G^{n-1}_{(1)}} K_{n-1}^M(F(x)) & \xrightarrow{s^*} & \bigoplus_{x\in G^{n}_{(1)}} K_{n-1}^M(F(x)) \\
& & & & \uparrow \partial \\
& & & & K_n^M(F(G^n))
\end{array} \quad (4.2)$$

For each partition $j_0+\dots+j_r = m-n$ we define a codimension $(n-r)$ irreducible subvariety $D(j_0, \dots, j_r) \in G^r_{(n-r)}$ and an element $\omega(j_0, \dots, j_r) \in K_r^M(D(j_0, \dots, j_r))$ such that a collection of all these elements forms a cocycle in (4.2).

Recall that $A^{m-n+1}(m)$ is the manifold of affine $(m-n+1)$-flags in V_m. For a partition $j_0 + \dots + j_r = m - n$ define a codimension $n - r$ manifold

$$\widetilde{D}_{j_0, \dots, j_r} \subset \underbrace{A^{m-n+1}(m) \times \dots \times A^{m-n+1}(m)}_{r+1 \text{ times}}$$

as follows: $(L_0^\bullet, \dots, L_r^\bullet) \in \widetilde{D}_{j_0, \dots, j_r}$ if and only if

$$\dim\Big(\mathop{+}_{p=0}^{r} L_p^{j_p+1}\Big) = r + \sum_{p=0}^{r} j_p = \dim\Big(\bigoplus_{p=0}^{r} L_p^{j_p+1}\Big) - 1.$$

Note that for generic $(L_0^\bullet, \dots, L_r^\bullet) \in \widetilde{D}_{j_0, \dots, j_r}$ the sum $+_{p=0}^{r} L_p^{j_p}$ is direct and the configurations of $r+1$ vectors

$$\Big(\bigoplus_{p=0}^{r} L_p^{j_p} \mid l_0^{j_0+1}, \dots, l_r^{j_r+1}\Big) \tag{4.3}$$

in $V_m / \bigoplus_{p=0}^{r} L_p^{j_p}$ generates a subspace of dimension r. Recall that there is a homomorphism (see (3.2))

$$\overline{f}_r(r)\colon C_r(r) \to \Lambda^r F^* \to K_r^M(F).$$

Applying it to the configuration of $r+1$ vectors (4.3) we obtain an element

$$\widetilde{\omega}_{j_0, \dots, j_r} \in K_r^M(F(\widetilde{D}_{j_0, \dots, j_r})). \tag{4.4}$$

Now choose $a \in A^{m-n+1}(m)$. Set

$$D_{j_0, \dots, j_r; a} := \{(g_1, \dots, g_r) \in G^r \mid (a, g_1 a, \dots, g_r a) \in \widetilde{D}_{j_0, \dots, j_r}\}.$$

Then $D_{j_0, \dots, j_r; a} \in G^r_{(n-r)}$ and $\widetilde{\omega}_{j_0, \dots, j_r}$ induces an element

$$\omega_{j_0, \dots, j_r; a} \in K_r^M(F(D_{j_0, \dots, j_r; a})). \tag{4.5}$$

Set

$$\widetilde{\omega}_r := \sum_{j_0+\dots+j_r=m-n} \widetilde{\omega}_{j_0, \dots, j_r} \in \bigoplus_{j_0+\dots+j_r=m-n} K_r^M(F(\widetilde{D}_{j_0, \dots, j_r})),$$

$$\omega_r := \sum_{j_0+\dots+j_r=m-n} \omega_{j_0, \dots, j_r} \in \bigoplus_{j_0+\dots+j_r=m-n} K_r^M(F(D_{j_0, \dots, j_r})).$$

THEOREM 4.3. *Collection of elements ω_r defines a cocycle in the bicomplex* (4.2).

PROOF. Choose a partition $i_0 + \dots + i_r = m - n + r$. Let $\widetilde{\mathscr{E}}_{i_0, \dots, i_r}$ be a subvariety in the manifold of $(r+1)$-tuples of affine $(m-n+1)$-flags in V^m

defined as follows:

$$\widetilde{\mathscr{E}}_{i_0,\dots,i_r} := \Big\{(L_0^\bullet, \dots, L_r^\bullet) \mid \dim\Big(\overset{r}{\underset{p=0}{+}} L_p^{i_p}\Big) = \Big(\sum_{p=0}^{r} i_p\Big) - 1\Big\}.$$

This is an irreducible subvariety of codimension $n - r + 1$.

PROPOSITION 4.4. *The component of $\partial\widetilde{\omega}_r$ on $\widetilde{\mathscr{E}}_{i_0,\dots,i_r}$ can be nonzero only if $i_k = 0$ for some k but $i_p > 0$ for $p \neq k$. In this case it is equal to*

$$\overline{f}_{r-1}(r-1)\Big(\bigoplus_{p\neq k} L_p^{i_p-1} \mid l_0^{i_0}, \dots, \widehat{l_k^{i_k}}, \dots, l_r^{i_r}\Big). \tag{4.6}$$

PROOF. Let $j_0 + \dots + j_r = m - n$ and

$$(l_0^1, \dots, l_0^{m-n+1}; \dots; l_r^1, \dots, l_r^{m-n+1}) \equiv (L_0^\bullet, \dots, L_r^\bullet) \in \widetilde{D}_{j_0,\dots,j_r}.$$

Choose a volume form in subspace $\langle l_0^1, \dots, l_0^{j_0+1}, \dots, l_r^1, \dots, l_r^{j_r+1}\rangle$ of codimension n. Then we can compute the determinant $\Delta(v_1, \dots, v_{m-n+r})$ for any $m - n + r$ vectors in this subspace, Set

$$\Delta(j_{k+1}) := \Delta(l_0^1, \dots, l_0^{j_0+1}, \dots, \widehat{l_k^{j_k+1}}, \dots, l_r^1, \dots, l_r^{j_r+1}).$$

By definition,

$$\widetilde{\omega}_{j_0,\dots,j_r} = \sum_{k=0}^{r}(-1)^k\{\Delta(j_0+1), \dots, \widehat{\Delta}(j_k+1), \dots, \Delta(j_r+1)\}. \tag{4.7}$$

The coboundary $\partial\widetilde{\omega}_{j_0,\dots,j_r}$ can be nonzero on divisors $\Delta(j_k+1) = 0$ in $\widetilde{D}_{j_0,\dots,j_r}$ only. The component of $\partial\widetilde{\omega}_{j_0,\dots,j_r}$ on the divisor $\Delta(j_k+1) = 0$ is equal to

$$\overline{f}_{r-1}(r-1)\Big(\bigoplus_{p=0}^{r} L_p^{j_p} \oplus l_k^{j_{k+1}} \mid l_0^{j_0+1}, \dots, \widehat{l_k^{j_{k+1}}}, \dots, l_r^{j_r+1}\Big). \tag{4.8}$$

This formula implies immediately that the component of $\partial\widetilde{\omega}_r$ on $\widetilde{\mathscr{E}}_{i_0,\dots,i_r}$ is zero if $i_{k_1} = i_{k_2} = 0$ for some $k_1 \neq k_2$.

It follows from (4.8) that if $i_p > 0$ for all p, the component of $\partial\widetilde{\omega}_r$ on $\widetilde{\mathscr{E}}_{i_0,\dots,i_r}$ is

$$\overline{f}_{r-1}(r-1)\Big(\sum_{k=0}^{r}(-1)^k\Big(\bigoplus_{p=0}^{r} L_p^{i_p-1} + l_k^{i_k} \mid l_0^{i_0}, \dots, \widehat{l_k^{i_k}}, \dots, l_r^{i_r}\Big)\Big). \tag{4.9}$$

Note that $(\bigoplus_{p=0}^{r} L^{i_p-1} \mid l_0^{i_0}, \dots, l_r^{i_r})$ is a configuration of $r+1$ vectors in an r-dimensional space. Therefore (4.9) is equal to

$$\overline{f}_{r-1}(r-1) \circ d'\Big(\bigoplus_{p=0}^{r} L_p^{i_p-1} \mid l_0^{i_0}, \dots, l_r^{i_r}\Big).$$

But this vanishes according to Lemma 3.2.

Now suppose that $i_k = 0$, $i_p \neq 0$ for $p \neq k$. Then (4.8) implies that the component of $\partial(\widetilde{\omega}_r)$ on $\widetilde{\mathscr{E}}_{i_0, \dots, i_r}$ is exactly (4.6). Proposition 4.4 is proved.

Theorem 4.3 follows immediately from Proposition 4.4.

4.3. Relation to the classical construction of Chern cycles. Assume that a vector bundle E over X has sufficiently many sections. Consider first of all the case when $\dim E = n$ and we are interested in $c_n(E) \in CH^n(X)$. Choose a section $s_0(x) \in \Gamma(X, E)$ that is transversal to the zero section of E. Then the subvariety

$$D_0 := \{x \in X \mid s_0(x) = 0\}$$

has codimension n and represents the class $c_n(E) \in CH^n(X)$. Now let $s_1(x)$ be another generic section of E (i.e., it is transversal to the zero section of E too). Then

$$D_1 := \{x \in X \mid s_1(x) = 0\}$$

should represent the same class in $CH^n(X)$. To see this let us consider a codimension $(n-1)$ subvariety

$$D_{01} := \{x \in X \mid \lambda_0 s_0(x) + \lambda_1 s_1(x) = 0 \text{ for some } \lambda_0, \lambda_1 \in \mathbb{C}\}.$$

There is a canonical rational function

$$\lambda_{01} := \frac{\lambda_0}{\lambda_1} \in F(D_{01}) \quad \text{and} \quad \operatorname{Div}(\lambda_{01}) = D_0 - D_1.$$

So D_0 and D_1 are canonically rationally equivalent cycles. Now let $s_2(x)$ be a third generic section of E. Put

$$D_{012} = \{x \in X \mid \dim\langle s_0(x), s_1(x), s_2(x)\rangle = 2\}.$$

Then $\operatorname{codim} D_{012} = n-2$ and there is a canonical element

$$\lambda_{012} := f_2(2)(s_0, s_1, s_2) \in K_2(F(D_{012})),$$
$$\partial(\lambda_{012}) = \lambda_{01} - \lambda_{02} + \lambda_{12},$$

where $\partial\colon K_2(F(Y)) \to \coprod_{y \in Y_{(1)}} F(y)^*$ is the tame symbol. Continuing this process we get, for a generic $(r+1)$-tuple of sections $s_0(x), \dots, s_r(x)$ of E, a codimension $(n-r)$ subvariety

$$D_{01\cdots r} := \{x \in X \mid \dim\langle s_0(x), \dots, s_r(x)\rangle = r\},$$

and a canonical element

$$\lambda_{01\cdots r} := \overline{f}_r(z)(s_0, \dots, s_r) \in K_r^M(F(D_{01\cdots r})),$$

satisfying the relation

$$\partial(\lambda_{01\cdots r}) = \sum_{i=0}^{r} (-1)^i \lambda_{01\cdots \widehat{i} \cdots r}$$

(∂ is the differential in complex (4.1)).

Now let E be a vector bundle of dimension $m > n$ and $p = m - n + 1$. Let

$$L_0^\bullet(x) = (l_0^1(x), \dots, l_0^p(x))$$

be a *generic* section of the bundle of affine p-flags of X. Put

$$D_0 := \{x \in X \mid l_0^1(x) \wedge \dots \wedge l_0^p(x) = 0 \text{ but } l_0^1(x) \wedge \dots \wedge l_0^{p-1}(x) \neq 0\}.$$

It is well known (see, for example, § 3 of Chapter III in [GH]) that the image of the cycle D_0 in the Chow group $CH^n(X)$ is just $c_n(E)$. Let $L_0^\bullet, \dots, L_r^\bullet$ be $r+1$ generic sections of the bundle of affine p-flags. For any partition $j_0 + \dots + j_r = p - 1$, $j_k \geqslant 0$, put

$$D(j_0, \dots, j_r) := \Big\{x \in X \mid \text{an } (r+1)\text{-tuple of vectors } (L_0^{j_0} + \dots + L_r^{j_r} \mid l_0^{j_0+1}, \dots, l_r^{j_r+1}) \text{ generates an } r\text{-dimensional vector space and } \dim \mathop{+}_{p=0}^{r} L_k^{j_k} = \sum_{k=0}^{r} j_k\Big\}. \tag{4.10}$$

Then $D(j_0, \dots, j_r)$ is a codimension $n-r$ cycle in X. There is a canonical element

$$\overline{f}_r(r)((L_0^{j_0} \oplus \dots \oplus L_r^{j_r} \mid l_0^{j_0+1}, \dots, l_r^{j_r+1})) \in K_r^M(F(D(j_0, \dots, j_r))). \tag{4.11}$$

Let us define an element

$$\lambda_{01\cdots r} \in \coprod_{j_0+\dots+j_r=p-1} K_r^M(F(D(j_0, \dots, j_r))) \subset \coprod_{x \in X_{(n-r)}} K_r^M(F(x)),$$

as the sum of elements (4.11):

$$\lambda_{01\cdots r} := \sum_{j_0+\dots+j_r=p-1} \overline{f}_r(r)\Big(\bigoplus_{k=0}^{r} L_k^{j_k} \mid l_0^{j_0+1}, \dots, l^{j_r+1}\Big).$$

THEOREM 4.5. $\partial(\lambda_{01\cdots r}) = \sum_{i=0}^{r}(-1)^i \lambda_{01\cdots\widehat{i}\cdots r}$.

PROOF. Follows immediately from the proof of Proposition 4.4.

4.4. An algebraic construction of the ring $H^*(GL_m(\mathbb{C}))$. I will construct a nonzero class in $W_0H^{2n-1}(GL_m(\mathbb{C}), \mathbb{Q}(n))$. This vector space is one-dimensional for $m \geqslant n$. Let us define for any $0 \leqslant j \leqslant m-n$ a subvariety $\widetilde{D}_j \subset A^{m-n+1}(m) \times A^{m-n+1}(m)$ as follows:

$$\widetilde{D}_j := \{(L_1^\bullet, L_2^\bullet) \text{ such that } (L_1^j + L_2^{m-n-j} \mid l_1^{j+1}, l_2^{m-n-j+1}) \text{ is a pair of collinear nonzero vectors}\}. \tag{4.12}$$

The ratio $\dfrac{\text{first vector}}{\text{second vector}}$ in (4.12) gives canonical invertible function $\widetilde{f}_j$ on $\widetilde{D}_j$. Now we choose an affine $(m-n+1)$-flag $L^\bullet$ in V_m. Set

$$GL(V_m) \supset D_j := \{g \in GL(V_m) \mid (gL^\bullet, L^\bullet) \in \widetilde{D}_j\}.$$

There is a canonical function $f_j \in \mathcal{O}(D_j)^*$.

THEOREM 4.6. *The current* $\sum_j d\log f_j$ *represents a nonzero class in the space*

$$W_0H^{2n-1}(GL_m(\mathbb{C}), \mathbb{Q}(n)).$$

PROOF. Let us prove that $\sum_j \operatorname{div} f_j = 0$, where $\operatorname{div} f_j$ is the divisor of f_j on $\overline{D}_j$ considered as a codimension n cycle on $GL(V_m)$. Note that $\operatorname{div} \widetilde{f}_j = Z_j^+ - Z_j^-$, where

$$Z_j^+ = \{(L_1^\bullet, L_2^\bullet) \mid \langle L_1^{j+1}, L_2^{m-n-j}\rangle = \langle L_1^j, L_2^{m-n-j}\rangle \text{ and } L_1^j \cap L_2^{m-n-j} = 0\},$$
$$Z_j^- = \{(L_1^\bullet, L_2^\bullet) \mid \langle L_1^j, L_2^{m-n-j+1}\rangle = \langle L_1^j, L_2^{m-n-j}\rangle \text{ and } L_1^j \cap L_2^{m-n-j} = 0\}.$$

Therefore it is easy to see that $\sum_j \operatorname{div} \widetilde{f}_j = 0$ and hence $\sum_j \operatorname{div} f_j = 0$. So the current $\sum_j d\log f_j$ represents a class in $W_0H^{2n-1}(GL_m(\mathbb{C}), \mathbb{Q}(n))$. It remains to prove that it is nontrivial.

Let $\mathrm{Gr}(N-m, N)$ be the Grassmannian of codimension m subspaces in V_N. There is a canonical m-dimensional bundle E over it: the fiber over plane h is V_N/h. Let us choose an affine $(m-n+1)$-flag $L^1 \subset \cdots \subset L^{m-n+1}$ in V_N. It determines a Chern cycle $c_m(E; L^\bullet) \subset \mathrm{Gr}(N-m, N)$. Let $\pi\colon \widetilde{E} \to \mathrm{Gr}(N-m, N)$ be the bundle of frames $(e_1, \dots, e_n)$ in fibers of E. This is a principal GL_m-bundle. Let us construct a cycle $B_m \subset \widetilde{E}$ together with a rational function $g_m \in F(B_m)$ such that

$$\operatorname{div} g_m = \pi^{-1}(c_m(E; L^\bullet)), \tag{4.13}$$

and for generic $h \in \mathrm{Gr}(N-m, N)$ the intersection

$$(B_m, g_m) \cap \pi^{-1}(h) \text{ coincides with } \sum_j (D_j, f_j), \tag{4.14}$$

constructed using the projection of the flag $L^\bullet$ onto V_N/h. (More precisely, a frame $(e_1, \dots, e_m)$ defines an affine $(m-n+1)$-flag $(e_1; \dots; e_{m-n+1})$ and this flag together with the projection of $L^\bullet$ should satisfy 4.12.) Conditions (4.13) and (4.14) just mean that the cohomology class of the current $\sum_j d\log f_j$ is the transgression of the mth Chern class of the universal bundle. Moreover, they give a precise description of the cycle B_m: it is closure of the union of cycles $\sum D_j \subset \bigcup_h \pi^{-1}(h)$ constructed using the projection of $L^\bullet$; here h runs through an open part in $\mathrm{Gr}(N-m, N)$. It is easy to see that for the natural invertible function g_m on B_m (4.13) holds.

§5. Explicit formulas for the universal motivic Chern classes $c_n \in H^{2n}(BGL_{m\bullet}, \mathbb{Q}(n))$ for $n \leqslant 3$

First of all let us recall what the motivic complexes are. So for convenience of the reader I will reproduce in 5.1–5.3 basic definitions and results from [G1, G2].

5.1. Motivic complexes. Let F be an arbitrary field. Denote by $\mathbb{Z}[P_F^1]$ a free abelian group generated by symbols $\{x\}$, where x runs through all F-points of P^1. Let us define subgroups $R_n(F) \subset \mathbb{Z}[P_F^1]$ $(n \leqslant 3)$ as follows:

- $R_1(F)$ is a subgroup generated by $\{xy\} - \{x\} - \{y\}$, where x, y run through all elements of F^*.
- $R_2(F)$ is a subgroup generated by $\sum_{i=0}^4 (-1)^i \{r(x_0, \dots, \widehat{x}_i, \dots, x_4)\}$, where $(x_0, \dots, x_4)$ run through all configurations of five distinct points of P_F^1 and $r(x_1, \dots, x_4) := \frac{(x_1-x_3)(x_2-x_4)}{(x_1-x_4)(x_2-x_3)}$ is the cross-ratio.
- $R_3(F)$ is a subgroup generated by $\sum_{i=0}^6 (-1)^i \{r_3(l_0, \dots, \widehat{l}_i, \dots, l_0)\}$, where $(l_0, \dots, l_6)$ run through all configurations of seven points in P_F^2 in general position and $r_3(l_1, \dots, l_6) \in \mathbb{Z}[P_F^1]$ is the *generalized cross-ratio*:

$$r_3(l_1, \dots, l_6) := \operatorname{Alt}\left\{\frac{\Delta(\widetilde{l}_1\widetilde{l}_2\widetilde{l}_4)\cdot\Delta(\widetilde{l}_2\widetilde{l}_3\widetilde{l}_5)\cdot\Delta(\widetilde{l}_3\widetilde{l}_1\widetilde{l}_6)}{\Delta(\widetilde{l}_1\widetilde{l}_2\widetilde{l}_5)\cdot\Delta(\widetilde{l}_2\widetilde{l}_3\widetilde{l}_6)\cdot\Delta(\widetilde{l}_3\widetilde{l}_1\widetilde{l}_4)}\right\}, \tag{5.1}$$

where $\operatorname{Alt} f(l_1, \dots, l_6) := \sum_{\sigma\in S_6} (-1)^{|\sigma|} f(l_{\sigma(1)}, \dots, l_{\sigma(6)})$.

Here $\widetilde{l}_i$ are vectors in $V^3 \setminus 0$ that project to the points $l_i \in P(V_3)$. The right-hand side of (5.1) does not depend on the volume form in V_3 and on lengths of vectors l_i. So the cross-ratio of six points in P_F^2 is well defined. Put

$$B_n(F) := \frac{\mathbb{Z}[P_F^1]}{R_n(F), \{0\}, \{\infty\}}.$$

There is a canonical isomorphism $B_1(F) \xrightarrow{\sim} F^*$ given by the map $\{x\} \mapsto x$; $\{0\}, \{\infty\} \mapsto 1$. Let us consider the following complexes $B_F(n)$:

$$\begin{aligned} &B_F(1)\colon F^*, \\ &B_F(2)\colon B_2(F) \xrightarrow{\delta_2} \Lambda^2 F^*, \\ &B_F(3)\colon B_3(F) \xrightarrow{\delta_3} B_2(F)\otimes F^* \xrightarrow{\delta_3} \Lambda^3 F^*. \end{aligned} \tag{5.2}$$

Here

$$\begin{aligned} &\delta_2(\{x\}) = (1-x)\wedge x, \\ &\delta_3(\{x\}) = \{x\}\otimes x; \qquad \delta_3(\{x\}\otimes y) = (1-x)\wedge x\wedge y, \end{aligned} \tag{5.3}$$

and by definition $\delta_n(\{0\}) = \delta_n(\{\infty\}) = 0$, $n = 2, 3$. Note that $\delta_3 \circ \delta_3(\{x\}) = (1-x)\wedge x \wedge x = 0$ modulo 2-torsion, so $B_F(3)$ is a complex.

THEOREM 5.1. $\delta_n(R_n(F)) = 0$.

PROOF. See § 3 in [G2]; see also §5.3 below.

In complexes (5.2) the groups $B_n(F)$ are placed at degree 1 and δ_n has degree $+1$.

The complex $B_F(2)$ is the well-known Bloch-Suslin complex.

5.2. The motivic complexes $\Gamma(X; n)$ for a regular scheme X $(n \leqslant 3)$. Let F be a field with a discrete variation v and the residue class field $\overline{F}_v$. Let us construct a canonical homomorphism of complexes

$$\partial_v : B_F(n) \to B_{\overline{F}_v}(n-1)[-1].$$

There is a homomorphism $\theta : \Lambda^n F^* \to \Lambda^{n-1}\overline{F}_v^*$ uniquely defined by the following properties ($u_i \in U$, $u \mapsto \overline{u}$ is the natural homomorphism $U \to \overline{F}_v^*$, and π is a prime, $\operatorname{ord}_v \pi = 1$):

1. $\theta(\pi \wedge u_1 \wedge \cdots \wedge u_{n-1}) = \overline{u}_1 \wedge \cdots \wedge \overline{u}_{n-1}$;
2. $\theta(u_1 \wedge \cdots \wedge u_n) = 0$.

It clearly does not depend on the choice of π.

Let us define a homomorphism $s_v : \mathbb{Z}[P_F^1] \to \mathbb{Z}[P_{\overline{F}_v}^1]$ as follows:

$$s_v\{x\} = \begin{cases} \{\overline{x}\} & \text{if } x \text{ is a unit,} \\ 0 & \text{otherwise.} \end{cases} \tag{5.4}$$

PROPOSITION 5.2. *Homomorphism* (5.4) *induces a homomorphism*

$$s_v : B(F) \to B_n(\overline{F}_v), \qquad n = 2, 3.$$

PROOF. Straightforward but tedious computations using formula (3.17) from [G2] for generators of the subgroup $R_3(F)$.

To avoid these computations one can consider subgroups $\mathscr{R}_n(F) \subset \mathbb{Z}[P_F^1]$ defined in § 1.4 of [G2]. Then, essentially by definition, $s_v(\mathscr{R}(F)) = \mathscr{R}_n(\overline{F}_v)$ and $\delta(R_n(F)) = 0$. So we get the groups $\mathscr{B}_n(F) := \mathbb{Z}[P_F^1]/\mathscr{R}_n(F)$ together with homomorphisms $s_v : \mathscr{B}_n(F) \to \mathscr{B}_n(\overline{F}_v)$.

Set

$$\partial_v := s_v \otimes \theta : B_k(F) \otimes \Lambda^{n-k} F^* \to B_k(\overline{F}_v) \otimes \Lambda^{n-k-1}\overline{F}_v^*. \tag{5.5}$$

LEMMA 5.3. *The homomorphism ∂_v commutes with the coboundary δ and hence defines a homomorphism of complexes* (5.3).

PROOF. Straightforward computation. See also § 1.14 in [G1], where the corresponding fact is proved for groups $\mathscr{B}_n(F)$.

Now let X be an arbitrary regular scheme, $X_{(i)}$ the set of all codimension i points of X, $F(x)$ the field of functions corresponding to a point $x \in X_{(i)}$. We define the motivic complexes $\Gamma(X, n)$ as the total complexes associated with the following bicomplexes:

$$\Gamma(X, 1): \quad F(X)^* \xrightarrow{\partial_1} \coprod_{x \in X_{(1)}} \mathbb{Z}$$

$$\begin{array}{llllll} \Gamma(X, 2): & \Lambda^2 F(X)^* & \xrightarrow{\partial_1} & \coprod_{x \in X_{(1)}} F(x)^* & \xrightarrow{\partial_2} & \coprod_{x \in X_{(2)}} \mathbb{Z} \\ & \uparrow \delta & & & & \\ & B_2(F(X)) & & & & \end{array}$$

$$\begin{array}{ccccccc}
\Gamma(X,3): & \Lambda^3 F(X)^* & \xrightarrow{\partial_1} & \coprod_{x\in X_{(1)}} \Lambda^2 F(X)^* & \xrightarrow{\partial_1} & \coprod_{x\in X_{(2)}} F(x)^* \xrightarrow{\partial_2} & \coprod_{x\in X_{(3)}} \mathbb{Z} \\
& \uparrow \delta & & \uparrow \delta & & & \\
& B_2(F(X))\otimes F(X)^* & \xrightarrow{\partial_1} & B_2(F(X)) & & & \\
& \uparrow \delta & & & & & \\
& B_3(F(X)) & & & & &
\end{array}$$

where $B_n(F(X))$ is placed at degree 1 and coboundaries have degree $+1$.

The coboundaries ∂_i are defined as follows. $\partial_1 := \coprod_{x\in X_{(i)}} \partial_{v_x}$. The others are a little bit more complicated. Let $x \in X_{(k)}$ and $v_1(y), \dots, v_m(y)$ be all discrete valuations of the field $F(x)$ over a point $y \in X_{(k+1)}$, $y \in \overline{x}$. Then $\overline{F(x)_i} := \overline{F(x)}_{v_i(y)} \supset F(y)$. (If $\overline{x}$ is nonsingular at the point y, then $\overline{F(x)_i} = F(y)$ and $m = 1$.) Let us define a homomorphism $\partial_2 : \Lambda^2 F(x) \to F(y)^*$ as the composition

$$\Lambda^2 F(x) \xrightarrow{\oplus \partial_{v_i(y)}} \bigoplus_{i=1}^{m} \overline{F(x)_i}^* \xrightarrow{\oplus N_{F(x)_i/F(y)}} F(y)^*, \tag{5.6}$$

and $F(x)^* \xrightarrow{\oplus \partial_{v_i}} \bigoplus_{i=1}^m \mathbb{Z} \xrightarrow{\Sigma} \mathbb{Z}$.

5.3. Motivic Chern classes $c_n \in H^{2n}_{\mathcal{M}}(BGL_m(F)_\bullet, \mathbb{Z}(n))$, $n \leqslant 3$. Recall that

$$BG_\bullet := \mathrm{pt} \overset{s_0}{\underset{s_1}{\Leftarrow}} G \Lleftarrow G^2 \Lleftarrow G^3 \cdots .$$

We must construct a $2n$-cocycle c_n in the bicomplex

$$\Gamma(G;n) \xrightarrow{s^*} \cdots \xrightarrow{s^*} \Gamma(G^n;n) \xrightarrow{s^*} \cdots \xrightarrow{s^*} \Gamma(G^{2n-1};n), \tag{5.7}$$

where $s^* = \sum (-1)^i s_i$. Its components in

$$\Gamma(G;n) \xrightarrow{s^*} \cdots \xrightarrow{s^*} \Gamma(G^n;n),$$

should be in the following part of the bicomplex:

$$
\begin{array}{ccccc}
\bigoplus_{x\in G_{(n)}} \mathbb{Z} & & & & \\
\uparrow \partial & & & & \\
\bigoplus_{x\in G_{(n-1)}} F(x)^* & \xrightarrow{s^*} & \bigoplus_{x\in G^2_{(n-1)}} F(x)^* & & \\
 & & \uparrow \partial & & \\
 & & \bigoplus_{x\in G^2_{(n-2)}} \Lambda^2 F(x)^* & \xrightarrow{s^*} & \cdots \\
 & & & & \\
 & & \cdots & \xrightarrow{s^*} & \bigoplus_{x\in G^n_{(1)}} \Lambda^{n-1} F(x)^* \\
 & & & & \uparrow \partial \\
 & & & & \Lambda^n F(G^n)^*.
\end{array}
\tag{5.8}
$$

In fact the components of c_n in (5.8) were already constructed in § 4. Recall this construction. Let a be an affine $(m-n+1)$-flag in an m-dimensional vector space V^m. For each partition $j_0 + \cdots + j_r = m - n$, irreducible subvarieties

$$D_{j_0, \dots, j_r; a} \in G^r_{(n-r)},$$

together with elements

$$\widetilde{\omega}_{j_0, \dots, j_r; a} \in \Lambda^r F(D_{j_0, \dots, j_r; a})^*, \tag{5.9}$$

were constructed. More precisely, if

$$(L_0^\bullet, \dots, L_r^\bullet) := (a, g_1 a, \dots, g_r a),$$

where $(g_1, \dots, g_r) \in D_{j_0, \dots, j_r; a} \subset G^r$, then

$$\left(\bigoplus_{p=0}^{r} L_p^{j_p} \mid l_0^{j_0+1}, \dots, l_r^{j_r+1}\right)$$

is a configuration of $r+1$ vectors in an r-dimensional vector space. Applying to it the homomorphism $f_r(r)\colon C_r(r) \to \Lambda^r F^*$, we get the element (5.9). The collection of elements

$$
\begin{aligned}
\widetilde{\omega}_r := \sum_{j_0+\cdots+j_t=m-n} \widetilde{\omega}_{j_0, \dots, j_r; a} &\in \bigoplus_{j_0+\cdots+j_2=m-n} \Lambda^r F(D_{j_0, \dots, j_r; a})^* \\
&\in \bigoplus_{x\in G^r_{(n-r)}} \Lambda^r F(x)^*
\end{aligned}
\tag{5.10}
$$

forms a cocycle in the bicomplex (5.8). (The proof of this fact is similar to the proof of Theorem 4.3.) The components of c_n in the bicomplex

$$\Gamma(G^n\,;\,n) \xrightarrow{s^*} \Gamma(G^{n+1}\,;\,n) \xrightarrow{s^*} \cdots \xrightarrow{s^*} \Gamma(G^{2n-1}\,;\,n) \qquad (5.11)$$

are constructed as follows. There is a homomorphism of complexes (see (2.4), (2.5))

$$T\colon C_*(A^{m-n+1}(m)) \to BC_*(n)\,,$$

where $BC_*(n)$ is the total complex for the Grassmannian bicomplex (1.2).

We will construct morphisms of complexes

$$f(n)\colon BC^*(n) \to B_F(n) \qquad (n \leqslant 3) \qquad (5.12)$$

such that for $r \geqslant n+1$ the ∂-coboundaries of elements

$$f(n) \circ T(a\,,\, g_1 a\,,\, \ldots\,,\, g_r a) \qquad (5.13)$$

are equal to zero. The collection of elements (5.10) and (5.13) form a cocycle c_n in the bicomplex (5.7).

Let us describe the construction of the homomorphism (5.12).

a) $n = 1$. $f_1(1)\colon C_1(1) \to F^*$ is the only homomorphism we need. It is easy to check that both $f_1(1) \circ d'\colon C_2(2) \to F^*$ and $f_1(1) \circ d\colon C_2(1) \to F^*$ are zero homomorphisms, so that we get a homomorphism $f(1)\colon BC^*(1) \to F^*[-1]$.

b) $n = 2$. We must construct a homomorphism from the total complex associated with the bicomplex

$$\begin{array}{ccccc} & & \downarrow & & \downarrow \\ \longrightarrow & & C_4(3) & \xrightarrow{d} & C_3(3) \\ & & \downarrow{\scriptstyle d'} & & \downarrow{\scriptstyle d'} \\ \xrightarrow{d} & & C_3(2) & \xrightarrow{d} & C_2(2) \end{array}$$

to the complex

$$0 \to B_2(F) \to \Lambda^2 F^*.$$

A homomorphism $f_2(2)\colon C_2(2) \to \Lambda^2 F^*$ was defined by formula (3.2). Lemma 3.2 shows that one can take a map from $C_3(3)$ to $B_2(F)$ equal to zero. Let us define a homomorphism

$$f_3(2)\colon C_3(2) \to B_2(F)$$

setting

$$(l_0\,,\, \ldots\,,\, l_3) \mapsto \{r(\bar{l}_0\,,\, \ldots\,,\, \bar{l}_3)\}_2\,,$$

where $(\bar{l}_0\,,\, \ldots\,,\, \bar{l}_3)$ is a configuration of four points in P^1_F corresponding to the one $(l_0\,,\, \ldots\,,\, l_3)$ of four vectors in V^2. Then $f_3(2) \circ d\colon C_4(2) \to B_2(F)$ is zero by definition of the group $B_2(F)$.

LEMMA 5.4. $f_3(2) \circ d' = 0$.

PROOF. We have to prove that for $(l_0, \dots, l_4) \in C_4(3)$,

$$\sum_{i=0}^{4}(-1)^i\{r(\bar{l}_i \mid \bar{l}_0, \dots, \widehat{\bar{l}_i}, \dots, \bar{l}_4)\}_2 = 0 \quad \text{in } B_2(F). \tag{5.14}$$

There is a conic (a curve of order 2) passing through five points $\bar{l}_0, \dots, \bar{l}_4$ in P_F^2. Let us consider it as a projective line. Then (5.14) is just the 5-term relation for five points $\bar{l}_i$ on *this* projective line.

So we have defined a homomorphism $f(2)\colon BC^*(2) \to B_F(2)$. It is nonzero only on the Grassmannian subcomplex $C^*(2) \subset BC^*(2)$.

c) $n = 3$. We have to define a homomorphism from the total complex associated with the bicomplex

$$\begin{array}{ccccccc}
 & & \vdots & & \vdots & & \vdots \\
 & & \downarrow & & \downarrow & & \downarrow \\
\cdots & \xrightarrow{d} & C_6(4) & \xrightarrow{d} & C_5(4) & \xrightarrow{d} & C_4(4) \\
 & & \downarrow{\scriptstyle d'} & & \downarrow{\scriptstyle d'} & & \downarrow{\scriptstyle d'} \\
\cdots & \xrightarrow{d} & C_5(3) & \xrightarrow{d} & C_4(3) & \xrightarrow{d} & C_3(3)
\end{array}$$

to the complex

$$B_3(F) \xrightarrow{\delta_3} B_2(F) \otimes F^* \xrightarrow{\delta_3} \Lambda^3 F^*.$$

A homomorphism $f_3(3)\colon C_3(3) \to \Lambda^3 F^*$ was defined by formula (3.2). Set

$$\begin{aligned}
&f_4(3)\colon C_4(3) \to B_2(F) \otimes F^*, \\
&f_4(3)\colon (l_0, \dots, l_4) \mapsto (1/2)\operatorname{Alt}(\{r(\bar{l}_0 \mid \bar{l}_1, \dots, \bar{l}_u)\}_2 \otimes \Delta(l_0, l_1, l_2)).
\end{aligned} \tag{5.15}$$

PROPOSITION 5.5. *$f_4(3)$ does not depend on the choice of the volume form $\omega_3 \in \Lambda^3(V^3)^*$ in the definition of $\Delta(l_i, l_j, l_k)$.*

PROOF. The difference between the right-hand sides of (5.15) computed using $\lambda \cdot \omega_3$ and ω_3 is proportional to (left-hand side of (5.14)) $\otimes\lambda$. So it is zero by Lemma 5.4.

PROPOSITION 5.6. *We have $f_3(3) \circ d = \delta \circ f_4(3)$.*

PROOF. Direct computation using the formula

$$r(\bar{l}_1, \dots, \bar{l}_4) = \frac{\Delta(l_1, l_3) \cdot \Delta(l_2, l_4)}{\Delta(l_1, l_4) \cdot \Delta(l_2, l_3)}.$$

Now set

$$\begin{aligned}
&f_5(3)\colon C_5(3) \to B_3(F) \\
&f_5(3)\colon (l_0, \dots, l_5) \mapsto \operatorname{Alt}\left\{\frac{\Delta(l_0, l_1, l_3) \cdot \Delta(l_1, l_2, l_4) \cdot \Delta(l_2, l_0, l_5)}{\Delta(l_0, l_1, l_4) \cdot \Delta(l_1, l_2, l_5) \cdot \Delta(l_2, l_0, l_3)}\right\}_3.
\end{aligned} \tag{5.16}$$

THEOREM 5.7. *We have* $f_4(3) \circ d = \delta \circ f_5(3)$.

PROOF. See the proof of Theorem 3.10 in [G2].

PROPOSITION 5.8. $f_k(3) \circ d' = 0$ *for* $k = 3, 4, 5$.

PROOF. For $k = 3$ this is Lemma 3.2. For $k = 4, 5$ see Theorem 3.12 in [G2].

PROPOSITION 5.9. $f_5(3) \circ d = 0$ *in* $B_3(F)$.

PROOF. Follows immediately from the definition of the group $B_3(F)$.

So one can define a homomorphism $f(3)\colon BC^*(3) \to B_F(3)$ using homomorphisms $f_k(3)$ on the subcomplex $C^*(3) \subset BC^*(3)$ and zero homomorphisms otherwise.

Now consider an element

$$f_4(3) \circ T(a, g_1 a, \dots, g_4 a) \in B_2(F(G^4)) \otimes F(G^4)^*.$$

Then

$$\partial_1 \circ f_4(3) \circ T(a, g_1 a, \dots, g_4 a) \in \bigoplus_{x \in G^4_{(1)}} B_2(F(x)). \tag{5.17}$$

LEMMA 5.10. *The left-hand side of* (5.17) *is equal to zero.*

PROOF. It follows from the definition (5.5) of ∂_v and the following remark: the term $\Delta(l_0, l_1, l_2)$ appears in formula (5.15) with the coefficient

$$\{r(\bar{l}_3 \mid \bar{l}_0, \bar{l}_1, \bar{l}_2, \bar{l}_4)\}_2 - \{r(\bar{l}_4 \mid \bar{l}_0, \bar{l}_1, \bar{l}_2, \bar{l}_3)\}_2$$

which is obviously zero if $\Delta(l_0, l_1, l_2) = 0$.

So we have proved that the collection of elements (5.10) and (5.13) form a cocycle in the bicomplex (5.7). The cohomology class of this cocycle does not depend on the choice of an affine $(m - n + 1)$-flag a. (Different flags yield canonically cohomologous cocycles.)

5.4. Chern classes in Deligne cohomology. Let us assume that there exists a $2n$-cocycle $\mathbb{L}_n'$ from Conjecture 1.1′. (A precise construction of this cocycle for $n \leqslant 3$ can be found in § 9 of [G1], see also [G2] and 6.3 below.) The main construction of § 2 gives an explicit construction of Chern classes in bi-Grassmannian cohomology and hence, applying $\mathbb{L}_n'$, in real Deligne cohomology. We will see in the next section that these Chern classes coincide with the classical ones (see Theorem 5.11).

5.5. The universal Chern classes in Deligne cohomology. Assuming the existence of $\mathbb{L}_n'$ we construct

$$c_n \in H_D^{2n}(BGL_m(\mathbb{C}), \mathbb{R}(n)).$$

The Dolbeaux resolution of the complex associated with the bicomplex (1.13) provides us with a complex computing real Deligne cohomology of an algebraic manifold over $\mathbb{C}$. We will denote this complex by $\mathbb{R}(X, n)$. We have to construct a $2n$-cocycle in the bicomplex

$$\mathbb{R}(G, n) \xrightarrow{s^*} \cdots \xrightarrow{s^*} \mathbb{R}(G^n, n) \xrightarrow{s^*} \cdots \xrightarrow{s^*} \mathbb{R}(G^{2n-1}, n) \tag{5.18}$$

(compare with (4.7)). First of all let us construct its components in

$$\mathbb{R}(G, n) \xrightarrow{s^*} \cdots \xrightarrow{s^*} \mathbb{R}(G^n, n). \tag{5.19}$$

For a subvariety $Y \hookrightarrow X$ of codimension d, there is a canonical morphism of complexes $i_*: \mathbb{R}(Y, n) \to \mathbb{R}(X, n+d)[2d]$. In 5.3 we have constructed a chain (5.10) in the bicomplex (5.8) corresponding to an affine $(m-n+1)$-flag a in V_m. Each component of this chain lies in $\Lambda^r F(x)^*$, where x is a codimension $n-r$ point in G^r. There is a canonical map

$$\Lambda^r \mathbb{C}(x)^* \to \mathbb{R}(\operatorname{Spec} \mathbb{C}(x), r),$$

$$f_1 \wedge \cdots \wedge f_r \mapsto \Big(\pi_r \operatorname{Alt}\Big(\sum_{k=0}^{[(n-1)/2]} \frac{1}{(2k+1)!\,(n-2k-1)!} \log|f_1|\, d\log|f_1| \wedge \cdots \wedge d\log|f_{2k+1}| \wedge d\, i\, \arg f_{2k+2} \wedge \cdots \wedge d\, i \arg f_{2n}, \; d\log f_1 \wedge \cdots \wedge d\log f_n\Big)\Big)$$

commuting with residue homomorphisms. So we get a chain in (5.19).

The components of c_n in the bicomplex

$$\mathbb{R}(G^n, n) \xrightarrow{s^*} \cdots \xrightarrow{s^*} \mathbb{R}(G^{2n-1}, n)$$

are constructed as a composition of the homomorphism of complexes

$$T: C_*(A^{m-n+1}(m)) \to BC_*(n),$$

with the $2n$-cocycle $\mathbb{L}'_n$ that lives on $BC_*(n)$ (or, better, on the bi-Grassmannian $\widehat{\mathbb{G}}(n)$). More precisely, to construct the $\mathbb{R}(G^k, n)$-component of c_n we must restrict homomorphism T to elements $(a, g_1 a, \dots, g_k a)$, where a is a given affine $(m-n+1)$-flag in V_m.

THEOREM 5.11. a) *The constructed chain c_n is a cocycle in* (5.18).
b) *The class of c_n equals the usual Chern class in $H_D^{2n}(BGL_m(\mathbb{C}), \mathbb{R}(n))$.*

PROOF. Part a) follows from the definition and previous results.

b) (Compare with the proof of Theorem 5.10 in [G2].) Let $\pi: EG_\bullet \to BG_\bullet$ be the universal G-bundle. Then $EG_{(p)} = BG_{(p+1)}$ and so any i-cochain $c_{(\bullet)}$ for $BG_\bullet$ defines an $(i-1)$-cochain $\widetilde{c}_{(\bullet)}$ for $EG_\bullet$: $\widetilde{c}_{(p)} := c_{(p+1)}$. Moreover, if $c_{(0)} = 0$ and $c_{(\bullet)}$ is a cocycle then $d\widetilde{c}_{(\bullet)} = c_{(\bullet)}$. Therefore $c_{(1)} = \widetilde{c}|_G$ is the transgression of the cocycle $c_{(\bullet)}$.

Applying this result to the cocycle c_n, we get a cocycle c'_n in $H_D^{2n-1}(GL_m(\mathbb{C}), \mathbb{R}(n))$ The usual exact sequence for Deligne cohomology gives us

$$\dots \longrightarrow H_D^{2n-1}(GL_m(\mathbb{C}), \mathbb{R}(n)) \xrightarrow{\alpha} \\ \longrightarrow H^{2n-1}(GL_m(\mathbb{C}), \mathbb{R}(n)) \cap H^{2n-1}(GL_m(\mathbb{C}), \Omega^{\geqslant n}).$$

It follows from definitions that $\alpha(c'_n)$ coincides with the class constructed in 4.4. It is nontrivial according to Theorem 4.6.

This proves that the cocycle c_n coincides with the usual Chern class in $H_D^{2n}(BGL_m(\mathbb{C})_\bullet, \mathbb{R}(n))$ modulo decomposable elements. (Decomposable elements go to zero under transgression.)

To prove that c_n coincides with the Chern class exactly, consider the standard long exact sequence for Deligne cohomology:

$$\cdots \to H^{2n-1}(BGL_m(\mathbb{C})_\bullet, \mathbb{C}) \to H_D^{2n}(BGL_m(\mathbb{C})_\bullet, \mathbb{R}(n)) \to \\ \xrightarrow{\beta_1+\beta_2} H^{2n}(BGL_m(\mathbb{C})_\bullet, \mathbb{R}(n)) \oplus H^{2n}(BGL_m(\mathbb{C})_\bullet, \Omega^{\geqslant n}) \to \cdots .$$

It is known that $H^{2n-1}(BGL_m(\mathbb{C})_\bullet, \mathbb{C}) = 0$. So we get an inclusion

$$H_D^{2n}(BGL_m(\mathbb{C})_\bullet, \mathbb{R}(n)) \overset{\beta_1+\beta_2}{\hookrightarrow} H^{2n}(BGL_m(\mathbb{C})_\bullet, \mathbb{R}(n)) \oplus H^{2n}(BGL_m(\mathbb{C})_\bullet, \Omega^{\geqslant n}).$$

(In fact both β_1 and β_2 are isomorphisms because $H^{2n}(BGL_m(\mathbb{C}), \mathbb{R}(n))$ is a pure Hodge structure of type $(0, 0)$.) To prove b) it suffices to prove the following

LEMMA 5.11$'$. *The class $\beta_2(c_n)$ coincides with the usual nth Chern class in $H^{2n}(BGL_m(\mathbb{C})_\bullet, \Omega^{\geqslant n})$.*

This lemma will be proved in 5.8 below.

5.6. Explicit formulas for measurable cocycles of $GL(\mathbb{C})$. We assume that there exists a function P'_n on $\widehat{G}^n_{n-1}$ satisfying $(2n-1)$-term relations (1.14). Recall that such a function can be considered as a function on configurations of $2n$ vectors in general position in $\mathbb{C}^n$ satisfying the equations

$$\sum_{i=0}^{2n}(-1)^i P'_n(l_0, \dots, \widehat{l_i}, \dots, l_{2n}) = 0, \tag{5.20a}$$

$$\sum_{i=0}^{2n}(-1)^i P'_n(l_i \mid l_0, \dots, \widehat{l_i}, \dots, l_{2n}) = 0. \tag{5.20b}$$

We assume also that P'_n is a component of a $2n$-cocycle $\mathbb{L}'_n$ from Conjecture 1.1$'$.

THEOREM 5.12. *Let a be an affine $(m-n+1)$-flag in V_m. Then $P_n(T(g_0 a, \dots, g_{2n-1}a))$ is a $2n$-cocycle of $GL_m(\mathbb{C})$. Its cohomology class coincides with the Borel class in $H^{2n-1}_{(m)}(GL_m(\mathbb{C}), \mathbb{R})$ ($m \geqslant n$).*

Recall that here $T\colon C_*(A^{m-n+1}(n)) \to BC_*(n)$ is a homomorphism of complexes. The cocycle condition follows from this fact and $(2n+1)$-terms equations (5.20).

Let G^δ be a Lie group G made discrete. The morphism of groups $GL_m(\mathbb{C})^\delta \to GL_m(\mathbb{C})$ provides a morphism

$$e\colon BGL_m(\mathbb{C})^\delta_\bullet \to BGL_m(\mathbb{C})_\bullet.$$

Therefore,

$$\begin{aligned} e^*\colon H^{2n}_D(BGL_m(\mathbb{C})_\bullet, \mathbb{R}(n)) &\to H^{2n}_D(BGL_m(\mathbb{C})^\delta_\bullet, \mathbb{R}(n)) \\ &= H^{2n-1}(BGL_m(\mathbb{C})_\bullet, S^0) \equiv H^{2n-1}_{(m)}(GL_m(\mathbb{C}), \mathbb{R}(n-1)). \end{aligned}$$

Here S^0 is a sheaf of smooth functions. It is known that e^* maps the indecomposable class in $H^{2n}_D(BGL_m(\mathbb{C}), \mathbb{Z}(n))$ just to the Borel class in $H^{2n-1}_{(m)}(GL_m(\mathbb{C}), \mathbb{R}(n-1))$ (see [B2, DMZ]). The arguments in the proof of Theorem 5.11 show that the constructed class $c_n \in H^{2n}_D(BGL_m(\mathbb{C})_\bullet, \mathbb{R}(n))$ lies in

$$\operatorname{Im} H^{2n}_D(BGL_m(\mathbb{C})_\bullet, \mathbb{Z}(n)) \to H^{2n}_D(BGL_m(\mathbb{C})_\bullet, R(n)),$$

and in fact coincides with the image of the standard class in

$$H^{2n}_D(BGL_m(\mathbb{C})_\bullet, \mathbb{Z}(n)).$$

In our case $e^*(c_n)$ coincides with $P_n(T(g_0 a, \dots, g_{2n-1}a))$ just by definition. Theorem 5.12 is proved.

REMARK 5.13. Explicit formulas for functions P_n are known for $n \leqslant 3$:

$$\begin{aligned} P_2(l_1, \dots, l_4) &:= \mathscr{L}_2(r(l_1, \dots, l_4)), \\ P_3(l_1, \dots, l_6) &:= \mathscr{L}_3(r_3(l_1, \dots, l_6)). \end{aligned} \tag{5.21}$$

5.7. Chern classes $c^M_n(E) \in H^n(X, \underline{\underline{K}}^M_n)$ constructed in §§ 3–4 coincide with standard classes. Gersten resolution (4.1) (see § 6 in [So] or [NS]) provides us with an isomorphism

$$i\colon H^n(X, \underline{\underline{K}}^M_n) \otimes \mathbb{Q} \xrightarrow{\sim} CH^n(X) \otimes \mathbb{Q}.$$

Our next goal is to show that $i(c^M_n(E))$ coincides with the usual Chern class $c_n(E) \in CH^n(X)$.

Let us call a vector bundle E "nice" if there is a section $l = (l^1; \dots; l^p)$ of the bundle of affine p-flags ($p = \dim E - n + 1$) such that the cycle

$$c_n(E; l) := \{x \in X \mid l^1(x) \wedge \dots \wedge l^p(x) = 0 \text{ but } l^1(x) \wedge \dots \wedge l^{p-1}(x) \neq 0\}$$

has codimension n. Then it represents the usual Chern class in $CH^n(X)$.

PROPOSITION 5.14. *For a "nice" vector bundle E one has*

$$i(c_n^M(E)) = c_n(E).$$

PROOF. Let $X = \bigcup U_i$ be a Zariski covering such that $E|_{U_i}$ is trivial and there are sections $s_i \in \Gamma(U_i, A^p(E))$ satisfying the condition that for any point x of $U_{i_0 \cdots i_n}$ sections $s_{i_0}(x), \dots, s_{i_n}(x)$ are in general position. Then, according to Theorem 3.4,

$$c_n^M(E; s_i) := \overline{f}_n(n) \circ T(s_{i_0}, \dots, s_{i_n}) \in K_n^M(\mathscr{O}_{U_{i_0 \cdots i_n}})$$

is a cocycle in the Čech complex $\check{C}(U_\bullet, \underline{\underline{K}}_n^M)$.

Let us consider the bicomplex computing the Čech hypercohomology for this covering with coefficients in the Gersten resolution for $\underline{\underline{K}}_n^M$.

For example in the case $n = 2$ it looks as follows:

$$\begin{array}{ccccc}
\bigoplus_i \coprod_{x \in U_{i(2)}} \mathbb{Z} & & \bigoplus_{i_0<i_1} \coprod_{x \in U_{i_0 i_1 (2)}} \mathbb{Z} & & \\
\partial \uparrow & & \partial \uparrow & & \\
\bigoplus_i \coprod_{x \in U_{i(1)}} F(x)^* & \xrightarrow{\delta} & \bigoplus_{i_0<i_1} \coprod_{x \in U_{i_0 i_1 (1)}} F(x)^* & \xrightarrow{\delta} & \cdots \\
\partial \uparrow & & \partial \uparrow & & \\
\bigoplus_i K_2(F(U_i)) & \xrightarrow{\delta} & \bigoplus_{i_0<i_1} K_2(F(U_{i_0 i_1})) & \xrightarrow{\delta} & \bigoplus_{i_0<i_1<i_2} K_2(F(U_{i_0 i_1 i_2})) \to \cdots
\end{array}$$

Here $X_{(n)}$ is the set of codimension n irreducible subvarieties in X, δ is Čech coboundary, ∂ and δ have degree $+1$, and the group $\bigoplus_i K_2^M(F(U_i))$ in the bottom left corner are placed in degree $(0, 0)$. Note that $c_n(E; l)$ can be considered as an element in $\coprod_{x \in U_{i(n)}} \mathbb{Z}$.

Let us construct a chain $b = \sum b_{k, n-1-k}$ of degree $n - 1$ in the total complex associated with the bicomplex such that $(\partial + \delta)b = c_n(E; l) - c_n^M(E; s_i)$.

To explain the idea let me first construct b in the case $\dim E = n = 2$. Then

$$b_{1,0} := \sum_{i_0<i_1} \overline{f}_2(r)(l, s_{i_0}, s_{i_1}) \in \bigoplus_{i_0<i_1} K_2(F(U_{i_0 i_1})),$$

and

$$b_{1,0} := \sum_{i;x} \frac{l(x)}{s_i(x)} \in \bigoplus_i \coprod_{x \in U_{i(1)}} F(x)^*.$$

In the last formula summation is over all codimension 1 subvarieties of U_i such that l is collinear to s_i at points of this subvariety. In this situation there is a rational function $l/s_i \in F(x)^*$.

In the general case set

$$b_{n-1,0} = \sum_{i_0<\cdots<i_{n-1}} \overline{f}_n(n) \circ T(l, s_{i_0,\dots,s_{i_{n-1}}}) \in \bigoplus_{i_0<\cdots<i_{n-1}} K_n^M(F(U_{i_0\cdots i_{n-1}})),$$

and

$$b_{n-k,k-1} := \sum_{i_0<\cdots<i_{n-k-1}} \overline{f}_{n-k+1}(n-k+1) \circ \widetilde{T}(l(x), \widetilde{s}_{i_0}(x), \dots, \widetilde{s}_{i_{n-k}})$$
$$\in \bigoplus_{i_0<\cdots<i_{n-k}} \coprod_{x\in U_{i_0\cdots i_{n-k}}} K_{n-k+1}^M(F(x)).$$

To explain the meaning of the right-hand side of this formula we need several definitions. Let $(L_0^\bullet, \dots, L_r^\bullet)$ be a configuration of $r+1$ affine p-flags, $J = (j_0, \dots, j_r)$, and $j_0 + \cdots + j_r = p-1$. We say that this configuration is J-weakly degenerate if the sum $L_0^{j_0} + \cdots + L_r^{j_r}$ is direct, the configuration of $r+1$ vectors

$$(L_0^{j_0} + \cdots + L_r^{j_r} \mid l_0^{i_0+1}, \dots, l_r^{j_r+1}) \tag{5.22}$$

generates an r-dimensional vector space, and every r of these vectors generate the same vector space.

In the right-hand side of the formula for $b_{k,n-k-1}$ summation is over all codimension $k-1$ subvarieties of $U_{i_0\cdots i_{n-k}}$ such that the configuration of $(n-k+1)$-flags $(l, s_{i_0}, \dots, s_{i_{n-k}})$ at the generic point of x is J-weakly degenerate for some partition J. In this case $\widetilde{T}(l(x), s_{i_0}(x), \dots, s_{i_{n-k}}(x))$ is the sum over all such partitions J of configurations of vectors (5.22). Then it follows immediately from the proof of Proposition 4.4, that

$$(\partial + \delta)b = c_n(E; l) - c_n^M(E; s_i).$$

Proposition 5.14 is proved.

THEOREM 5.15. *For any vector bundle E over X the characteristic class*

$$c_n^M(E) \in H^n(X, \underline{\underline{K}}_n^M)$$

constructed in §§3–4 coincides with the usual one, i.e.,

$$i(c_n^M(E)) = c_n(E).$$

PROOF. We need the following result which is a particular case of Proposition 2.4 from a paper of V. V. Schechtman [Sch].

PROPOSITION 5.16. *Let*

$$\overline{c}_i = \overline{c}_i(E_m) \in H^i(BGL_{m^\bullet}, \underline{\underline{K}}_i), \qquad 1 \leqslant i \leqslant m,$$

be the Chern classes of the canonical m-dimensional vector bundle over $BGL_{m^\bullet}$. Then one has an isomorphism

$$\bigoplus_m H^m(BGL_{m^\bullet}, \underline{\underline{K}}_m) = \mathbb{Z}[\overline{c}_1, \dots, \overline{c}_m].$$

We have constructed classes

$$c_i^M \in H^i(BGL_{m^\bullet}, \underline{\underline{K}}_i^M),$$

and proved (Proposition 5.14) that for "nice" vector bundles E over X one has

$$f_E^*(\overline{c}_i) = f_E^*(c_i^M).$$

Here $f_E^*: H^i(BGL_{m^\bullet}, \underline{\underline{K}}_i) \to H^i(X, \underline{\underline{K}}_i)$ is the homomorphism induced by vector bundle E. This implies Theorem 5.15. Indeed, consider as an example of a "nice" vector bundle, for example, the canonical m-dimensional vector bundle over $\mathrm{Gr}(N-m, N)$, N is big. Then the corresponding homomorphism f_E^* is injective. Theorem 5.15 is proved.

5.8. Proof of Lemma 5.11′. Recall that there is canonical map of sheaves

$$d\log: \underline{\underline{K}}_n^M \to \Omega^n_{\log} \hookrightarrow \Omega^n_{cl},$$
$$\{f_1, \dots, f_n\} \mapsto d\log f_1 \wedge \dots \wedge d\log f_n.$$

It follows from the definitions that

$$\beta_2(c_n(E)) = d\log(c_n^M(E)).$$

This together with Theorem 5.15 implies Lemma 5.11′.

§6. Conjecture

I will formulate a much stronger and more precise version of Conjectures 1.1 and 1.1′. It makes sense for any field F, implies these conjectures when $F = \mathbb{C}$, and explains how (and why) multivalued forms appear in Conjecture 1.1. For this I have to recall the definition of the Hopf algebra $A(F)_\bullet$ [BGSV, BMS] and its analytic counterpart, Aomoto polylogarithms [Ao].

6.1. The motivic Hopf algebra $A(F)_\bullet$. First of all let us reproduce from [BGSV] the definition of groups $A_n(F)$, $n = 0, 1, 2, \dots$. An n-simplex is a family of $n+1$ hyperplanes $L = (L_0, \dots, L_n)$ in P^n_F. An n-simplex is said to be nondegenerate if the hyperplanes are in general position. A face of an n-simplex is any nonempty intersection of hyperplanes from L. A pair of n-simplexes (L, M) is said to be admissible if L and M have no common faces.

Define the group $A_n(F)$ as the group with generators $(L; M)$, where $(L; M)$ runs over all admissible pairs of simplexes, and with the following relations:

(A1) If one of the simplexes L or M is degenerate, then $(L; M) = 0$.

(A2) *Skew symmetry.* For every permutation $\sigma: (0, 1, \dots, n) \longrightarrow (0, 1, 2, \dots, n)$ one has

$$(\sigma L; M) = (L; \sigma M) = \mathrm{sgn}(\sigma)(L; M),$$

where $\mathrm{sgn}\,\sigma$ is the parity of σ and $\sigma L = (L_{\sigma(0)}, \dots, L_{\sigma(n)})$.

(A3) *Additivity in* L. For every family of hyperplanes $(L_0, \dots, L_{n+1})$ and any n-simplex M such that all pairs $(\widehat{L^j}; M)$, where $\widehat{L^j} = (L_0, \dots, \widehat{L_j}, \dots, L_{n+1})$, are admissible

$$\sum_{j=0}^{j=n+1} (-1)^j (\widehat{L^j}; M) = 0.$$

Additivity in M. For every family $(M_0, \dots, M_{n+1})$ and any simplex L such that all $(L; \widehat{M^j})$ are admissible

$$\sum_{j=0}^{j=n+1} (-1)^j (L; \widehat{M^j}) = 0.$$

(A4) *Projective invariance.* For any $g \in PGL_{n+1}(F)$ one has $(gL; gM) = (L; M)$.

There is a canonical isomorphism

$$r\colon A_1(F) \longrightarrow F^*, \quad r\colon (L_0, L_1; M_0, M_1) \longrightarrow r(L_0, L_1, M_0, M_1).$$

By definition, $A_0 = \mathbb{Z}$. Set

$$A(F)_\bullet := \bigoplus_{n=0}^{n=\infty} A_n(F).$$

Then there is a multiplication $m\colon A_n \otimes A_m \longrightarrow A_{n+m}$ and a comultiplication $\Delta\colon A_n \longrightarrow \bigoplus_{k+l=n} A_k \otimes A_l$ that make $A(F)_\bullet$ a graded commutative Hopf algebra.

In the case $F = \mathbb{C}$ there is a canonical holomorphic differential form w_L with logarithmic singularities on the hyperplanes L_i. If $z_i = 0$ is the homogeneous equation of L_i, then $w_L = d\log(z_1/z_0) \wedge \cdots \wedge d\log(z_n/z_0)$. Let Δ_M be an n-cycle representing a generator of the group $H_n(P^n_{\mathbb{C}}, \bigcup M_j)$. Then

$$\alpha_n(L; M) := \int_{\Delta_M} w_L$$

is a multivalued analytic function called the Aomoto polylogarithm [Ao]. This integral depends on the choice of Δ_M but does not change under continuous deformation.

Let us denote by $\operatorname{Bar}\widetilde{A}(F)$ the complex that we would get applying the reduced bar construction to this Hopf algebra. Recall that

$$\operatorname{Bar}\widetilde{A}(F)\colon \widetilde{A}_\bullet \xrightarrow{\delta} \widetilde{A}_\bullet \otimes \widetilde{A}_\bullet \xrightarrow{\delta} \widetilde{A}_\bullet \otimes \widetilde{A}_\bullet \otimes \widetilde{A}_\bullet \xrightarrow{\delta} \cdots,$$

where

$$\delta(a_{i_1} \otimes a_{i_2} \otimes \cdots \otimes a_{i_n}) = \widetilde{\Delta} a_{i_1} \otimes a_{i_2} \otimes \cdots \otimes a_{i_n} - a_{i_1} \otimes \widetilde{\Delta} a_{i_2} \otimes \cdots \otimes a_{i_n} + \cdots.$$

Here $\widetilde{A} := \bigoplus_{i=1}^{\infty} A_i$, $\widetilde{\Delta}(a) := \Delta(a) - (a \otimes 1 + 1 \otimes a)$ is the reduced coproduct, A placed in degree 1 and the differential has degree $+1$. Let $\mathrm{Bar}_{(n)}\widetilde{A}(F)$ be the degree n part of this complex:

$$\mathrm{Bar}_{(n)}\widetilde{A}(F)\colon A_n \xrightarrow{\delta} \bigoplus_{i_1+i_2=n} A_{i_1} \otimes A_{i_2} \xrightarrow{\delta} \bigoplus_{i_1+i_2+i_3=n} A_{i_1} \otimes A_{i_2} \otimes A_{i_3} \xrightarrow{\delta} \cdots .$$

6.2. Recall that there is a "cohomological" complex $BC^*(n)$ with a differential ∂ of degree $+1$:

$$\cdots \xrightarrow{\partial} BC^0(n) \xrightarrow{\partial} BC^1(n) \xrightarrow{\partial} \cdots \xrightarrow{\partial} BC^n(n).$$

By definition, $BC^i(n) := BC_{2n-i}(n)$ and $BC^i(n)$ is placed at degree i.

CONJECTURE 6.1. *There is a canonical homomorphism of complexes*

$$h_n\colon BC^*(n) \longrightarrow \mathrm{Bar}_{(n)}\widetilde{A}(F)$$

such that the homomorphism

$$h_n(n)\colon BC^n(n) \longrightarrow \bigotimes^n A_1(F) = \bigotimes^n F^*$$

is given by the formula

$$h_n(n)(l_0, \ldots, l_n) := \mathrm{Alt} \bigotimes_{i=1}^{i=n} \Delta(l_0, \ldots, \widehat{l_i}, \ldots, l_n)$$

(*compare with formula* (3.2)).

6.3. Relation with Conjecture 1.1. Let $a \in A_n(\mathbb{C}(X))$ and $\alpha(a)$ be the corresponding Aomoto n-logarithm considered as a function on (an open part of) X.

PROPOSITION 6.2. *Let* $\Delta^{n-1,1}(a) = \sum b_i \otimes c_i \in A_{n-1} \otimes A_1$ *be the* $(n-1, 1)$ *component of the coproduct. Then*

$$d\alpha(a) = \sum \alpha(b_i) d \log \alpha(c_i).$$

PROOF. An exercise.

Now let us define a map of complexes

$$\mathscr{D}\colon \mathrm{Bar}_n \widetilde{A}(\mathbb{C}(X)) \longrightarrow \widetilde{\Omega}^\bullet(\mathrm{Spec}\,\mathbb{C}(X)),$$

as follows ($\widetilde{\Omega^\bullet}$ is the De Rham complex of multivalued forms). Let $a_1 \otimes \cdots \otimes a_n \in A_{i_1} \otimes \cdots \otimes A_{i_n}$. Then

$$\mathscr{D}(a_1 \otimes \cdots \otimes a_{i_n}) := \alpha(a_1) d \log \alpha(a_2) \wedge \cdots \wedge d \log \alpha(a_n)$$

if $i_2 = i_3 = \cdots = i_n = 1$ and zero otherwise.

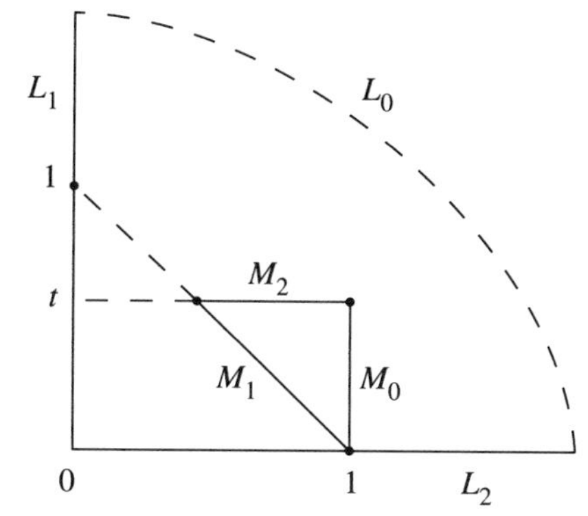

FIGURE 2. $L_2\{t\}$

LEMMA 6.3. *$\mathscr{D}$ is a homomorphism of complexes.*

PROOF. Follows immediately from Proposition 6.2.

The composition $\mathscr{D} \circ h_n$ gives us a multivalued n-logarithm L_n on the bi-Grassmannian satisfying all conditions of Conjecture 1.1.

Let me emphasize that the constructed n-logarithm L_n catches only a small part of the homomorphism h_n, because $\mathscr{D}$ is zero almost everywhere. Further, multivalued functions appear only at the last moment.

I know an explicit construction of homomorphism h_n for $n \leqslant 4$. For $n \leqslant 3$ it is the composition of homomorphism of complexes (5.12)

$$f(n)\colon BC_*(n) \longrightarrow B_F(n),$$

constructed in § 5.3 with canonical homomorphism of complexes

$$l(n)\colon B_F(n) \longrightarrow \mathrm{Bar}_{(n)}\widetilde{A}(F).$$

To give the definition of $l(n)$, let us first define a homomorphism

$$L_n\colon \mathbb{Z}[P^1_F] \longrightarrow A_n(F).$$

Let $x_1, \dots, x_n$ be coordinates in the affine space $P^n \backslash L_0$ such that L_i is the hyperplane $x_i = 0$. Consider the hyperplanes

$$0 = 1 - x_1; \quad 1 - x_1 = x_2; \quad x_2 = x_3; \dots ; x_{n-1} = x_n; \quad x_n = t, \quad t \in F^*,$$

by $M_0, M_1, \dots, M_{n-1}, M_n(t)$. Set

$$(L, M(t)) = (L_0, \dots, L_n; M_0, \dots, M_{n-1}, M_n(t)),$$
$$L_n(\{t\}) := (L; M(t)), \qquad L_n(\{0\}) = L_n(\{\infty\}) = 0$$

(see Figure 2). Then $\alpha(L_n(\{t\})) = \mathrm{Li}_n(t)$.

I will also use a notation $z = L_1(\{1-z\})$. Now let us define a homomorphism of complexes $l(2)$ as follows:

$$\begin{array}{ccc} B_2(F) & \longrightarrow & \Lambda^2 F^* \\ \downarrow{\scriptstyle l_1(2)} & & \downarrow{\scriptstyle l_2(2)} \\ A_2 & \xrightarrow{\delta} & A_1 \otimes A_1 \end{array}$$

$$l_2(2)\colon x \otimes y \mapsto (1/2)(x \otimes y - y \otimes x),$$
$$l_1(2)\colon \{x\} \mapsto L_2(\{x\}) - (1/2)\,\mathrm{Li}_1(\{x\}) \cdot x.$$

THEOREM 6.4 [BGSV]. *$l(2)$ is a homomorphism of complexes.*

Let us define a homomorphism of complexes

$$\begin{array}{ccccc} B_3(F) & \longrightarrow & B_2(F) \otimes F^* & \longrightarrow & \Lambda^3 F^* \\ \downarrow{\scriptstyle l_1(3)} & & \downarrow{\scriptstyle l_2(3)} & & \downarrow{\scriptstyle l_3(3)} \\ A_3 & \xrightarrow{\delta} & A_2 \otimes A_1 \oplus A_1 \otimes A_2 & \xrightarrow{\delta} & \otimes^3 A_1 \end{array}$$

by the following formulas:

$$l_3(3)\colon x_1 \otimes x_2 \otimes x_3 \mapsto (1/6)\,\mathrm{Alt}(x_1 \otimes x_2 \otimes x_3),$$
$$\begin{aligned} l_2(3)\colon \{x\} \otimes y \mapsto {} & (1/2)(l_1(2)(\{x\}) \otimes y - y \otimes l_1(2)(\{x\})) \\ & - (1/12)(L_1(\{x\}) \cdot y \otimes x + x \otimes L_1(\{x\}) \cdot y) \\ & + (1/12)(L_1(\{x\}) \otimes x \cdot y + x \cdot y \otimes L_1(\{x\})), \end{aligned}$$
$$l_1(3)\colon \{x\} \mapsto L_3(\{x\}) - (1/12) L_2(\{x\}) \cdot x + (1/2) L_1(x) \cdot x^2.$$

In these formulas the dot denotes multiplication in the Hopf algebra $A(F)_\bullet$.

THEOREM 6.5. *$l(3)$ is a homomorphism of complexes.*

There are similar formulas for homomorphism $l(n)$ for all n.

REFERENCES

[A] M. Atiyah, *Complex analytic connections in fibre bundles*, Trans. Amer. Math. Soc. **85** (1957), 181–207.

[Ao] K. Aomoto, *Addition theorem of Abel type for hyperlogarithms*, Nagoya Math. J. **88** (1982), 55–71.

[BT] H. Bass and J. Tate, *The Milnor ring of the global field*, Lecture Notes in Math., vol. 342, Springer-Verlag, Berlin and New York, 1973, pp. 349–446.

[B1] A. A. Beilinson, *Height pairings between algebraic cycles*, Lecture Notes in Math., vol. 1289, Springer-Verlag, Berlin and New York, 1987, pp. 1–26.

[B2] ———, *Higher regulators and values of L-functions*, Itogi Nauki i Tekhniki. Sovremennye Problemy Matematiki. Noveĭshie Dostizheniya, vol. 24, VINITI, Moscow, 1984, pp. 181–138; English transl. in J. Soviet Math. **30** (1985), 2036–2070.

[B3] ———, *Letter to P. Deligne*, September 1991.

[BMS] A. A. Beilinson, R. MacPherson, and V. V. Shechtman, *Notes on motivic cohomology*, Duke Math. J. **54** (1987), 679–710.

[BGSV] A. A. Beilinson, A. B. Goncharov, V. V. Shechtman, and A. N. Varchenko, *Projective geometry and algebraic K-theory*, Algebra i Analiz **3** (1990), 78–131; English transl. in Leningrad Math. J. **3** (1991).

[Bl1] S. Bloch, *Higher regulations, algebraic K-theory and zeta functions of elliptic curves*, Lecture Notes, Univ. of California, Irvine, CA, 1977.

[Bl2] ———, *Applications of the dilogarithm function in algebraic K-theory and algebraic geometry*, Proc. Int. Symp. Alg. Geometry, Kyoto, 1977, pp. 1–14.

[Bott] R. Bott, *On the Chern-Weil homomorphism and the continuous cohomology of the Lie groups*, Adv. Math. **11** (1973), no. 2, 285–303.

[BM1] J.-L. Brylinski and D. A. McLaughlin, *A geometric construction of the first Pontryagin class*, preprint (1991).

[BM2] ———, *Čech cocycles for characteristic classes*, preprint (1991).

[DS] J. Dupont and C.-H. Sah, *Scissors congruences*. II, J. Pure Appl. Algebra **25** (1982), 159–195.

[D] J. Dupont, *The dilogarithm as a characteristic class for flat bundles*, J. Pure Appl. Algebra **44** (1987), 137–164.

[DHZ] J. Dupont, R. Hain, and S. Zucker, *Regulators and characteristic classes of flat bundles*, preprint (1992).

[EGA] A. Grothendieck, *Elements de géometrie algébrique*. III. *Étude cohomologique des faisceaux coherents*. I, Inst. Hautes Études Sci. Publ. Math. **11** (1961).

[GGL] A. M. Gabrielov, I. M. Gelfand, and M. V. Losik, *Combinatorial computation of characteristic classes*, Funktsional Anal. i Prilozhen. **9** (1975); English transl. in Functional Anal. Appl. **9** (1975), 103–115.

[GM] I. M. Gelfand and R. MacPherson, *Geometry in Grassmannians and a generalisation of the dilogarithm*, Adv. Math. **44** (1982), 279–312.

[GM2] ———, *A combinatorial formula for the Pontryagin classes*, Bull. Amer. Math. Soc. (N.S.) **26** (1992), no. 2, 304–309.

[G1] A. B. Goncharov, *Geometry of configurations, polylogarithms, and motivic cohomology*, Adv. Math. (to appear).

[G2] ———, *Polylogarithms and motivic Galois groups*, Submitted to the proceedings of the AMS Research Summer conference "Motives".

[GH] Ph. Griffiths and J. Harris, *Principles of algebraic geometry*, Wiley, New York, 1978.

[Gu] A. Guichardet, *Cohomologie des groupes topologiques et des algébres de Lie*, CEDIC, Paris, 1980.

[Gil] H. Gillet, *Riemann-Roch theorem in higher K-theory*, Adv. Math. **40** (1981), 203–289.

[H] B. Harris, *Group cohomology classes with differential form coefficients*, Algebraic K-theory, Lecture Notes in Math., vol. 551, Springer-Verlag, Berlin and New York, 1976, pp. 278–282.

[HM] R. Hain and R. MacPherson, *Higher logarithms*, Illinois J. Math. **34** (1990), 392–475.

[HaM] M. Hanamura and R. MacPherson, *Geometric construstion of polylogarithms*, preprint (1991).

[Ka] K. Kato, *Milnor K-theory and the Chow group of zero cycles*, Contemp. Math. **55** (1985), 241–255.

[L] L. Lewin, *Dilogarithms and associated functions*, North Holland, New York and Amsterdam, 1981.

[L1] S. Lichtenbaum, *Values of zeta functions at non-negative integers*, Journées Arithmetiques (Noorwykinhoot, Netherlands, 1983), Lecture Notes in Math., vol. 1068, Springer-Verlag, Berlin and New York, 1984, pp. 127–138.

[M] R. MacPherson, *The combinatorial formula of Gabrielov, Gelfand and Losik for the first Pontryagin class*, Sem. Bourbaki **497** (1977).

[M1] J. Milnor, *Algebraic K-theory and quadratic forms*, Invent. Math. **9** (1970), 318–340.

[NS] Yu. P. Nesterenko and A. A. Suslin, *Homology of the full linear group over a local ring and Milnor K-theory*, Izv. Akad. Nauk SSSR Ser. Mat. **53** (1989), 121–146; English transl. in Math. USSR-Izv. **34** (1990).

[Q1] D. Quillen, *Higher algebraic K-theory*. I, Algebraic K-theory, Lecture Notes in Math., vol. 341, Springer-Verlag, Berlin and New York, 1973, pp. 85–197.

[Q2] ———, *A course on algebraic K-theory*, MIT (unpublished).

[Sa] C.-H. Sah, *Homology of classical groups made discrete*. III, J. Pure Appl. Algebra **56** (1989), 269–312.

[Sch] V. V. Schechtman, *On the delooping of Chern character and Adams operations*, Lecture Notes in Math., vol. 1289, Springer-Verlag, Berlin and New York, 1987, pp. 265–319.

[So] C. Soulé, *Operations en K-thèorie Algébrique*, Canad. J. Math **27** (1985), 488–550.

[S1] A. A. Suslin, *Homology of* GL_n, *characteristic classes and Milnor's K-theory*, Trudy Mat. Inst. Steklov. (1985), no. 3, 207–226; English transl. in Lecture Notes in Math., vol. 1046, Springer-Verlag, Berlin and New York, 1989, pp. 357–375.

[S2] ———, *Algebraic K-theory of fields*, Proc. International Congress of Mathematicians (Berkeley, California, 1986), vol. 1, Amer. Math. Soc., Providence, RI, 1988, pp. 222–243.

[S3] ———, K_3 *of a field and Bloch's group*, Trudy Mat. Inst. Steklov. **189** (1990), 180–199; English transl. in Proc. Steklov Inst. Math. **1990**, no. 4.

[Y1] J. Yang, *The Hain-MacPherson's trilogarithm, the Borel regulators and the value of Dedekind zeta function at* 3, preprint (1990).

[You] B. V. Youssin, *Sur les formes* $S^{p,q}$ *apparaissant dans le calsul combinatorie de la deuxieme class de Pontryagin par la methode de Gabrielov, Gelfand, et Losik*, C. R. Acad. Sci. Paris Sér. I Math. **292** (1981), 641–649.

[Z] D. Zagier, *Dedekind zeta functions and the algebraic K-theory of fields*, Arithmetic Algebraic Geometry (Texel, 1989), Birkhäuser, Boston, 1991, pp. 391–430.

DEPARTMENT OF MATHEMATICS, MASSACHUSETTS INSTITUTE OF TECHNOLOGY, CAMBRIDGE, MASSACHUSETTS 02139

ADVANCES IN SOVIET MATHEMATICS
Volume 16, Part 1, 1993

The Riemann-Roch Theorem for Elliptic Operators

MIKHAEL GROMOV AND MIKHAIL A. SHUBIN

§1. Introduction

The classical Riemann-Roch theorem for nonsingular complex algebraic curves has been generalized in different ways to multidimensional situations. The best known generalizations are the Riemann-Roch-Hirzebruch theorem and the Riemann-Roch-Grothendieck theorem, both of them motivated by algebraic geometry. In this paper we suggest a generalization that is motivated by classical analysis of solutions of general elliptic equations with point singularities. Namely, for any real C^∞-manifold we introduce point divisors which are elements of a free abelian group generated by the points of this manifold. Then we define spaces of "meromorphic" solutions of an elliptic equation depending on a given divisor: these solutions are allowed to have some poles (at points that enter to the divisor with positive degrees) and are required to have zeros (at the points that enter to the divisor with negative degrees). The main theorem gives a formula that connects dimensions of two such spaces of "meromorphic" solutions corresponding to a given divisor and its dual (inverse) divisor. This formula contains also the index of the given elliptic operator (that can be calculated by the Atiyah-Singer formula) and a degree of the divisor, which is written in terms of binomial coefficients and depends on the dimension of the manifold and on the order of the operator. It is a direct generalization of the classical Riemann-Roch theorem, but on the other hand it can be considered as an extension of the Atiyah-Singer index formula. We consider first the case of compact closed manifolds and then give a generalization to noncompact manifolds with a compact boundary. In this case we need appropriate boundary conditions and conditions at infinity are imposed in order to ensure that the given elliptic operator defines a Fredholm operator in appropriate spaces.

A special case of the scalar Laplacian on a Riemannian manifold was considered in a beautiful paper by Nadirashvili [1], which we took as a starting point. But our proof is based on different ideas and is even simpler than the

1991 *Mathematics Subject Classification*. Primary 58G03, 58G10.

proof in [1] due to its generality that allows us a more transparent use of duality arguments.

The main theorem is proved simultaneously with a duality theorem that gives a solvability condition for the given elliptic equation if the solution is allowed to have poles and is required to have zeros at prescribed points.

We also give a number of applications including those similar to well-known applications of the classical Riemann-Roch theorem. The first of them is an analogue of the Riemann inequality, and it gives an existence result for nontrivial meromorphic solutions that have zeros at a given finite set with multiplicities bounded from below by given numbers. This can be achieved provided that a sufficient number of poles (or a pole of sufficiently high order) is allowed. We also supply a more elementary proof of the inequality (this proof does not use our main theorem).

Next we prove that if the adjoint operator has a unique continuation property, then sometimes even the equality can be proved; however, this result is not as effective as in the classical case of ample divisors. On the other hand, we present simple arguments showing that no effective general result is possible in the same terms as in the classical case. We also show that a condition of unique continuation type is necessary for the equality to be true.

As an application of the duality theorem we prove a local solvability result for an elliptic equation with a right-hand side and with an additional condition on the order of zero of the solution at a given point.

Then we investigate simplest properties of sheaves of solutions, naturally associated with point divisors, i.e., sheaves of meromorphic solutions in the sense described above. The mentioned local solvability result leads to a construction of a fine resolution for such a sheaf. Applying this resolution we obtain triviality of cohomologies of these sheaves in dimensions 2 and higher and also a Serre type duality between cohomologies in dimensions 0 and 1.

Using the Riemann-Roch theorem for the Laplacian in the Euclidean space, Nadirashvili [1] proved an estimate for the maximal possible multiplicity of zeros for the Coulomb potential of a finite system of point charges in $\mathbb{R}^3$. At the end of the last section of this paper we reproduce his arguments in more detail, extending them to the Coulomb potentials in $\mathbb{R}^n$. We put basic arguments of Nadirashvili in a more general context (Riemannian manifolds) and also give a more elementary proof of his estimate (a proof that does not use a Riemann-Roch-type theorem). After that we study a similar problem for Riemann surfaces and prove estimates for the maximal possible multiplicity of zeros for harmonic functions with simple (logarithmic) singularities at points from a given finite subset of a Riemann surface. For the case of the Riemann sphere we also prove a precise (though elementary) estimate for this maximal multiplicity and give necessary and sufficient conditions for the existence of a harmonic function with a given configuration of several poles and a zero, so that the zero has the maximal possible multiplicity. This condition is naturally formulated in terms of cross ratios in this configuration.

It is also possible to prove an L^2-version of the main result (on covering manifolds) but we will do it in a separate paper.

Another possible generalization of the main result deals with the case of multidimensional divisors. More exactly, one can introduce the so called "rigged divisors" that consist of a pair of disjoint closed subsets in the manifold and a finite-dimensional space of distributions supported on each of them. Then a finite-dimensional set of singularities of solutions on the first subset may be allowed and finitely many vanishing conditions on the second subset imposed; both singularities and vanishing conditions are defined by the given distribution spaces. Then the Riemann-Roch-type theorem holds in this case too and we will also treat it in a separate paper.

The results of this paper were partially announced in [5], but the reader should keep in mind that a different sign convention was used there. Here we put the sign conventions in agreement with the classical Riemann-Roch theorem.

The second author is grateful to Forschungsinstitut für Mathematik ETH Zürich for the support during the time when this paper was being written, and also to Rahel Boller who partially typed the manuscript.

§2. Notations and main results

A. Let X be a compact closed C^∞-manifold, $\dim X = n \geqslant 2$, E and F complex vector bundles over X, $q = \dim_{\mathbb{C}} E_x = \dim_{\mathbb{C}} F_x$ the dimension of their fibres, and $\Gamma(U, E)$ the space of all C^∞-sections of E over an open set $U \subset X$. Let

$$A\colon \Gamma(X, E) \longrightarrow \Gamma(X, F) \tag{2.1}$$

be an elliptic differential operator of order d and $\operatorname{ind} A$ its index that can be calculated by the Atiyah-Singer index formula. Let $\Omega(X)$ be the bundle of complex densities on X, so for every $\omega \in \Gamma(X, \Omega(X))$ the integral $\int_X \omega \in \mathbb{C}$ is naturally defined. If E is a complex vector bundle over X then the bundle $E^* = \operatorname{Hom}_{\mathbb{C}}(E, \Omega(X))$ is called the dual bundle and there is a natural pairing of bundles $E \otimes E^* \to \Omega(X)$ which gives the pairing in sections

$$\langle\cdot, \cdot\rangle\colon \Gamma(X, E) \times \Gamma(X, E^*) \longrightarrow \mathbb{C}, \qquad \langle u, v\rangle = \int_X \langle u(x), v(x)\rangle, \tag{2.2}$$

where $\langle u(x), v(x)\rangle \in \Omega_x(X)$ is obtained by use of the pairing between E_x and E_x^*. Now using the pairing (2.2) define the transposed operator $A^t\colon \Gamma(X, F^*) \to \Gamma(X, E^*)$ by the formula

$$\langle Au, v\rangle = \langle u, A^t v\rangle, \qquad u \in \Gamma(X, E), \ v \in \Gamma(X, F^*). \tag{2.3}$$

Then A^t is again an elliptic differential operator. Note that $\operatorname{ind} A^t = -\operatorname{ind} A$.

Actually it is often convenient to use more general "dual" (adjoint or transpose) operators to a given differential operator (2.1). Namely, suppose that $\widetilde{E}$, $\widetilde{F}$ are vector bundles given together with nondegenerate bilinear or sesquilinear C^∞-pairings of bundles

$$E \times \widetilde{E} \longrightarrow \Omega(X), \qquad F \times \widetilde{F} \longrightarrow \Omega(X).$$

We denote these pairings at a point $x \in X$ by $\langle\cdot, \cdot\rangle_x$.

An example of such a situation arises e.g. if positive definite hermitian C^∞-metrics h_E, h_F are given in E and F and a positive smooth density $d\mu$ on X is fixed. Then we can take $\widetilde{E} = E$, $\widetilde{F} = F$, defining the pairings by

$$\langle u, \widetilde{u}\rangle_x = h_E(u, \widetilde{u})\, d\mu(x), \qquad u, \widetilde{u} \in E_x,$$

and similarly for F.

In the general case we have bilinear or sesquilinear pairings

$$\Gamma(X, E) \times \Gamma(X, \widetilde{E}) \longrightarrow \mathbb{C}, \qquad \Gamma(X, F) \times \Gamma(X, \widetilde{F}) \longrightarrow \mathbb{C}$$

obtained by the fiberwise pairing and integration. Denoting these pairings $\langle\cdot, \cdot\rangle_E$ and $\langle\cdot, \cdot\rangle_F$, we can define the operator $A^*\colon \Gamma(X, \widetilde{F}) \longrightarrow \Gamma(X, \widetilde{E})$ by the formula

$$\langle Au, v\rangle_F = \langle u, A^*v\rangle_E, \qquad u \in \Gamma(X, E), \ v \in \Gamma(X, \widetilde{F}).$$

Let us describe a connection between the general adjoint operator A^* and the canonical transpose A^t. To do this define a (linear or antilinear) bundle isomorphism $J_E\colon \widetilde{E} \longrightarrow E^*$ by

$$J_E\widetilde{e} = \langle\cdot, \widetilde{e}\rangle_x, \qquad \widetilde{e} \in \widetilde{E}_x,$$

where $\langle\cdot, \cdot\rangle_x$ denotes the given pairing $E_x \times \widetilde{E}_x \longrightarrow \Omega(X)_x$. Similarly define a bundle isomorphism $J_F\colon \widetilde{F} \longrightarrow F^*$. We denote the corresponding isomorphisms on sections by the same letters.

LEMMA 2.1. $A^* = J_E^{-1} A^t J_F$.

PROOF. For any $u \in \Gamma(X, E)$, $\widetilde{v} \in \Gamma(X, \widetilde{F})$ we have

$$\begin{aligned}\langle Au, \widetilde{v}\rangle_E &= \langle Au, J_F\widetilde{v}\rangle = \langle u, A^t J_F \widetilde{v}\rangle \\ &= \langle u, J_E(J_E^{-1} A^t J_F)\widetilde{v}\rangle = \langle u, J_E^{-1} A^t J_F \widetilde{v}\rangle_F,\end{aligned}$$

hence $A^* = J_E^{-1} A^t J_F$ as required.

It follows in particular that $\dim \operatorname{Ker} A^* = \dim \operatorname{Ker} A^t$, hence $\dim \operatorname{Ker} A^*$ does not depend on the choice of arbitrary elements in the definition of A^*.

We use point divisors, which are elements of free abelian groups generated by points $x \in X$. For brevity we refer to them simply as divisors. We write the operation in the group of the divisors as multiplication. Hence such a divisor can be written in the form

$$\mu = x_1^{p_1} x_2^{p_2} \cdots x_m^{p_m}, \tag{2.4}$$

where $x_i \neq x_j$ if $i \neq j$ and $p_i \in \mathbb{Z} - \{0\}$. We fix such a divisor for some time. The dual divisor is defined as

$$\mu^{-1} = x_1^{-p_1} x_2^{-p_2} \dots x_m^{-p_m}$$

and we introduce also the positive and negative parts of μ as divisors

$$\lambda = x_1^{p_1^+} \dots x_m^{p_m^+}, \qquad \nu = x_1^{p_1^-} \dots x_m^{p_m^-}$$

with $p^+ = \max(p, 0)$, $p^- = \min(p, 0)$. Here all factors of the form x_i^r with $r = 0$ have to be omitted. We evidently have $\mu = \lambda\nu$ and $\mu^{-1} = \nu^{-1}\lambda^{-1}$, where ν^{-1} and λ^{-1} are the positive and negative part of μ^{-1} respectively.

For the divisor (2.4) (with $p_i \neq 0$) we use the notation

$$\operatorname{supp} \mu = \{x_1, \dots, x_m\}.$$

Therefore, $\operatorname{supp} \mu$ is a finite subset in X such that $\operatorname{supp} \mu = \operatorname{supp} \lambda \cup \operatorname{supp} \nu$, $\operatorname{supp} \lambda \cap \operatorname{supp} \nu = \varnothing$.

DEFINITION 2.1. The *degree* of the divisor μ is defined as

$$d(\mu) = \sum_{1 \leqslant i \leqslant m} \operatorname{sign} p_i \left[\binom{|p_i| + n - 1}{n} - \binom{|p_i| + n - 1 - d}{n} \right], \tag{2.5}$$

where $\binom{N}{n} = \frac{N!}{n!(N-n)!}$ if $N \geqslant n$ and 0 otherwise.

Note that in this definition $d(\mu)$ depends also on the order d of the operator A. Actually if $|p_i|$ is sufficiently large then the expression in the square brackets can be written as $f_{n,d}(|p_i|)$ where $f_{n,d}$ is a polynomial of degree $n-1$ with the coefficient $d/(n-1)!$ by x^{n-1}:

$$f_{n,d}(x) = \frac{d}{(n-1)!} x^{n-1} + \cdots. \tag{2.6}$$

In particular

$$f_{2,d}(x) = dx - \frac{d(d-1)}{2}.$$

It is easy to check that if $n = 2$, $d = 1$ then

$$d(\mu) = \sum_{1 \leqslant i \leqslant m} p_i, \tag{2.7}$$

which gives the usual degree of the divisor on a (2-dimensional) surface.

Now we introduce solutions with poles and zeros. Let $x_0 \in X$, $p \in \mathbb{Z} - \{0\}$, $u \in \Gamma(U - \{x_0\}, E)$ where U is a neighbourhood of x_0 in X, and $Au = 0$ in $U - \{x_0\}$.

DEFINITION 2.2. (i) Suppose $p > 0$. Then we write $\operatorname{ord}_{x_0} u \geqslant -p$ iff $u = u_s + u_r$, where $u_r \in \Gamma(U, E)$ and

$$u_s(x) = o(|x - x_0|^{d-n-p}) \quad \text{as } x \longrightarrow x_0. \tag{2.8}$$

(ii) Suppose $p < 0$. Then we write $\operatorname{ord}_{x_0} u \geqslant -p$ iff u can be extended to a section $u \in \Gamma(U, E)$ and

$$u(x) = o(|x - x_0|^{|p|-1}) \quad \text{as } x \longrightarrow x_0. \tag{2.9}$$

Equivalently we could also write in this case $u(x) = O(|x-x_0|^{|p|})$ as $x \to x_0$, or $j^{|p|-1}_{x_0} u = 0$ where $j^{|p|-1}_{x_0} u$ is the jet of order $|p|-1$ of the section u at x_0.

We write $\operatorname{ord}_{x_0} u = -p$ if $\operatorname{ord}_{x_0} u \geqslant -p$ but it is not true that $\operatorname{ord}_{x_0} u \geqslant -p+1$.

The given definition is compatible with the standard definition of the order of a pole or a zero for a meromorphic function on a Riemann surface (this corresponds to the case $A = \bar\partial$, $n = 2$, $d = 1$). A motivation of (2.8) in the general case becomes clear if we consider the well-known structure of the fundamental solution of the elliptic operator A (see, e.g., [2]). If x_0 is an isolated singularity of a solution of the equation $Au = 0$ (i.e., u is a solution in $U - \{x_0\}$ where U is a neighbourhood of x_0) and $u(x) = o(|x - x_0|^{-N})$ with $N \in \mathbb{R}$ then we can find a "regularization" of u, a distributional section $\tilde u$ of E over U such that $\tilde u = u$ on $U - \{x_0\}$. Then $A\tilde u$ is a distributional section of F supported at x_0, hence locally

$$A\tilde u(x) = \sum_{|\alpha| \leqslant M} c_\alpha \partial_x^\alpha \delta(x - x_0)$$

with $c_\alpha \in F_{x_0}$. Here $\delta(x-x_0)$ is the standard Dirac measure. Now if B is a local parametrix of the operator A (so B is a classical or polyhomogeneous pseudodifferential operator of order $-d$) then we can conclude that near x_0 section $\tilde u$ has the form

$$\tilde u(x) = \sum_{|\alpha| \leqslant M} (-\partial_{x_0})^\alpha K_B(x, x_0) c_\alpha + u_r(x),$$

where $u_r \in \Gamma(U, E)$, K_B is the Schwartz kernel of B. Now K_B is a Fourier transform of the symbol b of $B = b(x, D_x)$, i.e.,

$$K_B(x, x_0) = (2\pi)^{-n} \int e^{i(x-x_0)\cdot\xi} b(x, \xi)\, d\xi.$$

But since b is polyhomogeneous, K_B is the sum of homogeneous functions (with respect to $x - x_0$) up to logarithmic terms and up to a sufficiently smooth remainder. The orders of homogeneity of terms in K_B are actually $d-n, d-n+1, d-n+2, \dots$, hence in $\partial^\alpha_{x_0} K_B$ we have terms of homogeneity degree $d-n-|\alpha|$, $d-n-|\alpha|+1, \dots$. Hence the condition $u(x) = o(|x - x_0|^{d-n})$ eliminates all singularities, the condition $u(x) = o(|x-x_0|^{d-n-1})$ allows the simplest singularity of the form $K_B(x, x_0)c$ with $c \in F_{x_0}$, etc.

REMARK. If $d - n - p < 0$ then the condition on u in Definition 2.2(i) can be replaced by a simpler condition (given in terms of u instead of u_s):

$$u(x) = o(|x - x_0|^{d-n-p}) \quad \text{as } x \to x_0.$$

In the general case the condition can be written in terms of derivatives of u:

$$\partial^\alpha u(x) = o(|x - x_0|^{d-n-p-|\alpha|}) \quad \text{if } |\alpha| \geqslant d - n - p + 1,$$

and in fact, equivalently, it is sufficient to consider only α with $|\alpha| = d - n - p - 1$.

Now we introduce the space of solutions with poles and zeros subordinated to the divisor μ (of the form (2.4)) and the dimension of this space.

DEFINITION 2.3. Denote

$$L(\mu, A) = \{u \mid u \in \Gamma(X - \operatorname{supp}\mu, E),\ Au = 0 \text{ on } X - \operatorname{supp}\mu,\ \operatorname{ord}_{x_i} u \geqslant -p_i \text{ for each } i = 1, \dots, m\},$$

$$r(\mu, A) = \dim_{\mathbb{C}} L(\mu, A).$$

Our first main result is

THEOREM 2.1. *We have*

$$r(\mu, A) = \operatorname{ind} A + qd(\mu) + r(\mu^{-1}, A^*). \tag{2.10}$$

Note that $r(\mu^{-1}, A^*) = r(\mu^{-1}, A^t)$ due to Lemma 2.1, hence $r(\mu^{-1}, A^*)$ does not depend on the choice of the arbitrary elements involved in the definition of A^*.

Now let us mention two particular cases of Theorem 2.1.

EXAMPLE 2.1 (Riemann-Roch). Let X be a compact complex manifold, $\dim_{\mathbb{C}} X = 1$, hence $n = 2$. (In other words we might say that X is a compact Riemannian surface or a nonsingular algebraic curve over $\mathbb{C}$.) Consider the standard splitting of the cotangent bundle

$$T^*X = T^{1,0}(X) \oplus T^{0,1}(X),$$

where $T^{1,0}(X)$ and $T^{0,1}(X)$ consist of $(1,0)$- and $(0,1)$-forms respectively, the spaces of their C^∞-sections being denoted $\Lambda^{1,0}(X)$ and $\Lambda^{0,1}(X)$. Now let us take

$$A = \bar\partial : C^\infty(X) \longrightarrow \Lambda^{0,1}(X).$$

Using the standard orientation on X we can identify $\Omega(X)$ with $\Lambda^2(T^*X)$, hence $\Gamma(X, \Omega(X))$ with $\Lambda^2(X)$. We have the canonical pairing (given by the wedge product)

$$T^{0,1}(X) \times T^{1,0}(X) \longrightarrow \Lambda^2(T^*X).$$

Now if $\mathbb{C}_X$ is the trivial vector bundle (with the fiber $\mathbb{C}$) over X then $(\mathbb{C}_X)^* = \Lambda^2(T^*X)$. Using the Stokes formula it is easy to check that

$$A^* = \bar\partial : \Lambda^{1,0}(X) \longrightarrow \Lambda^{1,1}(X) = \Lambda^2(X)$$

for the given dualities.

It is well known [3] that $\operatorname{ind} A = 1 - g$, where g is the genus of X (actually $\dim \operatorname{Ker} A = 1$, since $\operatorname{Ker} A = \{\text{const}\}$ and $\dim \operatorname{Ker} A^* = g$, since

$\operatorname{Ker} A^*$ is the space of all holomorphic forms on X). Now $L(\mu, A)$ and $L(\mu^{-1}, A^*)$ become spaces of meromorphic functions and forms respectively with the restriction on orders of poles and zeros exactly as in the classical Riemann-Roch theorem.

Now note that in this case $q = 1$ and (2.7) is true. Hence (2.10) leads in this example to the classical Riemann-Roch theorem:

$$r(\mu) = 1 - g + d(\mu) + r'(\mu^{-1})$$

where

$$r(\mu) = r(\mu, \bar\partial), \quad \bar\partial: C^\infty(X) \to \Lambda^{0,1}(X),$$
$$r'(\mu^{-1}) = r(\mu^{-1}, \bar\partial^*), \quad d(\mu) = \sum p_i.$$

EXAMPLE 2.2 (Nadirashvili [1]). Let X be a compact Riemannian manifold and Δ_g the scalar Laplace-Beltrami operator, $\Delta_g: C^\infty(X) \to C^\infty(X)$. Using the standard Riemannian density we can identify $\Omega(X)$ with $\mathbb{C}_X$ and Δ_g^* with Δ_g. Evidently, $\operatorname{ind} \Delta_g = 0$, so (2.10) here has the form

$$r(\mu, \Delta_g) = d(\mu) + r(\mu^{-1}, \Delta_g), \tag{2.11}$$

and this was proved in [1]. Actually Nadirashvili considers some noncompact Riemannian manifolds with boundary introducing elliptic boundary conditions and conditions at infinity satisfying appropriate conditions (to make the corresponding closure of Δ_g a Fredholm selfadjoint operator). As a corollary of (2.11) he proves that the Coulomb potential of m point charges cannot have a zero of order $\geqslant m$ (unless it is identically zero). We shall make further comments about the noncompact case later, and will also consider estimates of multiplicities of zeros in Section 4.

B. Now we shall introduce new spaces that play an important role in the proof of Theorem 2.1 but on the other hand allow us to formulate a duality theorem which is important by itself.

First we introduce the space $\Gamma(X, \mu, A)$ of sections $u \in \Gamma(X - \operatorname{supp} \lambda, E)$ having prescribed poles and zeros but not necessarily solutions of $Au = 0$:

$$\begin{aligned}\Gamma(X, \mu, A) = \{u \mid u \in \Gamma(X - \operatorname{supp}\lambda, E);\ & j_{x_i}^{|p_i|-1} u = 0 \text{ if} \\ & x_i \in \operatorname{supp} \nu;\ \text{for every } x_i \in \operatorname{supp}\lambda \text{ there exist a} \\ & \text{neighbourhood } U \text{ of } x_i \text{ and a representation} \\ & u = u_s + u_r, \text{ where } u_s \in \Gamma(U - \{x_i\}, E),\ Au_s = \\ & 0 \text{ in } U - \{x_i\},\ \operatorname{ord}_{x_i} u_s \geqslant -p_i, \text{ and } u_r \\ & \text{can be extended to a section } \bar u_r \in \Gamma(U, E)\}.\end{aligned} \tag{2.12}$$

So actually we allow only singularities that can occur as singularities of solutions of $Au = 0$. The space $\Gamma(X, \mu, A)$ includes sections with the same zeros and singularities as allowed in the definition of $L(\mu, A)$, but on the other hand (2.12) contains no global restrictions, so all possible local singularities and zeros can be present at every point $x_i \in \operatorname{supp}\mu$ independently from what happens at other points.

Now introduce the reduced divisor

$$\widetilde{\mu} = x_1^{\widetilde{p}_1} x_2^{\widetilde{p}_2} \cdots x_m^{\widetilde{p}_m} \tag{2.13}$$

where $\widetilde{p}_i = \operatorname{sign} p_i(|p_i| - d)^+$ and the factors $x_i^{\widetilde{p}_i}$ with $\widetilde{p}_i = 0$ have to be omitted. So compared with μ, the absolute value of every exponent decreases by d (or becomes 0 if it initially was less than d). Note that $\widetilde{\mu^{-1}} = (\widetilde{\mu})^{-1}$.

For every divisor μ of the form (2.4) and every vector bundle E over X we define the space

$$\Gamma_\mu(X, E) = \{u \mid u \in \Gamma(X, E),\ j_{x_i}^{|p_i|-1} u = 0 \text{ if } p_i < 0\}. \tag{2.14}$$

So $\Gamma_\mu(X, E) \subset \Gamma(X, E)$ and $\Gamma_\mu(X, E)$ depends actually on ν only, hence $\Gamma_\mu(X, E) = \Gamma_\nu(X, E)$.

Now define a linear operator

$$\begin{aligned} \widetilde{A}\colon \Gamma(X, \mu, A) &\longrightarrow \Gamma_{\widetilde{\mu}}(X, F), \\ \widetilde{A}u &= \operatorname{ext}(Au), \end{aligned} \tag{2.15}$$

where $\operatorname{ext}(Au)$ is the extension by continuity of the section Au (defined a priori over $X - \operatorname{supp}\lambda$, but extendable to a C^∞-section on X due to the definition of $\Gamma(X, \mu, A)$ in (2.12)).

Now we introduce an important duality of the spaces introduced before.

DEFINITION 2.4. Bilinear or sesquilinear pairings

$$\begin{aligned} &\Gamma_{\widetilde{\mu}}(X, F) \times \Gamma(X, \mu^{-1}, A^*) \longrightarrow \mathbb{C}, \\ &\Gamma(X, \mu, A) \times \Gamma_{\widetilde{\mu}^{-1}}(X, \widetilde{E}) \longrightarrow \mathbb{C} \end{aligned} \tag{2.16}$$

are defined by integration over $X - \operatorname{supp}\mu$ and denoted by $\langle \cdot , \cdot \rangle$.

Note that the integrals that appear here actually converge. Indeed let $f \in \Gamma_{\widetilde{\mu}}(X, F)$, $v \in \Gamma(X, \mu^{-1}, A^*)$. Then near a point x_i with $p_i < 0$ we have $f(x) = O(|x - x_i|^{|p_i|-d})$, $v(x) = O(|x - x_i|^{d-n-|p_i|+\varepsilon})$ with $\varepsilon > 0$ (actually the last estimate is satisfied for any $\varepsilon < 1$), hence

$$\langle f(x), v(x) \rangle_x = O(|x - x_i|^{-n+\varepsilon})$$

and the integral

$$\langle f, v \rangle = \int_{X - \operatorname{supp}\mu} \langle f(x), v(x) \rangle$$

converges near all points x_i with $p_i < 0$ which are the only possible singularities.

Now let $\langle \cdot , \cdot \rangle \colon \mathscr{H} \times \mathscr{H}' \to \mathbb{C}$ be a bilinear or sesquilinear pairing between two complex linear spaces $\mathscr{H}$ and $\mathscr{H}'$. For any linear subspace $L \subset \mathscr{H}'$

we define its *annihilator* or *orthogonal complement* L° with respect to the pairing $\langle\cdot\,,\,\cdot\rangle$ as follows:

$$L^\circ = \{f \mid f \in \mathcal{H}\,,\ \langle f\,,\,v\rangle = 0 \text{ for every } v \in L\}.$$

Hence L° is a linear subspace in $\mathcal{H}$. Similarly, if L is a linear subspace in $\mathcal{H}$ then L° is defined and is a linear subspace in $\mathcal{H}'$.

In the following theorem L° will mean the annihilator with respect to the pairings (2.16). So if $L \subset \Gamma(X\,,\,\mu^{-1}\,,\,A^*)$ then $L^\circ \subset \Gamma_{\widetilde{\mu}}(X\,,\,F)$, etc.

THEOREM 2.2. (i) *$\langle \widetilde{A}u\,,\,v\rangle = \langle u\,,\,\widetilde{A}^*v\rangle$, where $u \in \Gamma(X\,,\,\mu\,,\,A)$, $v \in \Gamma(X\,,\,\mu^{-1}\,,\,A^*)$.*

(ii) *$\operatorname{Im}\widetilde{A} = (\operatorname{Ker}\widetilde{A}^*)^\circ$, i.e., $f \in \operatorname{Im}\widetilde{A}$ iff $f \in \Gamma_{\widetilde{\mu}}(X\,,\,F)$ and $\langle f\,,\,v\rangle = 0$ for all $v \in \operatorname{Ker}\widetilde{A^*}$.*

(iii) *$\dim\operatorname{Coker}\widetilde{A} = \dim\operatorname{Ker}\widetilde{A^*}$.*

Note that $\operatorname{Ker}\widetilde{A} = L(\mu\,,\,A)$, $\operatorname{Ker}\widetilde{A^*} = L(\mu^{-1}\,,\,A^*)$. Theorem 2.2 (ii) gives conditions of solvability of the equation $Au = f$ in the class $\Gamma(X\,,\,\mu\,,\,A)$ that consists of sections with prescribed orders of zeros and poles.

Theorem 2.2 will be proved simultaneously with Theorem 2.1. We shall apply it in Section 4 to prove a local solvability result when we look for a solution with a prescribed order of a zero at a given point.

C. Now let us describe generalizations of Theorems 2.1 and 2.2 to a special case of noncompact manifolds with boundary. This case will cover important situations when the operator A on an appropriate domain will still be a Fredholm operator in the usual sense. (Note that considering the case of elliptic operators on covering manifolds leads to operators which are Fredholm in the sense of Breuer in appropriate von Neumann algebras, but we will treat this case in a subsequent paper.)

So let X be a noncompact manifold with boundary ∂X (which need not be compact either).

Let $E\,,\,F$ be complex vector bundles over the open manifold of all interior points of X, which we denote $\operatorname{Int}X$. Let

$$A\colon \Gamma(\operatorname{Int}X\,,\,E) \to \Gamma(\operatorname{Int}X\,,\,F) \tag{2.17}$$

be an elliptic differential operator. As before, denote by $q = \dim_{\mathbb{C}} E_x = \dim_{\mathbb{C}} F_x$ the dimensions of fibres of E and F. Let $\widetilde{E}\,,\,\widetilde{F}$ be another pair of vector bundles over $\operatorname{Int}X$ given together with nondegenerate bilinear or sesquilinear C^∞ pairings of bundles

$$E \times \widetilde{E} \to \Omega(\operatorname{Int}X)\,, \quad F \times \widetilde{F} \to \Omega(\operatorname{Int}X).$$

For any vector bundle E over $\operatorname{Int}X$ denote by $\Gamma_c(\operatorname{Int}X)$ the space of all C^∞-sections of E over $\operatorname{Int}X$ with compact supports lying inside $\operatorname{Int}X$. Hence if $u \in \Gamma_c(\operatorname{Int}X)$ then $\operatorname{supp}u$ is a compact subset of $\operatorname{Int}X$; in particular, $\operatorname{supp}u$ does not intersect ∂X. In this case the adjoint elliptic operator

$A^*: \Gamma(\operatorname{Int} X, \widetilde{F}) \to \Gamma(\operatorname{Int} X, \widetilde{E})$ is again defined by the identity

$$\langle Au, \widetilde{v}\rangle_E = \langle u, A^*\widetilde{v}\rangle_F, \qquad u \in \Gamma_c(\operatorname{Int} X, E), \ \widetilde{v} \in \Gamma_c(\operatorname{Int} X, \widetilde{F}),$$

where the dualities $\langle\cdot, \cdot\rangle_E$, $\langle\cdot, \cdot\rangle_F$ are defined exactly as for the compact case.

Now suppose that the domains of A and A^* are distinguished as linear subspaces $\operatorname{Dom} A$ and $\operatorname{Dom} A^*$ such that

$$\Gamma_c(\operatorname{Int} X, E) \subset \operatorname{Dom} A \subset \Gamma(\operatorname{Int} X, E),$$
$$\Gamma_c(\operatorname{Int} X, \widetilde{F}) \subset \operatorname{Dom} A^* \subset \Gamma(\operatorname{Int} X, \widetilde{F}).$$

They may be defined, for example, by a choice of boundary conditions and appropriate conditions at infinity. Then let us define images of A, A^* as

$$\operatorname{Im} A = A(\operatorname{Dom} A), \qquad \operatorname{Im} A^* = A^*(\operatorname{Dom} A^*).$$

Suppose also that linear subspaces $\operatorname{Dom}' A$ and $\operatorname{Dom}' A^*$ are given such that

$$\Gamma_c(\operatorname{Int} X, \widetilde{E}) \subset \operatorname{Dom}' A \subset \Gamma(\operatorname{Int} X, \widetilde{E}), \quad \operatorname{Im} A^* \subset \operatorname{Dom}' A,$$
$$\Gamma_c(\operatorname{Int} X, F) \subset \operatorname{Dom}' A^* \subset \Gamma(\operatorname{Int} X, F), \quad \operatorname{Im} A \subset \operatorname{Dom}' A^*.$$

We shall suppose that the following integrability condition is satisfied:

$$\begin{gathered} x \mapsto \langle v, \widetilde{v}\rangle_x \text{ and } x \mapsto \langle u, \widetilde{u}\rangle_x \text{ are } L^1\text{-densities on } \operatorname{Int} X \\ \text{for any } v \in \operatorname{Dom}' A^*, \ \widetilde{v} \in \operatorname{Dom} A^*, \ u \in \operatorname{Dom} A, \ \widetilde{u} \in \operatorname{Dom}' A^* \end{gathered} \tag{2.18}$$

(i.e., these densities are absolutely integrable in Lebesgue sense over $\operatorname{Int} X$).

Integrating these densities over $\operatorname{Int} X$ we obtain bilinear or sesquilinear pairings

$$\operatorname{Dom}' A^* \times \operatorname{Dom} A^* \longrightarrow \mathbb{C}, \qquad \operatorname{Dom} A \times \operatorname{Dom}' A \longrightarrow \mathbb{C}$$

which will be denoted $\langle\cdot, \cdot\rangle_F$ and $\langle\cdot, \cdot\rangle_E$, the same way as for sections with compact support.

Our next requirement is

$$\langle Au, \widetilde{v}\rangle_F = \langle u, A^*\widetilde{v}\rangle_E, \qquad u \in \operatorname{Dom} A, \ \widetilde{v} \in \operatorname{Dom} A^*. \tag{2.19}$$

Now if we are given a linear subspace $L \subset \operatorname{Dom} A$, then its annihilator L° is naturally defined as a linear subspace in $\operatorname{Dom}' A$:

$$L^\circ = \{\widetilde{u} \mid \widetilde{u} \in \operatorname{Dom}' A, \langle u, \widetilde{u}\rangle_E = 0 \text{ for every } u \in L\}.$$

Similarly for a linear subspace $M \subset \operatorname{Dom} A^*$ its annihilator $M^\circ \subset \operatorname{Dom}' A^*$ is naturally defined.

Define also $\operatorname{Ker} A$ and $\operatorname{Ker} A^*$ as linear subspaces in $\operatorname{Dom} A$ and $\operatorname{Dom} A^*$ respectively, e.g.,

$$\operatorname{Ker} A = \{u \mid u \in \operatorname{Dom} A, Au = 0\}.$$

Now our next requirement is

$$\begin{array}{l} A \text{ and } A^* \text{ are Fredholm in the following sense:} \\ \text{(i) } \dim\operatorname{Ker} A < \infty, \quad \dim\operatorname{Ker} A^* < \infty; \\ \text{(ii) } \operatorname{Im} A = (\operatorname{Ker} A^*)^\circ, \quad \operatorname{Im} A^* = (\operatorname{Ker} A)^\circ. \end{array} \tag{2.20}$$

Hence $\operatorname{ind} A = \dim\operatorname{Ker} A - \dim\operatorname{Ker} A^*$ is well defined.

Let us introduce the (point) divisors on X by (2.4) with $x_i \in \operatorname{Int} X$, $i = 1, \dots, m$. Definition 2.1 for the degree is obviously applicable in this case as well as local Definition 2.2.

We write that $u \in \operatorname{Dom} A$ outside a compact subset $K \subset \operatorname{Int} X$ if $u \in \Gamma(\operatorname{Int} X - K, E)$ and there exists $\bar{u} \in \operatorname{Dom} A$ such that $\bar{u} = u$ on $X - K$. Now we can define

$$\begin{aligned} L(\mu, A) = \{u \mid\ & u \in \operatorname{Dom} A \text{ outside a neighbourhood of } \operatorname{supp}\mu, \\ & Au = 0 \text{ on } X - \operatorname{supp}\mu,\ \operatorname{ord}_{x_i} u \geqslant -p_i \text{ for each } i = 1, \dots, m\}, \\ r(\mu, A) &= \dim_{\mathbb{C}} L(\mu, A). \end{aligned}$$

$L(\mu, A^*)$ is defined similarly and $r(\mu, A^*) = \dim_{\mathbb{C}} L(\mu, A^*)$.

Now we can formulate a generalization of Theorem 2.1 to the noncompact case.

Theorem 2.3. *Let A be an elliptic differential operator* (2.17) *and let* (2.18)–(2.20) *be satisfied. Then for any divisor μ (with $\operatorname{supp}\mu \subset \operatorname{Int} X$)*

$$r(\mu, A) = \operatorname{ind} A + qd(\mu) + r(\mu^{-1}, A^*).$$

Let us give two examples when this theorem may be applied.

Example 2.3. Let X be a compact Riemannian manifold with a smooth boundary, $A = A^* = \Delta$ is the Laplacian of the given Riemannian metric. Then taking $E = F = \widetilde{E} = \widetilde{F} = \mathbb{C}_X$ (the trivial vector bundle with the fibre $\mathbb{C}$ over X) and defining the duality by the use of the Riemannian volume, we can take

$$\operatorname{Dom} A = \operatorname{Dom}' A = \operatorname{Dom} A^* = \operatorname{Dom}' A^* = \{u \mid u \in C^\infty(X),\ u|_{\partial X} = 0\},$$

i.e., define A as the Laplacian with the Dirichlet boundary condition. Then conditions (2.18)–(2.20) are satisfied and so Theorem 2.3 is applicable in this case with $\operatorname{ind} A = 0$, $q = 1$.

Similarly we can consider the Neuman condition.

Example 2.4. Let $X = \mathbb{R}^n$, $n \geqslant 3$, $A = A^* = \Delta$ (the standard Laplacian or the Laplacian of the flat metric), $E = F = \widetilde{E} = \widetilde{F} = \mathbb{C}_X$, and

$$\begin{aligned} \operatorname{Dom} A &= \operatorname{Dom} A^* = \operatorname{Dom}' A = \operatorname{Dom}' A^* \\ &= \{u \mid u \in C^\infty(\mathbb{R}^n),\ \Delta u \in C_c^\infty(\mathbb{R}^n), \text{ and } u(x) \to 0 \text{ as } |x| \to \infty\}. \end{aligned}$$

Here $C_c^\infty(\mathbb{R}^n) = C^\infty(\mathbb{R}^n) \cap \mathscr{E}'(\mathbb{R}^n)$ is the set of all C^∞-functions with compact support.

Note that the condition $u(x) \to \infty$ as $|x| \to \infty$ can be replaced by a formally stronger but in fact equivalent condition

$$u(x) = O(|x|^{2-n}) \quad \text{as } |x| \to \infty \tag{2.21}$$

because $u \in \operatorname{Dom} A$ is a harmonic function near infinity. It easily follows that all conditions (2.18)–(2.20) are satisfied because $\operatorname{Ker} A = \operatorname{Ker} A^* = \{0\}$ by the Liouville theorem and $\operatorname{Im} A = \operatorname{Im} A^* = C_c^\infty(\mathbb{R}^n)$ since the equation $\Delta u = f$ with any $f \in C_c^\infty(\mathbb{R}^n)$ can be solved by taking the convolution of f with the standard fundamental solution $c_n|x|^{2-n}$. Hence Theorem 2.3 is applicable to this situation.

We can also replace $\mathbb{R}^n$ by a closed subset $X \subset \mathbb{R}^n$ such that $\mathbb{R}^n - X$ is a bounded open subset with a C^∞-boundary, imposing the Dirichlet or Neuman boundary condition on ∂X.

D. Now let us formulate a generalization of the duality Theorem 2.2 to the noncompact case. Denote

$$\begin{aligned}\Gamma(X, \mu, A) = \{u \mid\ & u \in \Gamma(\operatorname{Int} X - \operatorname{supp} \mu, E),\ u \in \operatorname{Dom} A \text{ outside} \\ & \text{a neighbourhood of } \operatorname{supp} \lambda;\ j_{x_i}^{|p_i|-1} u = 0 \text{ if} \\ & x_i \in \operatorname{supp} \nu; \text{ for every } x_i \in \operatorname{supp} \lambda \text{ there exists} \\ & \text{a neighbourhood } U \text{ of } x_i \text{ and a representation} \\ & u = u_s + u_r, \text{ where } u_s \in \Gamma(U - \{x_i\}, E),\ Au_s = \\ & 0 \text{ in } U - \{x_i\},\ \operatorname{ord}_{x_i} u_s \geqslant -p_i, \text{ and } u_r \text{ can be} \\ & \text{extended to a section } \bar{u}_r \in \Gamma(U, E)\};\end{aligned} \tag{2.22}$$

$$\Gamma_\mu(X, A^*) = \{u \mid u \in \operatorname{Dom}' A^*,\ j_{x_i}^{|p_i|-1} u = 0 \text{ if } p_i < 0\}.$$

Similar definitions apply if we replace A by A^*.

The reduced divisor $\tilde{\mu}$ is defined as before and we have bilinear or sesquilinear pairings

$$\Gamma_{\tilde{\mu}}(X, A^*) \times \Gamma(X, \mu^{-1}, A^*) \to \mathbb{C}, \quad \Gamma(X, \mu, A) \times \Gamma_{\widetilde{\mu^{-1}}}(X, A) \to \mathbb{C} \tag{2.23}$$

as in Definition 2.4. Also the linear operator

$$\tilde{A}: \Gamma(X, \mu, A) \to \Gamma_{\tilde{\mu}}(X, A^*) \tag{2.24}$$

is defined as in (2.15).

In the next theorem L° means the annihilator with respect to the pairings (2.23).

THEOREM 2.4. *We have*

(i) $\langle \tilde{A}u, v \rangle = \langle u, \widetilde{A^*}v \rangle$; $u \in \Gamma(X, \mu, A)$, $v \in \Gamma(X, \mu^{-1}, A^*)$.

(ii) $\operatorname{Im} \tilde{A} = (\operatorname{Ker} \widetilde{A^*})^\circ$, *i.e.*, $f \in \operatorname{Im} \tilde{A}$ *iff* $f \in \Gamma_{\tilde{\mu}}(X, A^*)$ *and* $\langle f, v \rangle = 0$ *for all* $v \in \operatorname{Ker} \widetilde{A^*}$.

(iii) $\dim \operatorname{Coker} \tilde{A} = \dim \operatorname{Ker} \widetilde{A^*}$.

§3. Proofs of the main results

For simplicity we first prove Theorems 2.1 and 2.2. Then we shall indicate (minor) modifications that should be made to prove Theorems 2.3 and 2.4.

Theorems 2.1 and 2.2 will be proved simultaneously because these proofs intertwine.

Obviously we can assume without loss of generality that $A^* = A^t$, hence $\widetilde{E} = E^*$ and $\widetilde{F} = F^*$. So from now on the simplest canonical (bilinear) duality is used.

First we state and prove some preparatory results.

Denote by $\mathscr{D}'(X, F)$ the space of distribution sections of F (dual space to $\Gamma(X, F^*)$) and let $S(\mu, F)$ be the space of $s \in \mathscr{D}'(X, F)$ such that $\operatorname{supp} s \subset \operatorname{supp} \lambda$ and locally near the points $x_i \in \operatorname{supp} \lambda$ (i.e., the points entering in μ with positive exponents) s can be written as

$$s = \sum_{\{i|p_i>0\}} \sum_{|\alpha| \leqslant p_i - 1} c_{i\alpha} \partial_x^\alpha \delta(x - x_i) \tag{3.1}$$

where δ is the Dirac measure, $c_{i\alpha} \in F_{x_i}$. So actually $S(\mu, F) = S(\lambda, F)$. Similar spaces will be used for the bundle E and other divisors.

For every $u \in \Gamma(X, \mu, A)$ we can find a "regularization" $\widetilde{u} \in \mathscr{D}'(X, E)$ such that $\widetilde{u} = u$ on $X - \operatorname{supp} \lambda$ and $A\widetilde{u} = f + s$ with $f \in \Gamma(X, F)$ and $s \in S(\mu, F)$. Denote by $\widetilde{\Gamma}(X, \mu, A)$ the space of all such regularizations. Due to the standard elliptic regularity result and the structure of fundamental solutions described in Section 1, the space $\widetilde{\Gamma}(X, \mu, A)$ can be described as a set of $\widetilde{u} \in \mathscr{D}'(X, E)$ such that $\widetilde{u} \in C^\infty$ in a neighbourhood of $\operatorname{supp} \nu$, $j_{x_i}^{|p_i|-1} u = 0$ for every $x_i \in \operatorname{supp} \nu$, and $A\widetilde{u} = f + s$ with $f \in \Gamma(X, F)$, $s \in S(\mu, F)$. Now we need

LEMMA 3.1. *The sequence*

$$0 \longrightarrow S(\widetilde{\mu}, E) \xrightarrow{i_1} \widetilde{\Gamma}(X, \mu, A) \xrightarrow{r} \Gamma(X, \mu, A) \longrightarrow 0 \tag{3.2}$$

is exact. Here i_1 and r are natural inclusion and restriction maps.

PROOF. The surjectivity of r means the existence of a regularization as mentioned before, the injectivity of i_1 is evident. So we must only prove the exactness in the middle term which actually means that if $s \in \mathscr{D}'(X, E)$, $\operatorname{supp} s \subset \operatorname{supp} \lambda$, and $As \in S(\mu, F)$ then $s \in S(\widetilde{\mu}, E)$. This is a local question and it is sufficient to consider the case $\mu = x_1^{p_1}$ with $p_1 < 0$. But then the statement easily follows from ellipticity of A. □

LEMMA 3.2. *We have*

$$\dim S(\mu, E) = \dim(\Gamma(X, E)/\Gamma_{\mu^{-1}}(X, E)) = q \sum_{p_i>0} \binom{n + p_i - 1}{n}. \tag{3.3}$$

PROOF. Evidently,

$$S(\mu, E) = \bigoplus_{p_i>0} S(x_i^{p_i}, E), \qquad \Gamma(X, E)/\Gamma_{\mu^{-1}}(X, E) = \bigoplus_{p_i>0} J_{x_i}^{p_i-1}(E),$$

where $J_{x_i}^{p_i-1}(E)$ is the space of jets of the order $p_i - 1$ of sections of E at x_i. So it is sufficient to prove that for any $p_i > 0$

$$\dim S(x_i^{p_i}, E) = \dim J_{x_i}^{p_i-1}(E) = q\binom{n+p_i-1}{n}$$

which reduces to a well-known combinatorial exercise. □

REMARK. Note that the spaces $S(\mu, E)$ and $\Gamma(X, E^*)/\Gamma_{\mu^{-1}}(X, E^*)$ are dual to each other with respect to the natural duality induced by the duality between $\mathscr{D}'(X, E)$ and $\Gamma(X, E^*)$.

PROPOSITION 3.3. *We have*

$$\operatorname{ind}\widetilde{A} = \operatorname{ind} A + qd(\mu). \tag{3.4}$$

PROOF. Consider the following commutative diagram

$$\begin{array}{ccccccccc} 0 & \longrightarrow & S(\widetilde{\mu}, E) & \xrightarrow{i_1} & \widetilde{\Gamma}(X, \mu, A) & \xrightarrow{r} & \Gamma(X, \mu, A) & \longrightarrow & 0 \\ & & \downarrow{\scriptstyle A_S} & & \downarrow{\scriptstyle \hat{A}} & & \downarrow{\scriptstyle \widetilde{A}} & & \\ 0 & \longrightarrow & S(\mu, E) & \xrightarrow{i_2} & \Gamma_{\widetilde{\mu}}(X, F) \oplus S(\mu, F) & \xrightarrow{\pi_1} & \Gamma_{\widetilde{\mu}}(X, F) & \longrightarrow & 0 \end{array}$$

where the first row is as in Lemma 3.1, i_2 and π_1 are natural inclusion and projection respectively, and A_S, $\hat{A}$ are restrictions of A to the corresponding spaces of distributions. Both rows in the diagram are exact. Due to the well-known algebraic property of the Euler characteristic we have

$$\operatorname{ind}\widetilde{A} = \operatorname{ind}\hat{A} - \operatorname{ind} A_S.$$

But

$$\begin{aligned} \operatorname{ind} A_S &= \dim S(\widetilde{\mu}, E) - \dim S(\mu, F) \\ &= -q\sum_{p_i>0}\left[\binom{n+p_i-1}{n} - \binom{n+p_i-d-1}{n}\right] = -qd(\lambda). \end{aligned}$$

Hence

$$\operatorname{ind}\widetilde{A} = \operatorname{ind}\hat{A} + qd(\lambda). \tag{3.5}$$

Now consider the commutative diagram

$$\begin{array}{ccccccccc} 0 & \longrightarrow & \Gamma(\widetilde{\mu}, E) & \longrightarrow & \widetilde{\Gamma}(X, \mu, A) & \xrightarrow{\pi_2 \circ \hat{A}} & S(\mu F) & \longrightarrow & 0 \\ & & \downarrow{\scriptstyle A_\mu} & & \downarrow{\scriptstyle \hat{A}} & & \downarrow{\scriptstyle \mathrm{Id}} & & \\ 0 & \longrightarrow & \Gamma_{\widetilde{\mu}}(X, F) & \xrightarrow{i_2} & \Gamma_{\widetilde{\mu}}(X, F) \oplus S(\mu, F) & \xrightarrow{\pi_1} & S(\mu, F) & \longrightarrow & 0 \end{array}$$

where the rows again are exact, A_μ is the restriction of A. From this diagram we find

$$\operatorname{ind} \hat{A} = \operatorname{ind} A_\mu. \tag{3.6}$$

Finally, consider the commutative diagram

$$\begin{array}{ccccccccc} 0 & \longrightarrow & \Gamma_\mu(X, E) & \xrightarrow{i_\mu} & \Gamma(X, E) & \xrightarrow{p_\mu} & J_\mu(E) & \longrightarrow & 0 \\ & & \downarrow{\scriptstyle A_\mu} & & \downarrow{\scriptstyle A} & & \downarrow{\scriptstyle J_\mu(A)} & & \\ 0 & \longrightarrow & \Gamma_{\tilde{\mu}}(X, F) & \xrightarrow{i_{\tilde{\mu}}} & \Gamma(X, F) & \xrightarrow{p_{\tilde{\mu}}} & J_{\tilde{\mu}}(E) & \longrightarrow & 0 \end{array}$$

where $J_\mu(E) = \Gamma(X, E)/\Gamma_\mu(X, E) = \bigoplus_{p_i<0} J_{x_i}^{|p_i|-1}(E)$, i_μ and p_μ are natural inclusion and restriction maps, $J_\mu(A)$ is the natural quotient map. Then we find

$$\operatorname{ind} A_\mu = \operatorname{ind} A - \operatorname{ind} J_\mu(A).$$

But

$$\begin{aligned} \operatorname{ind} J_\mu(A) &= \dim J_\mu(E) - \dim J_{\tilde{\mu}}(F) \\ &= q \sum_{p_i<0} \left[\binom{n+|p_i|-1}{n} - \binom{n+|p_i|-d-1}{n} \right] = -qd(\nu), \end{aligned}$$

hence

$$\operatorname{ind} A_\mu = \operatorname{ind} A + qd(\nu)$$

and now using (3.5) and (3.6) we obtain

$$\operatorname{ind} \widetilde{A} = \operatorname{ind} A + qd(\nu) + qd(\lambda) = \operatorname{ind} A + qd(\mu). \quad \square$$

REMARK. Proposition 3.3 means that

$$\dim \operatorname{Ker} \widetilde{A} = \operatorname{ind} A + qd(\mu) + \dim \operatorname{Coker} \widetilde{A}, \tag{3.7}$$

so to prove Theorem 2.1 it suffices to prove the equality (iii) in Theorem 2.2. We start with the proof of (i) in Theorem 2.2.

LEMMA 3.4. *We have*

$$\langle \widetilde{A}u, v \rangle = \langle u, \widetilde{A^t}v \rangle, \qquad u \in \Gamma(X, \mu, A), \ v \in \Gamma(X, \mu^{-1}, A^t). \tag{3.8}$$

PROOF. Let us take a function $\chi \in C_0^\infty(\mathbb{R}^n)$, $\chi(x) = 1$ if $|x| < 1/2$, $\chi(x) = 0$ if $|x| > 1$. For every $\varepsilon > 0$ define $\chi_\varepsilon \in C_0^\infty(\mathbb{R}^n)$, $\chi_\varepsilon(x) = \chi(\varepsilon^{-1}x)$, so that $\operatorname{supp} \chi_\varepsilon \subset \{x \mid |x| \leqslant \varepsilon\}$, $\chi_\varepsilon(x) = 1$ if $|x| \leqslant \varepsilon/2$, and $|\partial^\alpha \chi_\varepsilon| \leqslant C_\alpha \varepsilon^{-|\alpha|}$.

For every point $x_i \in \operatorname{supp} \mu$ introduce local coordinates in a neighbourhood U_i of x_i and then using these local coordinates for small $\varepsilon > 0$ define

$$\psi_\varepsilon(x) = 1 - \sum_{1 \leqslant i \leqslant m} \chi_\varepsilon(x - x_i).$$

It follows that $\psi_\varepsilon(x) = 0$ in a neighbourhood of $\operatorname{supp}\mu$, $\psi_\varepsilon(x) = 1$ outside a small neighbourhood of $\operatorname{supp}\mu$, and $|\partial^\alpha \psi_\varepsilon| \leqslant C_\alpha \varepsilon^{-|\alpha|}$, where the derivative is taken in chosen local coordinates. Now using (2.3) and the convergence of the integrals defining both sides in (3.8) we obtain

$$\begin{aligned}\langle Au, v\rangle &= \lim_{\varepsilon\downarrow 0}\langle Au, \psi_\varepsilon v\rangle = \lim_{\varepsilon\downarrow 0}\langle u, A^t(\psi_\varepsilon v)\rangle \\ &= \lim_{\varepsilon\downarrow 0}[\langle u, \psi_\varepsilon A^t v\rangle + \langle u, [A^t, \psi_\varepsilon]v\rangle] \\ &= \langle u, A^t v\rangle + \lim_{\varepsilon\downarrow 0}\langle u, [A^t, \psi_\varepsilon]v\rangle.\end{aligned}$$

It remains to prove that the last limit vanishes. Denote $B_\varepsilon = [A^t, \psi_\varepsilon]$. Then B_ε is a differential operator of order $d-1$ with coefficients supported near $\operatorname{supp}\mu$ and near x_i:

$$B_\varepsilon = \sum_{|\alpha|\leqslant d-1} b_{\alpha,\varepsilon}(x) D^\alpha$$

with

$$\operatorname{supp} b_{\alpha,\varepsilon} \subset \{x \mid \varepsilon/2 \leqslant |x - x_i| \leqslant \varepsilon\}, \tag{3.9}$$

$$|b_{\alpha,\varepsilon}(x)| \leqslant C\varepsilon^{-d+|\alpha|}. \tag{3.10}$$

Now we must consider two cases: $p_i < 0$ and $p_i > 0$.

(a) Let $p_i < 0$. Then in U_i we have $u(x) = O(|x - x_i|^{|p_i|})$, $v(x) = o(|x - x_i|^{d-n-|p_i|})$, $D^\alpha v(x) = o(|x - x_i|^{d-n-|p_i|-|\alpha|})$. On $\operatorname{supp} b_{\alpha,\varepsilon}$ we have, due to (3.9):

$$u(x) = O(\varepsilon^{|p_i|}), \qquad D^\alpha v(x) = o(\varepsilon^{d-n-|p_i|-|\alpha|}).$$

Hence (3.10) gives

$$b_{\alpha,\varepsilon} D^\alpha v = o(\varepsilon^{-n-|p_i|}), \qquad \langle u(x), b_{\alpha,\varepsilon}(x) D^\alpha v(x)\rangle = o(\varepsilon^{-n}).$$

Now the volume of $\operatorname{supp} b_{\alpha,\varepsilon}$ is $O(\varepsilon^n)$, hence

$$\int_{U_i} \langle u(x), B_\varepsilon v(x)\rangle = o(1) \quad \text{as } \varepsilon \longrightarrow 0. \tag{3.11}$$

(b) Let $p_i > 0$. Then on $\operatorname{supp} b_{\alpha,\varepsilon}$ we have similarly $D^\alpha v(x) = O(\varepsilon^{p_i - |\alpha|})$,

$$u(x) = o(\varepsilon^{d-n-p_i}), \quad D^\alpha v(x) = O(\varepsilon^{p_i-|\alpha|}), \quad \langle u(x), b_{\alpha,\varepsilon}(x) D^\alpha v(x)\rangle = o(1),$$

so the conclusion (3.11) is again true, which proves the lemma. □

DEFINITION 3.5. Let $\mathscr{H}$, $\mathscr{H}'$ be two complex linear spaces and $\langle\cdot, \cdot\rangle : \mathscr{H} \times \mathscr{H}' \to \mathbb{C}$ is a bilinear or sesquilinear pairing. We say that this pairing is *nondegenerate* if

$$\{u \in \mathscr{H}, \langle u, v\rangle = 0 \text{ for all } v \in \mathscr{H}'\} \quad \text{implies} \quad u = 0$$

and

$$\{v \in \mathscr{H}', \langle u, v\rangle = 0 \text{ for all } u \in \mathscr{H}\} \quad \text{implies} \quad v = 0.$$

LEMMA 3.6. *The pairings* (2.16) *are nondegenerate.*

PROOF. The statement is evident since all spaces in (2.16) contain smooth sections of the corresponding bundles supported in $X - \operatorname{supp}\mu$ and on the other hand all elements of these spaces are uniquely defined by their (smooth) restrictions to $X - \operatorname{supp}\mu$. □

Now we need the following

LEMMA 3.7. *Let* $\langle\cdot,\cdot\rangle: \mathscr{H}\times\mathscr{H}' \to \mathbb{C}$ *be a nondegenerate bilinear or sesquilinear pairing between two complex linear spaces* $\mathscr{H}$, $\mathscr{H}'$. *Let* L *be a linear subspace in* $\mathscr{H}'$, L° *its annihilator in* $\mathscr{H}$, *and* $(L^\circ)^\circ$ *the annihilator of* L° *in* $\mathscr{H}'$. *Then*

$$L \subset (L^\circ)^\circ \tag{3.12}$$

and

$$\operatorname{codim} L \geqslant \dim L^\circ. \tag{3.13}$$

Furthermore, if F *is a linear subspace in* $\mathscr{H}'$ *then*

$$\operatorname{codim} F^\circ = \dim F. \tag{3.14}$$

PROOF. The inclusion (3.12) is obvious. It gives $\operatorname{codim} L \geqslant \operatorname{codim}(L^\circ)^\circ$. Hence (3.14) implies (3.13) and we have only to prove (3.14). Clearly $\operatorname{codim} F^\circ \leqslant \dim F$, so it remains to prove that

$$\operatorname{codim} F^\circ \geqslant \dim F.$$

It is sufficient to do it in the case when $\dim F < \infty$. Consider the natural map $j: F \to (\mathscr{H}/F^\circ)'$, where L' means the space of all complex linear maps from L to $\mathbb{C}$,

$$j(f)(x + F^\circ) = \langle f, x\rangle, \qquad x \in \mathscr{H}.$$

Then j is injective due to the nondegeneracy of the pairing. Hence

$$\operatorname{codim} F^\circ = \dim \mathscr{H}/F^\circ = \dim(\mathscr{H}/F^\circ)' \geqslant \dim F$$

as required. □

LEMMA 3.8. *In the first of the pairings* (2.16)

$$(\operatorname{Im}\widetilde{A})^\circ = \operatorname{Ker}\widetilde{A}^*.$$

PROOF. Clearly

$$\operatorname{Ker}\widetilde{A}^* = \{v \mid v \in \Gamma(X, \mu^{-1}, A^*),\ A^*v = 0 \text{ on } X - \operatorname{supp}\mu\}.$$

Lemma 3.8 follows because $\operatorname{Im}\widetilde{A}$ contains all sections $\widetilde{A}u$ if $u \in \Gamma(X, E)$ and $\operatorname{supp} u \subset X - \operatorname{supp}\mu$. □

PROOF OF THEOREMS 2.1 AND 2.2. Due to Lemmas 3.7 and 3.8 we have

$$\operatorname{Im}\widetilde{A} \subset (\operatorname{Ker}\widetilde{A}^t)^\circ, \tag{3.15}$$

$$\operatorname{codim}\operatorname{Im}\widetilde{A} \geqslant \dim\operatorname{Ker}\widetilde{A}^t, \tag{3.16}$$

and we must prove that both the inclusion and the inequality are actually equalities. Furthermore, (3.14) gives

$$\operatorname{codim}(\operatorname{Ker}\widetilde{A}^t)^\circ = \dim \operatorname{Ker}\widetilde{A}^t,$$

hence equality in (3.16) implies equality in (3.15). Since $\operatorname{codim}\operatorname{Im}\widetilde{A} = \dim\operatorname{Coker}\widetilde{A}$ we must only prove that

$$\dim\operatorname{Coker}\widetilde{A} = \dim\operatorname{Ker}\widetilde{A}^t, \tag{3.17}$$

which will immediately give the proofs of Theorems 2.1 and 2.2 due to (3.7). Clearly, (3.16) and (3.7) imply

$$\begin{aligned} \dim\operatorname{Ker}\widetilde{A} &= \operatorname{ind} A + qd(\mu) + \dim\operatorname{Coker}\widetilde{A} \\ &\geqslant \operatorname{ind} A + qd(\mu) + \dim\operatorname{Ker}\widetilde{A}^t. \end{aligned} \tag{3.18}$$

Now we can apply the same results to the divisor μ^{-1} (instead of μ) and the operator $\widetilde{A}^t$ (instead of $\widetilde{A}$). Then we obtain

$$\dim\operatorname{Ker}\widetilde{A}^t \geqslant \operatorname{ind} A^t + qd(\mu^{-1}) + \dim\operatorname{Ker}\widetilde{A} = -\operatorname{ind} A - qd(\mu) + \dim\operatorname{Ker}\widetilde{A}.$$

But this is the inequality opposite to (3.18). Hence in (3.18) and (3.17) we actually have equalities. Proofs of Theorems 2.1 and 2.2 are complete. □

REMARK. An easier proof of Theorem 2.1 can be given in the case when $\mu > 0$ or $\mu < 0$ (i.e., if $p_i > 0$ for all $i = 1, \dots, m$, or, vice versa, if $p_i < 0$ for all $i = 1, \dots, m$). Of course it is sufficient to consider the case $\mu > 0$. Denote then

$$\widetilde{L}(\mu, A) = \{u \mid u \in \mathscr{D}'(X, E),\ Au \in S(\mu, F)\}.$$

Then we have an exact sequence

$$0 \longrightarrow S(\widetilde{\mu}, E) \xrightarrow{\ i\ } \widetilde{L}(\mu, A) \xrightarrow{\ r\ } L(\mu, A) \longrightarrow 0,$$

where i and r are natural inclusion and restriction maps. Hence

$$\dim\widetilde{L}(\mu, A) = r(\mu, A) + \dim S(\widetilde{\mu}, E).$$

Now consider the following exact sequence

$$0 \to \operatorname{Ker} A \longrightarrow \widetilde{L}(\mu, A) \xrightarrow{\ A\ } S(\mu, F) \xrightarrow{\ \partial\ } (\operatorname{Ker} A^t)' \longrightarrow (\operatorname{Ker} A^t)' / \operatorname{Im}\partial \longrightarrow 0$$

where ∂ is the dualization map, $\partial(f)(v) = \langle f, v\rangle$, $v \in \operatorname{Ker} A^t$. Hence

$$\begin{aligned} 0 &= \dim\operatorname{Ker} A - \dim\widetilde{L}(\mu, A) + \dim S(\mu, F) - \dim\operatorname{Ker} A^t \\ &\quad + \dim\left[(\operatorname{Ker} A^t)'/\operatorname{Im}\partial\right] \\ &= \dim\operatorname{Ker} A - r(\mu, A) - \dim S(\widetilde{\mu}, E) \\ &\quad + \dim S(\mu, F) - \dim\operatorname{Ker} A^t + \dim\left[(\operatorname{Ker} A^t)'/\operatorname{Im}\partial\right] \\ &= \operatorname{ind} A + qd(\mu) - r(\mu, A) + \dim\left[(\operatorname{Ker} A^t)'/\operatorname{Im}\partial\right] \end{aligned}$$

and it remains to prove that

$$\dim[(\operatorname{Ker} A^t)'/\operatorname{Im}\partial] = r(\mu^{-1}, A^t).$$

But

$$\dim[(\operatorname{Ker} A^t)'/\operatorname{Im}\partial] = \dim(\operatorname{Im}\partial)^\circ,$$

where $(\operatorname{Im}\partial)^\circ$ is the annihilator of $\operatorname{Im}\partial$ in $\operatorname{Ker} A^t$. Clearly $(\operatorname{Im}\partial)^\circ = L(\mu^{-1}, A^t)$, which implies the required equality.

PROOF OF THEOREMS 2.3 AND 2.4. Only minor modifications are needed in the proofs of Theorems 2.1 and 2.2 to convert them into the proofs of more general Theorems 2.3 and 2.4.

In Section 2 we introduced the spaces $\Gamma(X, \mu, A)$ of smooth sections with singularities. In the noncompact case the definition takes into account domains (or boundary conditions and conditions at infinity) as well. In the compact case we used also the spaces $\Gamma_\mu(X, E)$. They should be replaced by the corresponding spaces $\Gamma_\mu(X, A)$. Also $\operatorname{Dom} A$ and $\operatorname{Dom} A^*$ should be used instead of $\Gamma(X, E)$ and $\Gamma(X, F^*)$ respectively. The space $\widetilde{\Gamma}(X, \mu, A)$ is defined as a space of distributions that are "regularizations" of functions from $\Gamma(X, \mu, A)$, so Lemmas 3.1 and 3.2 remain true. Then no change is needed in the proof of Proposition 3.3.

Using the duality, we must replace A^t with A^* and keep track of the dual domains $\operatorname{Dom}' A$ and $\operatorname{Dom}' A^*$. Lemma 3.4 is "by definition" true in our noncompact case for smooth sections from the corresponding domains (see (2.19)), and it is easily extended to sections with singularities, because further arguments are purely local and there is only a finite number of singularities. Then Lemmas 3.7 and 3.8 can be applied the same way to complete the proof. □

The easier proof given in the Remark after the end of the proofs of Theorems 2.1 and 2.2 for the case $\mu > 0$ or $\mu < 0$ can be easily extended as well.

§4. Applications

A. We begin with an obvious corollary of Theorems 2.1 or 2.3:

COROLLARY 4.1.

$$r(\mu, A) \geqslant \operatorname{ind} A + qd(\mu). \tag{4.1}$$

In particular, if $\operatorname{ind} A + qd(\mu) > 0$, then we have a nontrivial space $L(\mu, A)$ of solutions with poles allowed. So this space will be always nontrivial if we fix orders of zeros, but allow poles of sufficiently high order to make $d(\mu) > 0$ sufficiently large. For example if we fix any set of points $\{x_2, \dots, x_m\}$ and any negative integers $p_2, \dots, p_m$, but take $p_1 > 0$ sufficiently large, we obtain $r(\mu, A) > 0$, which means that there exists a nontrivial solution u of the equation $Au = 0$ on $X - \{x_1\}$ with a pole at x_1 and zeros at the points $x_2, \dots, x_m$ with multiplicities bounded from below by $|p_2|, \dots, |p_m|$. If A is a scalar operator (i.e., $q = 1$) this means that choosing a point x_1 and any finite set of points $x_2, \dots, x_m$ we can always include this set into the nodal set of a solution with a single pole at x_1. Note that the same is true for large q, where this result looks much stronger because for $q \geqslant n/2$ we can expect zeros to be generically isolated (or even absent).

REMARK. The inequality (4.1) can be proved without use of Theorem 2.1 (or 2.3). To do this let us use the space $\widetilde{L}(\lambda, A)$ that consists of all "regularizations" of functions from $L(\lambda, A)$, i.e.,

$$\widetilde{L}(\lambda, A) = \{u \mid u \in \mathscr{E}'(X, E) + \operatorname{Dom} A,\ Au \in S(\lambda, F)\}$$

(for the compact case it was used already in the Remark following the proof of Theorems 2.1 and 2.2).

Obviously, $\widetilde{L}(\lambda, A) \supset \operatorname{Ker} A$ and a solution $u \in \widetilde{L}(\lambda, A)$ of $Au = f \in S(\lambda, F)$ exists if and only if $f \in (\operatorname{Ker} A^*)^\circ$. Hence

$$\dim \widetilde{L}(\lambda, A) \geqslant \dim S(\lambda, F) + \dim \operatorname{Ker} A - \dim \operatorname{Ker} A^* = \dim S(\lambda, F) + \operatorname{ind} A.$$

But $L(\lambda, A) \cong \widetilde{L}(\lambda, A)/S(\widetilde{\lambda}, E)$, hence

$$\dim L(\lambda, A) \geqslant \dim S(\lambda, F) - \dim S(\widetilde{\lambda}, E) + \operatorname{ind} A = qd(\lambda) + \operatorname{ind} A.$$

Now

$$L(\mu, A) = \{u \mid u \in L(\lambda, A),\ j_{x_j}^{|p_j|-1} u = 0 \text{ if } x_j \in \operatorname{supp} \nu\}$$
$$= \{u \mid u \in L(\lambda, A) \cap (S(\nu, E^*))^\circ\}.$$

Note that $u \in [A^* S(\widetilde{\nu}, F^*)]^\circ$ for any $u \in L(\lambda, A)$, $A^* S(\widetilde{\nu}, F^*) \subset S(\nu, E^*)$, and $\dim A^* S(\widetilde{\nu}, F^*) = \dim S(\widetilde{\nu}, F^*)$. Hence

$$\dim L(\mu, A) \geqslant qd(\lambda) + \operatorname{ind} A - \dim S(\nu, E^*) + \dim S(\widetilde{\nu}, F^*)$$
$$= qd(\lambda) + qd(\nu) + \operatorname{ind} A = qd(\mu) + \operatorname{ind} A,$$

as required.

B. Now we indicate that sometimes even the equality in (4.1) can be claimed (if we have a sufficiently large number of poles, multiplicities counted).

DEFINITION 4.2. Let us say that a differential operator B on X has a *unique continuation property* if any local C^∞-solution u of the equation $Bu = 0$ having zero of infinite order at a point x_0, vanishes in a neighbourhood of x_0.

It is well known that the unique continuation property is not always true for elliptic operators. But it is true for elliptic operators B with real-analytic coefficients (because then all the solutions of $Bu = 0$ are analytic), for second-order elliptic operators that are either scalar or have a scalar principal symbol in case $n \geqslant 3$ (e.g., for the Laplacian on the differential forms on a Riemannian manifold), for elliptic operators with simple characteristics (see [4]).

PROPOSITION 4.3. *Suppose that the manifold X is connected, A^* has the unique continuation property, and the points $x_1, \dots, x_m$ are fixed. Then for every $N_0 > 0$ there exists $N > 0$ such that if $\sum_{p_i<0} |p_i| \leqslant N_0$ and $\sum_{p_i>0} p_i \geqslant N$ then*

$$r(\mu, A) = \operatorname{ind} A + q d(\mu). \tag{4.2}$$

PROOF. We have to prove that $r(\mu^{-1}, A^*) = 0$ if a positive integer N_0 is fixed and N is sufficiently large. Note first that there is only a finite number of possible choices of the points x_i with $p_i < 0$, so we can fix such a choice. After that consider the divisor $\nu_{\min} = \prod_{p_i<0} x_i^{-N_0}$. Then for any possible choice of numbers p_i with $\sum_{p_i<0} |p_i| \leqslant N_0$ we shall have

$$L(\mu^{-1}, A^*) \subset L(\nu_{\min}^{-1}, A^*).$$

Hence $L(\mu^{-1}, A^*)$ will always belong to a fixed finite-dimensional space of sections of $\widetilde{F}$ over $X - \operatorname{supp} \nu_{\min}$. We have to prove that for any point $x_0 \in X - \operatorname{supp} \nu_{\min}$ there exists an integer $N > 0$ such that if $v \in L(\nu_{\min}^{-1}, A^*)$ and $j_{x_0}^N v = 0$ then $v \equiv 0$. But if it is not true then using any Hilbert norm $\|\cdot\|$ in the finite-dimensional space $L(\nu_{\min}^{-1}, A^*)$ we can choose a sequence $v_k \in L(\nu_{\min}^{-1}, A^*)$ with $\|v_k\| = 1$ and $j_{x_0}^k v = 0$, $k = 1, 2, \dots$. Now using the compactness of the unit sphere in $L(\nu_{\min}^{-1}, A^*)$ we can even suppose that $v_k \to v$ in $L(\nu_{\min}^{-1}, A^*)$ and $\|v\| = 1$. But the convergence in $L(\nu_{\min}^{-1}, A^*)$ obviously implies the convergence in C^∞-topology of sections over $X - \operatorname{supp} \nu_{\min}$. Hence $j_{x_0}^k v = 0$ for all $k = 1, 2, \dots$, contradicting the unique continuation property. □

REMARK 1. Note that in the classical Riemann-Roch theorem (see Example 2.1 in Section 2) an effective result of this sort is possible:

$$r(\mu) = 1 - g + d(\mu) \quad \text{provided} \quad d(\mu) > 2g - 2.$$

In the general case the equality conditions cannot be given in terms of $d(\mu)$ alone. For example, let us take $A = \Delta$ (the scalar Laplacian) on a compact Riemannian manifold of dimension $n \geqslant 2$. Then we can find a nonconstant real-valued meromorphic solution v of $\Delta v = 0$, i.e., a nonzero element in $L(\mu_0^{-1}, \Delta)$ for a (point) divisor μ_0 (such solutions exist provided $d(\mu_0^{-1}, \Delta) \geqslant 2$ due to Corollary 4.1). By the Sard Lemma there exists $c \in \mathbb{R}$ such that $\{x \mid v(x) = c\}$ is a hypersurface, i.e., a submanifold of codimension 1 in X. Replacing v by $v - c$ we may suppose that the nodal set $X_0 = \{x \mid v(x) = 0\}$ is a hypersurface. Now let us take $\mu_N = \mu_0 x_1 \cdots x_N$ where $x_i \in X_0$, $x_i \notin \operatorname{supp} \mu_0$, $x_i \neq x_j$, $i, j = 1, \cdots, N$. Then obviously $d(\mu_N, \Delta) \to \infty$ as $N \to \infty$ but $r(\mu_N, \Delta) \neq d(\mu_N, \Delta)$ because $L(\mu_N^{-1}, \Delta) \ni v$, hence $r(\mu_N^{-1}, \Delta) \neq 0$.

REMARK 2. A condition of the unique continuation type is necessary for the statement of Proposition 4.3 to be true. Namely, suppose that there exists a solution $v \in \Gamma_c(X, F^*)$ of the equation $A^*v = 0$ such that $v \not\equiv 0$ and $\operatorname{supp} v \neq X$. Then for any divisor μ with $\operatorname{supp}\mu \subset X - \operatorname{supp} v$ we have

$$r(\mu, A) < \operatorname{ind} A + qd(\mu),$$

since $v \in L(\mu^{-1}, A^*)$, which implies that $r(\mu^{-1}, A^*) \neq 0$. Hence the conclusion of Proposition 4.3 is not true in this case.

C. Now we turn to a local application of Theorem 2.1. First introduce necessary notations. Let $\mathscr{E}_0$ be the linear space of germs at 0 of C^∞-functions $f\colon U \to \mathbb{C}$, where U is a neighbourhood of 0 in $\mathbb{R}^n$. So the elements of $\mathscr{E}_0$ are equivalence classes of such functions and $f_1 \sim f_2$ for two such functions $f_i\colon U_i \to \mathbb{C}$ iff there exists a neighbourhood U of 0 in $\mathbb{R}^n$ such that $U \subset U_1 \cap U_2$ and $f_1 \mid U = f_2 \mid U$. Now denote

$$\mathscr{E}_0^{(k)} = \{f \mid f \in \mathscr{E}_0,\ j_0^k f = 0\}.$$

Suppose that we have a $(q \times q)$-matrix elliptic differential operator A of order d defined in a neighbourhood of 0 in $\mathbb{R}^n$. Then it induces a linear map

$$A\colon \mathscr{E}_0^{(k+d)} \otimes \mathbb{C}^q \longrightarrow \mathscr{E}_0^{(k)} \otimes \mathbb{C}^q. \tag{4.3}$$

PROPOSITION 4.4. *The map* (4.3) *is surjective for each integer* $k \geqslant 0$.

PROOF. The idea is to use Theorem 2.2 (ii). Suppose that we have a C^∞-function $f\colon U \to \mathbb{C}^q$ such that $j_0^k f = 0$ (here U is a neighbourhood of 0 in $\mathbb{R}^n$). We want to find a smaller neighbourhood U_1 of 0 and a C^∞-function $u\colon U_1 \to \mathbb{C}^q$ such that $j_0^{(k+d)} u = 0$ and $Au = f \mid U_1$. To do this we extend A and f from a neighbourhood of 0 to a compact closed manifold X (e.g., $X = S^n$ or $T^n = \mathbb{R}^n/\mathbb{Z}^n$) so that 0 becomes a point $x_1 \in X$, A an elliptic operator over X, $A\colon \Gamma(X, E) \to \Gamma(X, F)$, and f a C^∞-section of the vector bundle F over X (actually bundles E and F can be chosen trivial). Such an extension is obviously possible.

Now let us choose a divisor $\mu = x_1^{-(k+d+1)}$. Then $f \in \Gamma_{\tilde{\mu}}(X, F)$. We want to modify f in such a way that the modified section (coinciding with f in a neighbourhood of x_1) belongs to $(\operatorname{Ker} \widetilde{A}^t)^\circ$. This will prove the existence of u due to Theorem 2.2.

Denote

$$N_{x_1}(X, F) = \{g \mid g \in \Gamma(X, F),\ g = 0 \text{ in a neighbourhood of } x_1\}.$$

Then we have to find $g \in N_{x_1}(X, F)$ such that $f - g \in (\operatorname{Ker} \widetilde{A}^t)^\circ$ (hence $f - g$ will be the desired modified section).

Consider the natural map $j\colon \Gamma_{\tilde{\mu}}(X, F) \to (\operatorname{Ker} \widetilde{A}^t)'$, $j(f)(v) = \langle f, v\rangle$. We want to prove that $j(f) \in j(N_{x_1}(X, F))$. But actually $j\colon N_{x_1}(X, F) \to (\operatorname{Ker} \widetilde{A}^t)'$ is surjective since the dual map $j'\colon \operatorname{Ker} \widetilde{A}^t \to (N_{x_1}(X, F))'$, which is defined similarly to j: $j'(v)(f) = \langle f, v\rangle$, is obviously injective. Indeed, j' is injective because $\operatorname{Ker}\widetilde{A}^t \subset \Gamma(X - \{x_1\}, F^*)$ and $N_{x_1}(X, F)$ includes all sections $g \in \Gamma(X, F)$ with $\operatorname{supp} g \subset X - \{x_1\}$. □

D. We make a few remarks about sheaves related with a given divisor and a given elliptic equation. These sheaves are actually natural localizations of the solution spaces described above.

In this subsection X is a closed compact C^∞-manifold, all other notations are the same as before. Denote by $\mathscr{L}(\mu, A)$ the sheaf of (local) solutions of $Au = 0$ with possible point singularities at the points x_j with $p_j > 0$ and zeros at the points x_j with $p_j < 0$, so that $\operatorname{ord}_{x_j} \geqslant -p_j$ for all $j = 1, \dots, k$ as in the definition of $L(\mu, A)$. So $L(\mu, A)$ becomes the space of all global sections of $\mathscr{L}(\mu, A)$, i.e.,

$$L(\mu, A) = \Gamma(X, \mathscr{L}(\mu, A)) = H^0(X, \mathscr{L}(\mu, A)).$$

Further information about the cohomologies of these sheaves is given by the following

THEOREM 4.1. (i) $H^p(X, \mathscr{L}(\mu, A)) = 0$ *for any* $p \geqslant 2$.

(ii) *The linear spaces* $H^1(X, \mathscr{L}(\mu, A))$ *and* $H^0(X, \mathscr{L}(\mu^{-1}, A^*))$ *are in a natural nondegenerate duality. In particular*

$$\dim H^1(X, \mathscr{L}(\mu, A)) = \dim H^0(X, \mathscr{L}(\mu^{-1}, A^*)).$$

PROOF. Let us introduce sheaves $\mathscr{E}(\mu, A)$ and $\mathscr{E}_{\widetilde{\mu}}(F)$ that are the localizations of the spaces $\Gamma(X, \mu, A)$ and $\Gamma_{\widetilde{\mu}}(X, F)$. So $\mathscr{E}(\mu, A)$ outside $\operatorname{supp}\mu$ coincides with the sheaf of all C^∞-sections of E. If $x_j \in \operatorname{supp}\mu$ and $p_j > 0$ then the local sections of $\mathscr{E}(\mu, A)$ near x_j have the form $u = u_s + u_r$, where u_s is a solution of $Au_s = 0$ defined in $\mathscr{U} - \{x_j\}$ ($\mathscr{U}$ is a neighbourhood of x_j) such that $\operatorname{ord}_{x_j} u_s \geqslant -p_j$, $u_r \in \Gamma(\mathscr{U}, E)$; if $p_j < 0$ then they are C^∞-sections u of E defined near x_j, such that $\operatorname{ord}_{x_j} u \geqslant -p_j$ (i.e., $j_{x_j}^{|p_j|-1} u = 0$). Similarly $\mathscr{E}_{\widetilde{\mu}}(F)$ is the sheaf of all C^∞-sections f of F, such that $j_{x_j}^{|\widetilde{p}_j|-1} f = 0$ if $x_j \in \operatorname{supp}\mu$ and $\widetilde{p}_j < 0$.

Now notice that both $\mathscr{E}(\mu, A)$ and $\mathscr{E}_{\widetilde{\mu}}(F)$ are fine sheaves because for any covering of X by open sets a partition of unity subordinated to this covering can be chosen on X in such a way that all functions that enter to this partition of unity are either identically zero or identically 1 near any point $x_j \in \operatorname{supp}\mu$. Hence due to Proposition 4.4 we get a fine resolution of $\mathscr{L}(\mu, A)$ as follows:

$$0 \longrightarrow \mathscr{L}(\mu, A) \longrightarrow \mathscr{E}(\mu, A) \xrightarrow{A} \mathscr{E}_{\widetilde{\mu}}(F) \longrightarrow 0.$$

Now (i) immediately follows. Also we get

$$H^1(X, \mathscr{L}(\mu, A)) = \operatorname{Coker}\{A\colon \Gamma(X, \mu, A) \longrightarrow \Gamma_{\widetilde{\mu}}(X, F)\}.$$

But the right-hand side here is in a nondegenerate duality with

$$\operatorname{Ker}\{A^* : \Gamma(X, \mu^{-1}, A^*) \longrightarrow \Gamma_{\widetilde{\mu}^{-1}}(X, E^*)\}$$

due to Theorem 2.2. This proves (ii). □

REMARK. Obviously (i) holds for noncompact manifolds (without boundary) as well.

The statement (ii) in Theorem 4.1 is an analogue of the Serre duality (see, e.g., [6, Chapter 3, Section 7]).

E. Nadirashvili [1] used the particular case $A = \Delta$ of Theorem 2.3, that he proved (see also Example 2.4 in Section 2) to give an estimate of multiplicity of possible zeros for the Coulomb potential of k point charges in $\mathbb{R}^3$. Namely, he proved that this multiplicity cannot be more than $k-1$ (unless all charges are 0). We reproduce here his arguments in more detail and make more use of them by considering Coulomb potentials in $\mathbb{R}^n$, $n \geqslant 3$. We shall also discuss similar questions on Riemannian surfaces.

We shall begin with a very general statement about meromorphic harmonic functions on general compact Riemannian manifolds.

Let X be a compact Riemannian manifold and $x_0, x_1, \dots x_k$ a collection of $k+1$ distinct points in X. We want to find out what is the maximum possible order of zero at x_0 for a nontrivial meromorphic harmonic function on X with possible simple poles at $x_1, \dots, x_k$ (and no other poles). Denote this maximal order by $l(X; x_0 \mid x_1, \dots, x_k)$. Obviously the estimate

$$l(X; x_0 \mid x_1, \dots, x_k) \leqslant l - 1 \tag{4.4}$$

is equivalent to the equality

$$r(x_0^{-l} x_1 \cdots x_k, \Delta) = 0 \tag{4.5}$$

LEMMA 4.5. *Let l be a positive integer. Then* (4.4) (*or* (4.5)) *is true if and only if there exist functions $u_1, \dots, u_k \in L(x_0^l, \Delta)$ such that $u_i(x_j) = \delta_{ij}$, $i, j = 1, \dots, k$.*

PROOF. Theorem 2.1 gives

$$\begin{aligned} r(x_0^{-l} x_1 \cdots x_k, \Delta) &= d(x_0^{-l} x_1 \cdots x_k) + r(x_0^l x_1^{-1} \cdots x_k^{-1}, \Delta) \\ &= -d(x_0^l x_1^{-1} \cdots x_k^{-1}) + r(x_0^l x_1^{-1} \cdots x_k^{-1}, \Delta) \\ &= -d(x_0^l) + k + r(x_0^l x_1^{-1} \cdots x_k^{-1}, \Delta), \end{aligned}$$

so (4.5) is equivalent to

$$r(x_0^l x_1^{-1} \cdots x_k^{-1}, \Delta) \leqslant d(x_0^l) - k.$$

Note that $d(x_0^l) = r(x_0^l, \Delta)$ because $r(x_0^{-l}, \Delta) = 0$ due to the maximum principle. Hence (4.5) is equivalent to

$$r(x_0^l x_1^{-1} \cdots x_k^{-1}, \Delta) \leqslant r(x_0^l, \Delta) - k. \tag{4.6}$$

Now note that

$$L(x_0^l x_1^{-1} \cdots x_k^{-1}, \Delta) = \{u \mid u \in L(x_0^l, \Delta),\ u(x_1) = \cdots = u(x_k) = 0\}.$$

Hence (4.6) is equivalent to linear independence of the conditions $u(x_i) = 0$, $i = 1, \ldots, k$, on $L(x_0^l, \Delta)$ which immediately proves Lemma 4.5. □

This lemma reduces the proof of the multiplicity estimate to the proof of existence of harmonic functions with a single (but not necessarily simple) pole such that their restrictions to the finite set $\{x_1, \ldots, x_k\}$ give linearly independent vectors.

Lemma 4.5 is easily extended to the noncompact case as discussed in Section 2. Namely, let X be a (noncompact) connected Riemannian manifold with a compact boundary and Δ the scalar Laplacian of the given Riemannian metric. Now let A, A^* be defined both with the help of Δ but possibly with different domains $\operatorname{Dom} A$, $\operatorname{Dom} A^*$; let $\operatorname{Dom}' A$, $\operatorname{Dom}' A^*$ be also chosen. Suppose that all the conditions of Theorem 2.3 are satisfied. Suppose also that

$$\operatorname{Dom} A \subset \{u \mid u|_{\partial X} = 0,\ u(x) \to 0 \text{ as } x \to \infty\}, \tag{4.7}$$

or

$$\operatorname{Dom} A \subset \{u \mid \partial u/\partial n|_{\partial X} = 0, u(x) \to 0 \text{ as } x \to \infty\}, \tag{4.8}$$

i.e., $\operatorname{Dom} A$ is defined by Dirichlet or Neuman boundary conditions and some conditions at infinity, that include the requirement that u vanishes at infinity. Then the following lemma holds:

Lemma 4.6. *$r(x_0^{-l} x_1 \cdots x_k, A) = 0$ if and only if there exist functions $u_1, \ldots, u_k \in L(x_0^l, A^*)$ such that $u_i(x_j) = \delta_{ij}$, $i, j = 1, \ldots, k$.*

The proof is obtained by the same arguments as in the proof of Lemma 4.5 except Theorem 2.3 should be applied instead of Theorem 2.1 and the maximum principle should be applied on $\operatorname{Dom} A$, which is possible due to (4.7) or (4.8).

In particular, Lemma 4.6 may be applied in the situation of Example 2.4. Here $X = \mathbb{R}^n$, $n \geqslant 3$, and $L(x_1 \cdots x_k, \Delta)$ consists of functions of the form

$$u(x) = \sum_{1 \leqslant i \leqslant k} \frac{q_i}{|x - x_i|^{n-2}}, \tag{4.9}$$

which are generalized Coulomb potentials of point charges $q_1, \ldots, q_k \in \mathbb{R}$, situated at the points $x_1, \ldots, x_k \in \mathbb{R}^n$, which are supposed to be distinct. In this case (4.5) is reduced to finding harmonic functions in $\mathbb{R}^n$ vanishing at infinity and having a single pole at x_0 of order at most l, such that their restrictions to the set $\{x_1, \ldots, x_k\}$ give linearly independent vectors. Nadirashvili could do this when $l = k$ (and $n = 3$, but his arguments are good for any $n \geqslant 3$), so his result is

THEOREM 4.2. *Let u have the form* (4.9) *and $u \not\equiv 0$. Then all zeros of u have orders less than or equal to $k-1$.*

PROOF. We have to check that $\operatorname{ord}_{x_0} u \leqslant k-1$ for any $x_0 \in \mathbb{R}^n$. Using the translation invariance we may assume without loss of generality that $x_0 = 0$. We shall use the Kelvin transform K, which acts on functions $u: \mathbb{R}^n - \{0\} \longrightarrow \mathbb{C}$ by the formula

$$Ku(x) = |x|^{2-n} u(x^*),$$

where $x \mapsto x^* = x/|x|^2$ is the inversion map on $\mathbb{R}^n - \{0\}$. The function Ku is harmonic on $\mathbb{R}^n - \{0\}$ if and only if u is harmonic there.

Now let us choose a 2-dimensional linear subspace in $\mathbb{R}^n$ such that all the points $0, x_1^*, \dots, x_k^*$ have distinct orthogonal projections to this plane. Using a rotation to change coordinates we may assume without loss of generality that this subspace is a coordinate plane. We will identify this plane with the complex line $\mathbb{C}$ and denote the orthogonal projection of $\mathbb{R}^n$ to this plane by π, so π^* will denote the corresponding map on functions.

Let us denote $z_i = \pi(x_i^*)$ and take the Lagrange interpolation polynomials

$$p_i: \mathbb{C} \longrightarrow \mathbb{C}, \quad \deg p_i = k-1, \quad p_i(z_j) = c_i \delta_{ij}, \quad i, j = 1, \dots, k,$$

where $c_i \in \mathbb{C} - \{0\}$, $i = 1, \dots, k$, are normalization constants to be chosen later. Now let us consider harmonic functions

$$u_i = K(\pi^* p_i), \qquad i = 1, \dots, k.$$

It is easy to check that they satisfy the requirements of Lemma 4.6 with $l = k$ after an appropriate choice of the normalization constants. □

REMARK 1. Note that the estimate of the multiplicity given by Theorem 4.2 is precise for $k = 1$ or 2. However we cannot expect it to be precise in general. In fact an easy parameter counting shows that the quantity

$$l(k) = \max\{l(\mathbb{R}^n; x_0 \mid x_1, \dots, x_k) \mid x_0, x_1, \dots, x_k \in \mathbb{R}^n\}$$

(the maximum possible order of zero over all possible configurations of charges and the zero) should be $O(k^{1/(n-1)})$. In particular we may expect an estimate by $O(\sqrt{k})$ if $n = 3$.

REMARK 2. Theorem 4.2 can be proved without use of Theorem 2.3. Namely, in this particular case Lemma 4.6 can be proved by elementary arguments from linear algebra. To do this consider (in $\mathbb{R}^n$) the following linear map

$$\mathbb{C}^k \cong L(x_1 \cdots x_k, \Delta) \longrightarrow J^l_{x_0}, \qquad u \mapsto j^l_{x_0} u,$$

where $J^l_{x_0}$ is the set of all l-jets at x_0 for scalar functions on $\mathbb{R}^n$. Here we identify the function (4.9) with the vector $(q_1, \dots, q_k) \in \mathbb{C}^k$. Now suppose that we want to prove that this map is injective. It is equivalent to the fact that the dual map is surjective. But it is easy to check that the dual map

$$S(x_0^l) \longrightarrow \mathbb{C}^k$$

maps $\sum_{|\alpha|\leqslant l} c_\alpha \delta^{(\alpha)}(x - x_0)$ to the vector, whose components are the values of the function $\sum_{|\alpha|\leqslant l} c_\alpha \partial_x^\alpha |x - x_0|^{2-n}$ at the points $x_1, \ldots, x_k$. (Here $S(x_0^l) = S(x_0^l, \mathbb{C}_{\mathbb{R}^n})$.) The necessary special case of Lemma 4.6 immediately follows.

F. Now we discuss multiplicity estimates for zeros of harmonic functions on compact Riemannian surfaces X with simplest (logarithmic) singularities. This amounts to investigating the numbers $l(X; x_0 \mid x_1, \ldots, x_k)$ in notations which were used in a more general context earlier in this section. Let us also denote

$$l(k; g) = \max\{l(X; x_0 \mid x_1, \ldots, x_k) \mid \operatorname{genus}(X) = g,\ x_0, x_1, \ldots, x_k \in X\},$$

i.e., $l(k; g)$ is the maximum of all possible multiplicities of zeros of nontrivial harmonic functions with k simple poles on a Riemannian surface of genus g over all compact Riemannian surfaces of genus g and over all possible configurations of poles.

THEOREM 4.3. $l(k; g) \leqslant k + 2g - 1$.

PROOF. For any $u \in L(x_0^{-l} x_1 \cdots x_k, \Delta)$ consider a meromorphic $(1, 0)$-form $\omega = \partial u = (\partial u/\partial z)\, dz$ (here z is a local complex parameter on X). Then clearly $\omega \in L(x_0^{-l+1} x_1 \cdots x_k, \bar\partial^*)$ where $\bar\partial^* = \bar\partial : \Lambda^{1,0}(X) \longrightarrow \Lambda^{1,1}(X) = \Lambda^2(X)$ is the $\bar\partial$-operator on $(1, 0)$-forms on X. Using the fact that the degree of every meromorphic differential should be equal to $2g - 2$ we immediately obtain the inequality $l - 1 - k \leqslant 2g - 2$ (with the equality if ω actually has poles in all points $x_1, \ldots, x_k$, has no other zeros except x_0, and the order of this zero is precisely $l - 1$). The desired result immediately follows. □

Now we will formulate a theorem that will give a more precise estimate but only for generic sets $\{x_0, x_1, \ldots, x_k\}$.

THEOREM 4.4. *There exists a dense open set $\mathfrak{A} \subset X \times \cdots \times X$ $(k + 1$ factors), such that for any $(x_0, x_1, \ldots, x_k) \in \mathfrak{A}$ and any nontrivial harmonic function with possible simple poles (with logarithmic singularities) at the points $x_1, \ldots, x_k$ only, the multiplicity of a possible zero of u at x_0 is less than or equal to $k/2$.*

PROOF. We shall use Lemma 4.5 and so the degree of the divisors will be always taken with respect to the second order operator Δ. If $l \geqslant 1$ then $\dim_{\mathbb{C}} L(x_0^l, \Delta) = d(x_0^l) = 2l - 1$. Now suppose that $2l - 1 \geqslant k$ or $l \geqslant (k + 1)/2$. Then fixing x_0 we can find linearly independent functions $u_1, \ldots, u_k \in L(x_0^l, \Delta)$. Hence we can find points $x_1, \ldots, x_k$ such that $\det(u_i(x_j)) \neq 0$. Passing to linear combinations of the chosen functions we may even assume that $u_i(x_j) = \delta_{ij}$. Moreover we shall obviously have then $\det(u_i(x_j)) \neq 0$ for $(x_1, \ldots, x_k)$ in a dense open set $\mathfrak{B}_{x_0} \subset X \times \cdots \times X$

(k factors). Note that locally $u_1, \dots, u_k$ can be made continuous with respect to x_0 because up to an additive constant they are determined by their "singular parts" at x_0, which are in a one-one correspondence with a set of distributions $f \in S(x_0^l, \mathbb{C}_X)$ (see Section 3). It follows that the set

$$\mathfrak{A} = \{(x_0, x_1, \dots, x_k) \mid (x_1, \dots, x_k) \in \mathfrak{B}_{x_0}\}$$

is also open in $X \times \cdots \times X$ ($k+1$ factors).

Now Lemma 4.5 implies that $\operatorname{ord}_{x_0} u < (k+1)/2$ or, which is equivalent, $\operatorname{ord}_{x_0} u \leqslant k/2$ for any $u \in L(x_1 \cdots x_k, \Delta)$ provided $(x_0, x_1, \dots, x_k) \in \mathfrak{A}$. □

Finally let us consider the simplest case $X = S^2 = \mathbb{CP}^1 = \mathbb{C} \cup \infty$, i.e., the case $g = 0$.

THEOREM 4.5. *Let* $g = 0$. *Then*

(i) $l(k; 0) = k - 1$, *i.e., the maximal possible multiplicity of a zero for a nontrivial harmonic function with* k *simple poles equals* $k - 1$.

(ii) *For any set of distinct points* $z_0, z_1, \dots, z_k \in X$ *and any positive integer* $l < k/2$ *there exists a nontrivial harmonic function* u *on* X *with possible simple poles at the points* $z_1, \dots, z_k$ *and a zero of order* l *at* z_0, *i.e.,* $l(X; z_0 \mid z_1, \dots, z_k) \geqslant l$ *for any integer* $l < k/2$ *and any points* $z_0, z_1, \dots, z_k$.

(iii) $l(X; z_0 \mid z_1, \dots z_k) = k - 1$ *if and only if*

$$\prod_{l \neq i, j} \left(\frac{z_i - z_0}{z_j - z_0} : \frac{z_i - z_l}{z_j - z_l} \right) \in \mathbb{R} \cup \infty, \qquad 1 \leqslant i, j \leqslant k, \quad i \neq j, \tag{4.10}$$

i.e., a nontrivial harmonic function with the only possible simple poles at the points $z_1, \dots, z_k$ *and with the maximal possible order of zero at* z_0 *exists if and only if all the products in* (4.10) *are real or infinite.*

PROOF. Note first that (iii) obviously implies (i) since the conditions (4.10) are satisfied, e.g., if $z_i \in \mathbb{R}$, $i = 0, \dots, k$.

Now let us prove (iii). We know already (Theorem 4.3) that $l(X; z_0 \mid z_1, \dots, z_k) \leqslant k - 1$. Using the conformal invariance of the statement (iii), we can suppose that $z_0, z_1, \dots, z_k \in \mathbb{C}$, i.e., that none of the points $z_0, z_1, \dots, z_k$ coincides with ∞. Suppose that $u \in L(z_0^{-k+1} z_1, \dots z_k, \Delta) - \{0\}$. Taking the real or imaginary part we may then suppose that u is real-valued. Then $\omega = \partial u$ will be a meromorphic (1, 0)-form on X with real residues, $\omega \in L(z_0^{-k+2} z_1 \cdots z_k, \bar{\partial}^*) - \{0\}$. Vice versa, having such a form, we can reconstruct u by the formula

$$u(z) = \operatorname{Re} \int_{z_0}^{z} \omega. \tag{4.11}$$

(The condition on the residues ensures that u is single-valued.)

So (iii) is equivalent to the statement that $\omega \in L(z_0^{-k+2} z_1 \cdots z_k, \bar{\partial}^*) - \{0\}$ with real residues exists if and only if the conditions (4.10) are satisfied. Since we suppose ω to be regular at infinity it should have the form $\omega = (P(z)/Q(z))\, dz$, where P and Q are polynomials, $\deg P \leqslant \deg Q - 2$. We can obviously assume that P and Q have no common zeros. Then we should have $\deg Q \leqslant k$ (otherwise ω will have more than k poles), hence $\deg P \leqslant k-2$. But then we should have (up to constant factors) $P(z) = (z - z_0)^{k-2}$, $Q(z) = (z - z_1) \cdots (z - z_k)$, hence

$$\omega = c \frac{(z - z_0)^{k-2}}{(z - z_1) \cdots (z - z_k)} dz$$

with $c \in \mathbb{C}$. The residues can be made real by a choice of the constant c here if and only if all the pairwise ratios of the residues of $(P(z)/Q(z))\, dz$ are real. These ratios are

$$\frac{(z_i - z_0)^{k-2}}{\prod_{l \neq i}(z_i - z_l)} : \frac{(z_j - z_0)^{k-2}}{\prod_{l \neq j}(z_j - z_l)} = - \prod_{l \neq i, j} \left(\frac{z_i - z_0}{z_j - z_0} : \frac{z_i - z_l}{z_j - z_l} \right),$$

which immediately leads to the desired result (iii).

To prove (ii) assume again that $z_0, z_1, \ldots, z_k \in \mathbb{C}$ and set

$$\omega = \frac{(z - z_0)^{l-1} g(z)\, dz}{(z - z_1) \cdots (z - z_k)},$$

where $g = g(z)$ is a polynomial, $\deg g \leqslant k - l - 1$. Obviously ω is then a meromorphic $(1, 0)$-form which is regular at infinity. Now the formula (4.11) will give us the desired function u provided ω does not vanish identically and has real residues. The resudues of ω have the form

$$\frac{(z_i - z_0)^{l-1} g(z_i)}{\prod_{l \neq i}(z_i - z_l)}, \qquad i = 1, \ldots, k.$$

Vanishing of their imaginary parts gives k real linear equations for $2(k-l)$ real numbers (the real and imaginary parts of the coefficients of g) that determine the polynomial g. If $2(k-l) > k$ or, equivalently, $2l < k$, then these equations have a nontrivial solution, and (ii) is proved. □

REFERENCES

1. N. S. Nadirashvili, *Harmonic functions with a given set of singularities*, Functional. Anal. i Prilozhen. **22** (1988), no. 1, 64–66; English transl. in Functional Anal. Appl. **22** (1988), no. 1.
2. L. Hörmander, *The analysis of linear partial differential operators*. III, Springer-Verlag, Belin, Heidelberg, New York, 1985.
3. R. Palais, *Seminar on the Atiyah-Singer index theorem*, Princeton Univ. Press, Princeton, NJ, 1965.
4. E. G. Sitnikova, *A strong zero theorem for a higher order elliptic equation*, Mat. Sb. **81** (1970), no. 3, 376–397; English transl. in Math. USSR–Sb. **10** (1970).

5. M. Gromov and M. A. Shubin, *The Riemann-Roch theorem for general elliptic operators*, C. R. Acad. Sci. Paris **314** (1992), no. 1, 363–367.

6. R. Hartshorne, *Algebraic geometry*, Springer-Verlag, Berlin, Heidelberg, New York, 1977.

INSTITUT DES HAUTES ÉTUDES SCIENTIFIQUES 91440, BURES-SUR-YVETTE, FRANCE

INSTITUTE OF NEW TECHNOLOGIES, 11, KIROVOGRADSKAYA MOSCOW 113587, RUSSIA and DEPARTMENT OF MATHEMATICS, NORTHEASTERN UNIVERSITY, BOSTON, MASSACHUSETTS 02115